The ecological implications of body size

CAMBRIDGE STUDIES IN ECOLOGY

The ecological implications of body size

ROBERT HENRY PETERS

McGill University

CAMBRIDGE UNIVERSITY PRESS

Cambridge
London   New York   New Rochelle
Sydney   Melbourne

Published by the Press Syndicate of the University of Cambridge
The Pitt Building, Trumpington Street, Cambridge CB2 1RP
32 East 57th Street, New York, NY 10022, USA
296 Beaconsfield Parade, Middle Park, Melbourne 3206, Australia

First published 1983

Printed in the United States of America

Library of Congress Cataloging in Publication Data
Peters, Robert Henry
The ecological implications of body size.
(Cambridge studies in ecology; 2)
1. Animal ecology.   2. Body size.   I. Title.   II. Series.
QH541.P44   1983   591.5   82–19755
ISBN 0 521 24684 9

To Ruth Gregory, for her assistance with the nets early in my career; Bob Bateman, who helped me to see Nature; and Frank Rigler, who taught me to study it

Contents

Preface

This book is an exercise in predictive ecology. It draws together a widely dispersed body of empirical relations that relate biological form and process to body size. These relationships are then applied to ecological problems; that is to say, to problems whose solutions require some knowledge about temporal and spatial patterns in the characteristics of organisms and in their abundance.

Although the book is largely based on the work of autecologists and environmental physiologists and is applied principally to the interests of ecologists, it should appeal to anyone interested in the development of a broad, quantitative science of organism function. The book was written for senior undergraduates, who should have some feeling for the goals and jargon of ecology. Nevertheless, the approach and information are original enough that many established scientists will find it interesting too. I hope the book is sufficiently clear that it is also open to the educated (if perseverant) layman.

In writing this book, I have used a simple design. I begin with a sketch of one typical body size relationship (Chapter 1) to show why the subject of this book deserves better recognition and wider use. In brief, these body size relations represent scientific theories and hypotheses, which include many examples of general, quantitative biological laws. Like all quantitative theories, body size relations require some basic numerical skills – the statistics of regression and the algebra of logarithms and powers – that are fortunately so few that they can be reviewed in the current context (Chapter 2). The terms of an animal's energy budget provide a convenient framework, around which most of the available body size relations can be organized; those relations describe rates of ingestion, respiration, growth, and defecation, plus a number of related phenomena, like predator–prey relations, resistance to starvation, and aspects of life history. Respiration receives the most attention because it has been disproportionately studied. This wider empirical base permits an examination of the basic relations between metabolism and size (Chapter 3), the confirmation of these relations by reference to the physiological components of respiration (Chapter 4) and the effect of other factors – temperature (Chapter 5) and activity (Chapter 7) – on the basic relations. Ingestion (Chapter 7) and growth (Chapter 8) are treated as extensively as available information allows. Not all relations that are important in ecology can be presented as terms in an individual energy budget. Thus, Chapter 9 deals with problems in material flow. Chapter 10 investigates the relations among size, population density, and community size structure; these allow the application of equations describing individual activities to questions in population or community ecology. Some peripheral topics – animal behavior, economic ecology, and evolution, which can be affected by individual size – are noted in Chapter 11. Chapter 12 examines simulation and modeling, by example, as one mechanism for the further elaboration and extension of body size relations. The book closes with a consideration of some explanations for the success of body size as a general predictor (Chapter 13) and with a brief prospectus (Chapter 14) for future investigations.

This sequence imposes a general progression from established empiricism to supposition. The same pattern is followed within each chapter and section. Each topic is addressed by first establishing the most fundamental of the relevant empirical relations, and then the discussion moves, by degrees, to less well-grounded theories and speculation. Provisional support for more tenuous conclusions is indicated whenever possible.

The ecological implications of body size are best examined as mathematical expressions of statistical trends. Such a prospect will scarcely encourage many potential readers. I am in sympathy with such sentiment, for I realize that equations and symbolism can render reading a chore and a paper indigestible. I have, therefore, tried to make the ideas as transparent as possible even for those who dislike mathematics. Very few equations and fewer mathematical arguments appear in the text. Those that do are particularly important. The bulk of existing body size relations are found in the appendixes, where they can be easily referenced and compared. Since there are hundreds of appropriate equations, these have been divided among 10 appendixes, which follow the sequence of topics in the chapters. Frequent reference to these appendixes in the text should direct the inquisitive reader to appropriate formulas. Some equations appear in figures or figure legends, where a graph makes the meaning clear. Mathematical deductions and arithmetic calculations are confined to "tables." If desired, these can be ignored because the conclusions are presented verbally and the argument is presented intuitively in the text.

I find most algebraic presentations in the ecological literature to be so terse that they are almost opaque. Often steps that are essential to the reader are omitted by the author as being self-evident. To reduce this source of irritation, tabled calculations are given in detail, particularly early in the book. These tables serve many purposes. First, they remove some difficulties in reading a quantitative text; second, their detail should make detection of any errors easier; and third, they should make subsequent calculations more familiar. Ideally, the tables can serve as patterns or paradigms (Kuhn 1977) in the sense that the tables are model solutions to sample problems. In the future, other students can adapt these models to deal with some of their own questions.

Some readers may be dismayed by my consistent use of the International System (S.I.) of units. We may mourn the passing of the calorie, but the advantages offered by standardization are too great to be denied. Without standardization, the amount of calculation and the number of errors in this book would be very significantly increased. Given that some standardization is required, it would be perverse to ignore the existing accord. In general, S.I. units become familiar with very little use, but when these seem highly inappropriate (as in the expression of life span in seconds), S.I. units are here supplemented by more meaningful units in figures and tables. Conversion factors are given in Appendix I.

These pages should not be read uncritically. I am most keenly aware of two faults. First, the scope of the book often takes me far beyond the limits of my personal expertise, and I have, no doubt, committed many errors of omission and ignorance. For these, I can only ask that readers take the time to correct me by indicating relations I have overlooked and tests or observations of which I am unaware. Second, this book is based on a single theme – the use of body size as a predictive tool – and the reader must appreciate that tunnel vision is characteristic of thematic presentations. I do not imply that size is the only source of predictive power in ecology; although I believe it is currently the best. I frequently mention the weakness of current relations and point to ways they could be improved. The book is a starting point, not a definitive summary.

Far more people have helped to develop this book than I can possibly mention here. I am grateful to the students in course 177–385B, "The

Environmental Physiology of Animals," at McGill University, who were so much a part of the book's early gestation. I am equally indebted to the faculty of McGill's Department of Biology, who have given me the freedom to explore new approaches, and to Livia Tonolli and Ettore Grimaldi and our colleagues at the Istituto Italiano di Idrobiologia, Pallanza, who provided the calm atmosphere of scholarship and concentration in which I could write. I particularly wish to thank those many friends who have read and reviewed parts of the text: John Downing, Don Kramer, Rolf Sattler, Valerie Pasztor, Jerry Pollack, Bob Carroll, Bill Leggett, Bob Lemon, Alain Vézina, Ed McCauley, David Currie, Robin Anderson, Jaap Kalff, Gertrud Nürnberg, Louis Lefebvre, Marty Lechowicz, Derek Roff, and Graham Bell. Cindy Sinclair and Karen Wassenberg provided technical help and Luba Harnaga, Mary Anne Pace, Sylvie Marchand, Celina Dolan, Jeanne Gould, Yvette Mark, and Joanne Smith typed the manuscript. Hélène Mailhot proved irreplaceable at all phases of preparation. I am grateful to all. Special thanks are due to my family, on both sides of the Atlantic, for their support and, when required, forbearance.

1

A philosophical introduction

The central tenet of this book is that there exists a set of theories that make simple, quantitative, and reasonably accurate predictions of biological phenomena. This book draws these theories together and explores their implications for ecology. These theories are the many equations that relate an animal's characteristics to its body size. Most *body size relations* take the form (Chapter 2)

$$Y = aW^b \tag{1.1}$$

where Y is the biological characteristic to be predicted, W is animal body mass, and a and b are empirically derived constants. Such equations are common in biology, but they are only rarely developed as a coherent body of theory. Nevertheless, the approach can be extended to yield basic descriptions of any animal. Body size relations, therefore, have a central role to play in animal ecology.

This book demonstrates the scope of body size relations. This introduction will establish criteria whereby scientific theories may be judged and then apply these criteria to body size relations. The purpose of this exercise is to show that these simple relationships are robust and powerful scientific theories.

The nature of scientific theory

One can imagine an infinite number of ways in which the parts of our universe might be related. This potential range leads to uncertainty about the way the world is constructed. Two classes of intellectual endeavor have developed to lessen this uncertainty. The first is logical analysis, which demonstrates the necessary relations among our concepts and terminologies thereby indicating phenomena that are not possible and will not be encountered. The second activity uses knowledge to select probable forms and connections from all logical possibilities. The tools that allow us to make this selection are called *scientific theories*. The process of

1

selection of the probable from the possible is called *prediction*. By definition, all scientific theories make predictions (Popper 1962, 1972). Better, more powerful theories exclude a greater range of possibilities as improbable.

Scientific theories predict by suggesting the probable nature of one thing, called the *dependent variable*, on the basis of some other thing, called the *independent variable*. Sometimes there is more than one independent variable, and usually the scope of the theory is restricted by stated limitations. However, these more complex theories present no difference in principle, and I will discuss theories as if they were all based on only one independent variable. To achieve prediction, a theory must describe or imply how to ascertain the value of the independent variable. The body of the theory then tells us how to use this value to arrive at a probable value or range of values for the dependent variable. When this is done, a prediction has been made.

The requirement that a theory make a prediction is often stated as a requirement that the theory be testable or (potentially) falsifiable. Attributes that make a theory more testable make it a better theory. Any increase in realism or operationalism, generality, precision, and quantification would increase the theory's testability. Because science is a human activity, anything that causes a theory to be studied more often also makes it more likely to be tested. Among these less formal characteristics are simplicity, utility, familiarity, and fertility.

Example: Daily sleep and body size in herbivorous mammals

An example may give substance to these criteria and simultaneously permit an analysis of body size relations in their light. The example is one of many, distinguished only because it is less well known than others and because a fuller statistical description is available. The relation between size and daily sleep is a typical body size relation.

Zepelin and Rechtschaffen (1974) list the average number of hours that a series of mammals sleep per day and the average body mass of each species. Scrutiny of this list suggests a pattern: Large herbivores sleep less than small ones. If these data are plotted on a double-logarithmic graph (Figure 1.1), the points seem to fall around a straight line. A statistical analysis (least-squares regression, Chapter 2) gives an algebraic description of this line and estimates the average variation or scatter of the points around the line (Figure 1.1B). If Y is the sleep time in hours of herbivorous mammals and W is average body mass in kilograms, then the

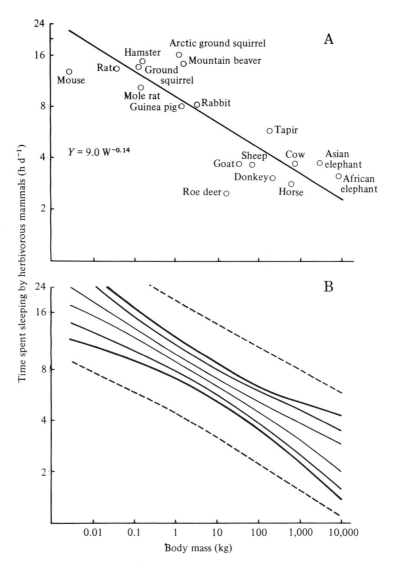

Figure 1.1. The effect of animal body size on the number of hours of daily sleep in herbivorous mammals. (A) Relevant data for 19 species reported by Zepelin and Rechtschaffen (1974). (B) A more general theory based on these data and the results of regression analysis (Appendix IIIc). The solid lines represent the 50, 95, and 99% confidence limits around the estimate of the mean. The dashed lines enclose 95% of all individual points.

regression line through the data is

$$Y = 9.0W^{-0.14}$$

(1.2)

This is the solid line in Figure 1.1A and serves as an average description of the observations. Some animals slept more and some less than this. Further statistical maneuvering allows a description of a zone that will probably enclose a given proportion of these individual values. Thus, 95% of all points will probably lie between the dashed lines in Figure 1.1B. In ecology, it is more common to report confidence limits around the regression line rather than around the individual points. For example, the 95% confidence limits around the regression line in Figure 1.1A imply that the average line through replicate samplings has a 95% probability of lying within these limits. By convention, confidence limits are normally drawn at the 99 or 95% level. However, ecology is a less precise science than some others, and it may often be necessary to make predictions where the chance of error is greater than 1 in 100 or 1 in 20 (Dillon & Rigler 1974). For example, there is a 50% probability that the average regression line lies between the innermost solid curves in Figure 1.1B. Confidence limits around the regression line describe a narrower envelope than those around individual points. In both types, increased levels of required accuracy result in decreased precision of the prediction (Figure 1.1B).

Theoretical status of body size relations

It should be clear why the lines in Figure 1.1 comprise a theory. Using only the definitions of hour and day, one can readily deduce that an animal must sleep between 0 and $24\,h\,d^{-1}$. This follows from logical necessity and excludes the impossibilities that an animal may sleep more than 24 or less than $0\,h\,d^{-1}$. No one would mistake this deduction for theory. A theory must exclude some value between 0 and 24 h as less probable; we can then claim to have some new knowledge about the external world. In fact, our present theory does much better than this minimum requirement, because it excludes a considerable range of values and allows us to distinguish between short and long sleepers on the basis of animal size.

Statistical curve-fitting provides an algebraic description of average trends in the data and therefore, allows us to make quantitative predictions like those in Figure 1.1B by assuming that these trends will describe future collections of data. Such data, once collected, can be used to test the theory by comparing the new data with the quantitative description. If the two are not statistically different, the theory is confirmed or corroborated. If they are different, then the theory has been disconfirmed or falsified. Since the new points will be independent from the old ones, logic does not guarantee that the theory will predict well, and there is always a possibility that the theory will fail some future test. Science never provides certain knowledge, only probabilistic theories.

Should the theory be falsified, one has a number of options. The simplest is to build a new statistical description based on all observations (old and new) and to begin the process again. Alternatively, one might look for factors that may have led to the discrepancy and might be important alternate or additional independent variables. On the other hand, one might despair of producing so general a theory and insist that sleep time be measured independently for each species. In the latter case, one has not abandoned theory but has replaced a general, but inaccurate, theory with a host of more specific theories that one hopes can predict better.

One of the most common charges against body size relationships is that they have sacrificed precision to achieve generality (Benedict 1938; May 1980). This is true; a statistical description is always less precise (and more general) than the individual members of the data set. In the case of body size relations, the error term that produces this imprecision reflects interspecific variation, other ecological variables, and errors in measurement and calculation associated with each datum. One cannot dismiss a theory because it is imprecise. Theories are rejected because they predict less well than their competitors. In most cases, there is no competing theory of equal generality, and so the criticism is empty unless it proposes that a number of specific theories will make the same predictions more accurately and more precisely. Certainly, if one wished to predict the average sleep time of the species listed by Zepelin and Rechtschatten (1974), one might use the average values, not Equation 1.2. This would not ensure success, because the listed averages of each species are subject to error. Such an error would also affect the regression analysis, but its effect would be modulated by the "good" points in the data set. More importantly, these tiny species-specific theories do not extend easily to the many mammalian herbivores unrepresented in the data set. A general theory based on body size or other variables, such as brain size or metabolic rate (Zepelin & Rechtschaffen 1974), must remain the method of choice whenever an appropriate specific theory does not exist. Since there are far more species on the face of the earth than there are biologists to study them, ecologists must take advantage of general relationships.

Body size relations, therefore, meet the formal requirements of specific theory. They predict the probable value of a dependent variable (like sleep time) from the measured value of the independent variable (body mass) under certain conditions (the animal must be a mammalian herbivore). Such theories are realistic because they are built empirically, using actual observations. They lend themselves to testing because they are quantitative. They are admittedly imprecise, but this is the price of generality.

The acceptability of body size relations

A theory must meet these formal requirements, but a good theory should also be relatively simple, useful, familiar and fertile. It must appeal to other scientists.

No theory could be simpler than a body size relation. The independent variable is probably the most easily determined of animal characteristics, the mathematical form of the equation is basic, and the dependent variable is well defined. Since no problem has been defined that would require an estimate of sleep time, this theory cannot seem very useful. Perhaps Equation 1.2 could suggest a probable schedule of activity for a little-known animal and so aid in the preliminary design of behavioral studies. The theory certainly suggests that large herbivores will be more interesting zoo exhibits than small ones. In general terms, body size relations provide a basis for action when no specific knowledge is at hand and a standard of comparison when such knowledge is available. Unfortunately, it is difficult to make the present example appear utilitarian. We will encounter relations that are much more useful (and others that are less) in subsequent chapters.

Theories should also be attractive to other scientists. One of the primary attractions of theories like Equation 1.2 lies in their extensive utilization of the quantitative literature. Scientific journals are crammed with data, the fruits of years of research in field and laboratory. Too often, this information is ignored, lost in the vast array of scientific facts. Body size relationships draw similar data together and so use these results of other scientists to build more general theories. Because each general theory is derived from the data of many scientists, these theories must meet the aims of these scientists. This should ensure a minimum audience for these theories.

Body size relations have a familiar form as well as familiar goals. For well over a century, animal body mass has been seen as an important independent variable, and power relations like Equation 1.1 have been used to describe size relations. The approach has been used in such diverse fields as paleontology (Gould 1966; Rensch 1960), animal morphology (Huxley 1972; Thompson 1961), physiology (Pedley 1977; Schmidt-Neilsen 1979), ecology (Kendeigh, Dol'nik, & Govrilov 1977), and animal behavior (Clutton-Brock & Harvey 1977a,b). The tempo at which such equations are produced seems to be increasing, and the tedium once associated with building and examining these equations has been immensely simplified by the development of computer-assisted analyses and pocket calculators. Of course, as these relations become more common, they also must appear less profound. As statements of trend – large animals eat more and live longer – they are often self-evident and trivial. Body size relations are more interesting if one wishes to see a quantitative biology that is capable of rigorous prediction and subject to rigorous tests. Fortunately, this need is now more keenly felt, and so these relations may seem familiar solutions to new problems.

Several properties of body size relations make them less attractive. Often regressions are produced to no purpose other than to demonstrate the existence of yet another curve. This risks the loss of the wheat in the chaff of the approach. One should bear in mind that the goal of science is prediction, not monotonous use of the same formula and variable. The predictions of these equations may be still more unattractive, because they deal with multispecific assemblages of animals and ignore differences among species. Much of traditional biology has, by contrast, dealt with differences among species and individuals. By glossing over these differences, these equations and other general theories sacrifice a part of their potential audience. Futhermore, biology has traditionally sought to explain living phenomena by reducing these to physicochemical laws or by demonstrating their evolutionary strategy and purpose. There is no good explanation for body size relations (Chapter 13), and those who crave explanations will find the approach wanting. Finally, although two major strengths of the body size approach lie in its potential for generality and for statistical analysis, many of the curves in this book apply only to limited taxa, and full statistical treatments of the data are rarely presented in the literature.

These weaknesses will discourage some scientists. Others hopefully will see these shortcomings as a challenge and an opportunity. Body size relations should seem a fertile field of research, because so much remains to be done (Chapter 14). Gaps in our knowledge must be filled by studying new groups of animals and many additional dependent variables. In addition, one could reanalyze the published data to test the adequacy of existing descriptions and to provide missing statistical information. One might also take advantage of the exponential growth of scientific information to collect new data to test existing relations. The relevance of curves based on laboratory observations to free-living animals should also be tested by examining the results of more recent field programs. The extensive summary of our present knowledge in the appendixes of this book is included to encourage this research. Although such a project must seem like simple accounting to some, its goal is improved general theories for ecology. This is surely as important as many of the specific problems to which ecologists now dedicate themselves.

An equally important prospect is the search for other independent variables that may be combined with body size to yield more precise predictions. Since the trend in Figure 1.1A shows considerable scatter, there must be other factors that contribute to the differences among animals. Allison and Cicchetti (1976) proposed the use of semiquantitative indexes of predation level and the security of the resting place to remove some of the scatter. Body size relations ought not be seen as ends but as starting points. Although they may have impressive predictive power, they can be greatly improved. The factors and forms used for each improvement are limited only by the imagination of the scientific community.

Scientific crisis in ecology

According to Thomas Kuhn's (1970) analysis of scientific change, new theories and new approaches are not accepted just because they are more powerful predictors or more useful tools. He argues that new approaches are accepted only after traditional avenues of research have led a field's researchers to a crisis. This crisis is precipitated by the realization that the old approach no longer yields interesting problems. It is difficult for an ecologist to read Kuhn without thinking of contemporary ecology.

I have argued elsewhere that much of modern ecology is scientifically weak and practically futile. Our theories are too frequently vague verbal statements, incapable of rigorous prediction, and, therefore, not subject to rigorous tests. Our most respected scientific constructs, like the "theory" of evolution by natural selection or the competitive exclusion principle, are deductive constructs identifying the possible but not the probable (Peters 1976, 1978a). Too often, variables are undefined, and sometimes they are unmeasurable (Peters 1977). We seem to follow an endless round of old concepts (Peters 1980a), marking our progress only with the latest fad in terminology. Saddest of all, most ecologists believe that mankind is racing blindly toward an ecological disaster that only the drastic application of ecological science could avert (Peters 1971). The emptiness of ecology is apparent to anyone who leafs through a contemporary ecology text searching for practical solutions to real problems. Over the past two decades, eloquent spokesmen have sensitized public and politician to environmental degradation. However, when asked for practical advice about agricultural policy, recreational development, wildlands management, or industrial siting, our "directions" were frequently little more than platitudes and aphorisms. If ecology has lost the ear of the public, it is our own fault. When we were asked, we had nothing to say.

There is a growing dissatisfaction with the old ecology. One of the most promising developments toward its replacement, toward a new ecology, is the growing interest in general empirical theories (Brown 1981; Peters 1980b). I hope this collection of body size relations will itself add to this promising field of predictive ecology and will stimulate others to make their own contributions.

2

A mathematical primer: Logarithms, power curves, and correlations

Only a small number of numerical skills are required to deal with body size relationships, and most of these are straightforward. They include a basic understanding of simple algebra, an ability to manipulate numbers expressed as powers and logarithms, a grasp of the implications of power formulas like Equation 1.1, and an appreciation of the strengths and weaknesses of regression analysis. This chapter is intended as a crutch for those who are nervous with the algebra of body size. Many readers will find this chapter a tedious exercise in the obvious and should pass over it.

Basic tools

Logarithms

Most analyses of body size relations begin by converting or "transforming" observed values to their logarithms. Logarithmic transformation is a simple device to ease and improve diagrammatic and statistical descriptions of the effect of body size on other attributes. This primer, therefore, begins by recalling the basic characteristics of logarithms.

Like any numbers, logarithms can be added, subtracted, multiplied, and divided, provided all logarithms in the calculation are converted to the same base. However, addition, subtraction, multiplication, and division performed with logarithms do not correspond to the same operations in "normal" arithmetic based on the antilogarithms. Because logarithms represent the power to which some base must be raised, a change of one logarithmic unit corresponds to an order of magnitude change in the antilogarithm. An arithmetic change (e.g., from 1 to 2 to 3) in the logarithm corresponds to a *geometric* or multiplicative change (e.g., from 10 to 100 to 1,000) in the antilogarithm. Although the rules describing the relations between operations involving logarithms and antilogarithms are elementary, Table 2.1 may still be a useful summary.

10

Table 2.1. *A summary of operations involving logarithms*

$$\log(N_1 N_2) = \log N_1 + \log N_2$$
$$\log(1/N_1) = -\log N_1$$
$$\log(N_1/N_2) = \log N_1 - \log N_2$$
$$\log N_1^c = c \log N_1$$
$$\log 1 = 0$$
$$\log \sqrt[c]{N_1} = \log N_1^{1/c} = (1/c) \log N_1$$

Note: The antilogarithms, N_1 and N_2, are any two positive numbers, c is a constant, and all logarithms are to the same base. There is no alternative to $\log(N_1 + N_2)$ or $\log(N_1 - N_2)$ and no logarithm exists for a negative number or 0.

Power formulas

Equations that describe the relationship between an animal's size (W) and another of its characteristics (Y) usually take the form of a power formula like Equation 1.1: $Y = aW^b$. This is called a *power formula* simply because the dependent variable changes as some power of the independent variable. Except in the special circumstances when $b = 0$ (and, thus, when $Y = a$) or when $b = 1$ (and, thus, when $Y/W = a$), this equation does not describe a straight line when Y is plotted against W. In fact, the relation between Y and W can have very different shapes depending on the value of b (Figure 2.1). Because these two phenomena, body size and whatever Y represents, increase at different rates, body size relations are often called *allometric relations*, and power formulas involving body mass are called *allometric equations* (*allos:* other; *metron:* measure). The change of Y with W is also called the *scaling* of that characteristic to body size: If one animal were built to the same design as another, but on a different scale, the various characteristics of the original would necessarily be scaled up or down to produce a working model. Allometric relationships are the rules by which this scaling is achieved.

One of the most frequent manipulations used in this book is the division or multiplication of one power equation by another. These calculations follow algebraic rules that are homologous to those used with logarithms (Table 2.1), but they are summarized in Table 2.2 because of their essential role in almost every chapter.

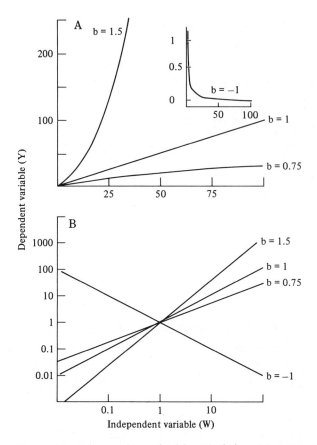

Figure 2.1 A comparison of arithmetic (A) and logarithmic (B) plots of power relationships, $Y = aW^b$ where $a = 1$ and b takes different values as indicated.

Transformation

Variations in the arithmetic rate of change of Y at different values of W make arithmetic plots of body size relationships hard to draw and interpret. Fortunately, power formulas yield straight lines if the logarithms of Y and W are plotted against one another instead. Since equal numbers must have equal logarithms, it follows from Equation 1.1 that

$$\log Y = \log (aW^b) \tag{2.1}$$

and from Table 2.1

$$\log Y = \log a + b \log W \tag{2.2}$$

Table 2.2. *A summary of operations involving powers*

$W^a W^b = W^{a+b}$	$W^a / W^b = W^{a-b}$
$(W^a)^b = W^{ab}$	$\sqrt[c]{W^a} = W^{a/c}$
$(W_1 W_2)^a = W_1^a W_2^a$	$(W_1 / W_2)^a = W_1^a / W_2^a$
$W^a / W = W^{a-1}$	$W_1^a W_2^b = W_1^a W_1^{b(\log_{w_1} W_2)}$
$1/W^b = W^{-b}$	
$W^0 = 1$	

Note: The first six expressions would correspond to those in Table 2.1 if the common base in Table 2.1 were W, if $\log_W N_1 = a$ and if $\log_W N_2 = b$. W_1 and W_2 are different bases.

Since log a is a constant, Equation 2.2 has the form of an equation for a straight line: $y = a + bx$. It is inconvenient to calculate the logarithm of each datum in order to achieve this linearity, but the same effect can be had by plotting the untransformed data on "double" or "full" logarithmic graph paper in which both axes are scaled logarithmically (Figure 2.1B).

Figure 2.2 illustrates this with a further description of mammalian sleep (Zepelin & Rechtschaffen 1974). The sleep of mammals cycles between two phases indicated by characteristic patterns of brain wave activity. In one phase, *paradoxical sleep* (also called *deep sleep* or *rapid eye movement sleep*), the brain shows periodic bursts of activity. The alternate phase of sleep involves brain waves with relatively low frequency and so is called *slow-wave sleep*. The length of time that an animal requires to pass through one period each of paradoxical and slow-wave sleep is called the *sleep cycle* (S_c, in h). This parameter increases with body size (W, in kg) in both herbivorous and carnivorous mammals as

$$S_c = 0.33 W^{0.19} \tag{2.3}$$

When these data are plotted arithmetically (Figure 2.2A), little pattern can be seen since the data for all species less than 10 kg in mass are confined to a narrow vertical band along the ordinate, and data for large species are comparatively rare. However, when the data are plotted on a logarithmic grid, the increasing trend is obvious (Figure 2.2B).

Since most of the relationships in this book require logarithmic transformation, the reader should be familiar with the effects of such

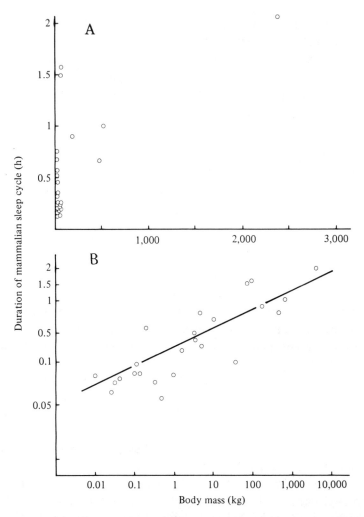

Figure 2.2. The effect of logarithmic transformation on the graphic presentation of scaling. An arithmetic plot (A) obscures the relation by limiting much of the data to a narrow zone close to the vertical axis. Logarithmic transformation (B) spaces the data more effectively revealing a regular increase in sleep cycle with body size. Sleep cycle is the amount of time required for a mammal to pass through one bout each of slow-wave and paradoxical sleep. Data from Zepelin and Rechtschaffen (1974).

transformation on the data. In very approximate terms, logarithmic transformation tends to diminish the difference among large numbers and to accentuate differences among small numbers. Points that lie 1 logarithmic unit above the curve in Figure 2.2B are ten times larger when expressed arithmetically, but those that lie 1 unit below the line are only one-tenth the size. It should be noted that since log 0 is undefined, there is no point on a double-logarithmic graph that corresponds to the arithmetic coordinates $(0, 0)$, and no negative values of the antilogarithms can be represented. The intercept of a double-logarithmic graph occurs when $\log W = 0$ or $W = 1$. Negative values of the logarithms correspond to fractions lying between 0 and 1.

These differences may lead to some suspicion of power curves and double-logarithmic plots (Benedict 1938; May 1980; Smith 1980). Much of this suspicion arises because logarithmic plots appear to reduce the error associated with larger values of Y. Certainly, data that lie well above the predicted value in arithmetic plots seem closer when logarithmic plots are used. However, this downward shift in the apparent position of high values is compensated by a downward shift in data lying below the predicted. Logarithmic transformation also tends to reduce the dispersion associated with high values, but this is again compensated by increased dispersion among low values. These changes affect the appearance of the data, but we have no basis to claim that either arithmetic or logarithmic transformation is more meaningful or correct.

Logarithmic transformations are only useful tools for the treatment of certain categories of data and are no more subject to abuse than any other system of computation (Stevens 1946). Once the peculiarities are appreciated, logarithms cannot be said to misrepresent the data. In fact, the data remain unchanged by transformation, and only the inflexibility of our minds makes one presentation more informative than another. Most of the problems associated with power curves and logarithms reflect unfamiliarity and disappear with use.

Regression analysis

A definitive review of regression analysis is beyond the scope of this book and that of its author. Instead, this section will examine only the least-squares method of regression analysis as it relates to body size relations. This is by far the most common approach to regression in biology. With few exceptions (Downing 1981; Ricker 1973; Zar 1968a,b) and despite some limitations, it is virtually the only technique used to determine allometric equations. This undoubtedly reflects the dominance of the least-squares method in statistics texts and commercial statistical packages for computer-assisted analysis. Regression analysis is applied to data sets in which one variable can be construed as a function of the other. The aim of regression is to find the line that, on average, describes the available data with the smallest errors. Least-squares regression achieves this by finding the straight line that passes through the point described by the mean of all X and Y values $(\bar{X}, \bar{Y})$ and minimizes the sum of the squares of the deviations $(\Sigma(Y_i - \hat{Y}_i)^2$, where Y_i is the observed value of the dependent variable and $\hat{Y}_i$ is the corresponding value predicted from the line at X_i). Table 2.3 summarizes the basic calculations used in least-squares regression. Data for allometric relations are usually transformed to logarithmic equivalents (i.e., log X_i,

Table 2.3. *Regression statistics calculated using the method of least square*

I. Basic calculations

Mean $X = \bar{X} = (\Sigma X_i)/n$,   mean $Y = \bar{Y} = (\Sigma Y_i)/n$

Sum of squares of deviations in $X = \Sigma x^2 = \Sigma(X_i - \bar{X})^2$

Sum of squares of deviations in $Y = \Sigma y^2 = \Sigma(Y_i - \bar{Y})^2$

Sum of products of deviations in X and $Y = \Sigma xy = \Sigma(X_i - \bar{X})(Y_i - \bar{Y})$

II. Sums of squares

Sum of squares attributable to regression $= \Sigma \hat{y}^2 = (\Sigma xy)^2/\Sigma x^2$

Sum of squares of deviations from regression $= d_{yx}^2 = \Sigma y^2 - \Sigma \hat{y}^2$

Mean square deviation from regression $= S_{xy}^2 = d_{yx}^2/(n-2)$

III. Regression line

Slope or regression coefficient of Y on $X = b = \Sigma xy/\Sigma x^2$

Intercept or elevation $= a = \bar{Y} - b\bar{X}$

Predicted value of Y from $X = \hat{Y} = a + bX$

IV. Indexes of fit

Coefficient of determinism $= r^2 = \Sigma \hat{y}^2/\Sigma y^2$

Correlation coefficient $= r$

F value $= \Sigma y^2/[(n-1)S_{xy}^2]$

Degrees of freedom $= df = (1, n-2)$

Table 2.3. (*cont.*)

V. Standard errors

Standard error of $\bar{Y}$ estimated from $\bar{X} = S_{\bar{Y}} = S_{xy}/\sqrt{n}$

Standard error of $b = S_b = S_{xy}/\sqrt{\sum x^2}$

Standard error of intercept $= S_a = S_{xy}\sqrt{1/n + \bar{X}^2/\sum x^2}$

VI. Confidence limits

$x_i = X_i - \bar{X}$

Population mean value of Y at some $X = \hat{\mu}$

Confidence limits for $\hat{\mu} = \pm t_\alpha S_{xy}\sqrt{1/n + (x_i^2/\sum x^2)}$

Confidence limits for an individual $Y = \pm t_\alpha S_{xy}\sqrt{1 + 1/n + (x_i^2/\sum x^2)}$

Confidence limits of $b = \pm t_\alpha S_b$

Confidence limits of $a = \pm t_\alpha S_a$

Note: This calculation determines the variation of some dependent variable Y with some independent variable X on the basis of n data pairs, X_i and Y_i, assuming $Y = a + bX$. The value of t_α depends on n and the confidence required for the prediction; these t values are read from an appropriate statistical table. For 95% confidence limits ($\alpha = 0.05$), the value of t_α is approximately 2 if $n > 20$. When $n > 20$, 50% confidence limits may be calculated assuming $t_\alpha = 0.68$. *Source:* Snedecor and Cochran (1967).

log Y_i) before calculations are begun. Consequently, the line that is actually fitted is not $Y = aW^b$ but log $Y = $ log $a + b$ log W. Results most often reported for allometric regressions are the slope of the regression line (b) and the intercept (a). This information is sufficient to predict mean values for the dependent variable, but it can scarcely be considered complete. No estimate of the data's variation around this mean has been presented, and it is, therefore, impossible to determine the precision of the prediction.

Indexes of fit

Three indexes of the goodness of fit, the effectiveness with which the regression describes the original data, are frequently used: r^2, r, and F.

Each reflects the improvement that the regression offers in describing the observed values of the dependent variable over that offered by their mean ($\bar{Y}$). If one used the mean to predict values of Y_i, an estimate of the error in prediction would be given by the total sums of squares of deviations in the dependent variable ($\Sigma y^2 = \Sigma(Y_i - \bar{Y})^2$). An analogous statistic, d_{yx}^2, describes the sum of squares of deviations from the regression line, so the reduction in the sum of squares as a result of regression can be calculated as $\Sigma \hat{y}^2 = \Sigma y^2 - d_{yx}^2$ although a seemingly different algebra is given in Table 2.3. If the regression significantly improves prediction, the sum of squares attributable to (or "explained by") regression ($\Sigma \hat{y}^2$) should be a significant fraction of the total deviation sum of squares of this dependent variable. This fractional value, called the *coefficient of determinism* ($r^2 = \Sigma \hat{y}^2 / \Sigma y^2$), is, therefore, a useful index to represent the improvement due to regression. The best regressions explain all variation in Y and, therefore, $r^2 = 1$. In the worst case, none of the variation is explained, and $r^2 = 0$. All values of r^2, therefore, lie between 0 and 1; an r^2 of 0.80 implies that the regression accounts for 80% of the variation around the mean, a quite substantial gain in predictive power. The *coefficient of correlation* ($r = \sqrt{r^2}$) conveys the same information although if $r = 0.80$, only ($r^2 = 0.80^2 = 0.64$) 64% of the variation is explained. If the slope is negative, r is often given as a negative fraction; thus, values of r range from -1 to 1. Perfect correlation is achieved if $r = -1$ or 1. The F value ($\Sigma y^2 / ((n - 1) S_{xy}^2)$) indicates if the average variation around the mean ($\Sigma y^2 / (n - 1)$ is significantly greater than the average deviation from the regression line ($S_{xy}^2 = (\Sigma y^2 - \Sigma \hat{y}^2 / (n - 2)$). The F value may be considered an estimate of the proportionate improvement in description made possible by regression. Thus, if the value of F is 20, the regression is 20 times more precise in describing the data than is the mean.

Both r^2 and F values compare variation around $\bar{Y}$ with residual variation around the regression line. High indexes of fit may be generated either by decreasing the error around the regression or by increasing the total variation around $\bar{Y}$, although the error around any section of the regression remains unchanged. The latter effect has two initially surprising corollaries. In terms of allometry, equally imprecise data will yield higher values of r^2 or F if one treats either a wider range of sizes or processes that vary more dramatically with size. Consequently "mouse-to-elephant" curves yield better indexes of fit than mouse-to-rat or horse-to-elephant relations, and relations between individual rate and size (which usually have slopes of about 0.75; Chapter 3) give better fits than those relating mass specific rate and size (which usually have slopes of about -0.25). These shifts in r^2 and F will occur even if the same data are employed. Facile comparisons of these indexes among relations might, therefore, mislead the unwary.

Limits of confidence

Each of the indexes of goodness of fit gives a simple summary of how well a regression line fits the data. It would also be useful to be able to predict the certainty or precision associated with predictions from the regression. This certainty can be interpreted as the confidence limits around any $\hat{Y}$.

The factors that affect the width of the confidence zone can be deduced from the appropriate equations in Table 2.4. Confidence limits may be very large if sample size (n) is small or if the deviations from regression (S_{xy}) are substantial. One can narrow the confidence band by using regressions based on a large range (Σx^2) of the independent variable or by permitting lower values of t_α and, therefore, accepting less accurate predictions. Finally, imprecision of the predictions is least when predictions are made close to $\bar{X}$, where x_i^2 is small. Figure 2.3A illustrates the principle of decreased precision (and increasing confidence limits) for predictions in which the value of the independent variable lies far from the mean. A special problem occurs among allometric relations based on power curves, because the symmetry of the confidence limits applies only to the logarithmic plot. When transformed back to antilogarithms, the confidence limits are highly skewed: The limits are wider at high values of Y than at low values, and the upper confidence limit always lies farther from the regression line than does the lower. Their effects are illustrated in Figure 2.3B.

The relevance of these comments on the calculation of confidence limits is limited, since statements of precision in the allometric literature are almost always so inadequate that confidence limits cannot be calculated. Among the most frequently omitted statistics are $\bar{X}$, n, S_{xy}, and Σx^2. Often, only the standard errors associated with the slope and intercepts are reported. Such values are useful only for rapid comparisons among slopes and elevations to gauge, approximately, whether apparent differences among regressions are likely to be important. These statistics are insufficient to calculate limits of confidence.

Comparison of isolated slopes, intercepts, and associated error is poor practice and may induce several problems that should not be overlooked.

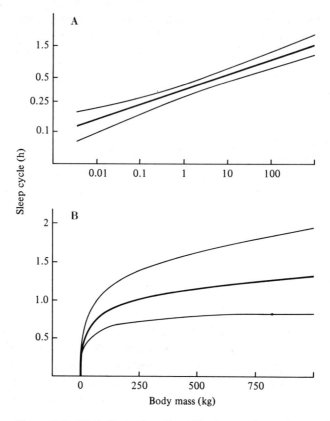

Figure 2.3. The effect of antilogarithmic transformation on the confidence limits around an allometric relation based on the logarithms of the dependent and independent variables. Both logarithmic (A) and antilogarithmic (B) (or arithmetic) plots are based on the relation between size and sleep cycle (Figure 2.2; Equation 2.3).

For example, the intercept at $X = 0$ may lie far from $\bar{X}$, frequently beyond the limits of the original data set. Confidence limits around the intercept can, therefore, give an erroneous impression of great imprecision. A better indicator of uncertainty associated with the elevation is the standard error of $\bar{Y}$, $S_{\bar{Y}}$, which can be extracted from S_a, but only if $\bar{X}$ is known. In addition, unless the intercept coincides with the mean, the intercept is not independent of the slope but depends upon it and $\bar{X}$, $\bar{Y}$ (Table 2.3). Essentially similar sets of data may produce quite different estimates of slope and intercept, because regression analysis, in a sense, reaches a compromise between the two: In allometric analyses, steep slopes may produce low intercepts and vice versa (see Chapter 4, Figure 4.2). Some authors report only the intercept or the slope of their regression lines (e.g., Gould 1979; Quiring 1941). This should be avoided, for the intercept depends on the slope and the applicability of the analysis is greatly limited if only one coefficient is given. Slope and intercept are parts of a single entity, the regression line, and both must be reported. Quick comparisons among regressions are better made by scrutinizing plots of the regression lines and their associated data. More adequate comparisons of regression lines use an *F* test (Snedecor & Cochran 1967) or sometimes a nonparametric procedure (Tsutakawa & Hewett 1977); these procedures will not be discussed here.

Problems in regression analysis

Like any statistical procedure, least-squares regression has limitations. Regrettably, these must sometimes be ignored since this analytical technique may be the investigator's only practical option. However, one should know when rules are broken and the effect of such transgressions on the statistical results.

Least-squares regression is characterized by its dependence on the squares of the deviations of X_i and Y_i from $\bar{X}$ and $\bar{Y}$ (Table 2.3). This squaring has important implications for it makes the regression very sensitive to extreme values. For example, a point lying 10 units from the mean has as much influence as 100 points lying 1 unit from the mean. If the data include one or a few extreme values and a swarm of other values, then least-squares regression essentially treats the swarm as a single point. If these extreme values are erroneous or exceptional, the regression curve may be misleading. Ideally, the data should be spread evenly along the range of the independent variable. If this is not the case, the extreme points should be excluded from the regression analysis. These extremes must, of course, be reported and may be used to examine extrapolations of the curve beyond the subset of more evenly distributed points.

In principle, least-squares regression assumes four properties of the data. These are:

1. There is a straight line (which the analysis approximates as $Y = a + bX$) that passes through the true mean value of *Y* at each value of *X;* thus, if enough estimates of the dependent variable were made at one value of the independent variable, the mean of the *Y* values would lie on this best regression line.

2. At each value of X, the corresponding values of Y are distributed normally around the true mean of Y.
3. The variance associated with such distributions of Y is the same at all levels of X.
4. The value of X must be determined without error.

Logarithmic transformation is performed to meet the first assumption by linearizing the curvilinear relationship between the antilogarithms. Although this is often successful, one cannot be sure that this transformation will work in any particular case. Consequently, the first step in any regression analysis is a plot of the data. If the data approach linearity, statistical treatment may be warranted. A further visual test for linearity can be performed after regression by plotting the residual deviations (the differences between predicted and observed values of the dependent variable $(Y_i - \hat{Y}_i)$ against the corresponding values of the independent variable (X_i). If Y is not linear with respect to X, a significant pattern in the residuals will emerge. A larger number of the residuals will fall to one side or the other of the horizontal line at $Y_i - \hat{Y}_i = 0$ in some region of X and the regression would be biased. In this case, the analyst would seek some other transformation that would remove any trend in the residuals.

By definition, when the variation around the regression line is not constant at all values of X, some parts of the data set show greater deviations than others. Since the analysis seeks to minimize the squares of all such deviations, these regions have a disproportionate effect on the regression statistics. A similar problem arises if the data are not normally distributed about the regression line but are skewed around the predicted value. The longer tail of the distribution will produce larger squared deviations of Y and will again have a disproportionate effect on the regression statistics. These biases are most easily found by scrutiny of the residual plots. They are not shown in the calculated confidence limits since those calculations assume a constant, normal variance. Confidence limits derived from skewed data will overestimate the precision of prediction for some values of $\hat{Y}$ and underestimate for others. Logarithmic transformation is frequently successful in normalizing and stabilizing variance in allometric data but the effectiveness of this transformation relative to other possibilities is rarely examined (Downing 1979, 1980; Zar 1968a,b).

Finally, least-squares regression analyses assume that the error associated with measuring the independent variable is negligible. Since an animal can be weighed with great accuracy, this assumption would appear inconsequential in allometry. However, many researchers do not report the body mass of their experimental animals, and regressions must frequently use an average size estimated from other sources. Given the variation in individual size within species, this substitution will introduce some error. Even when the size of an experimental animal is reported, only the average mass from each species is used in regression, and this mean body mass has an associated uncertainty. If a regression is applied over a very broad size range, this source of error is probably negligible. If the range of animal sizes is small, the error associated with estimation of body mass may become significant. In such cases, the estimate of the slope ($b = \Sigma xy/\Sigma x^2$; Table 2.3) will be biased downward by the increased variation in X, which leads to relatively higher values of Σx^2.

Despite these problems, almost all authors have used least-squares regressions to fit allometric data, and the data are rarely examined with respect to the assumptions of the analysis. This does not imply that existing relations are wrong; the high values for F and r associated with almost all body size relations indicate that much of the variation is explained and that the assumptions are not seriously violated. However, because the assumptions of least-squares regression may not be met, the existing relations could probably be improved. If the major aim of science is improved prediction, the amelioration of existing allometric relations by more appropriate statistical treatment is one area in which measurable scientific advance is readily achieved.

3

Metabolism

This chapter begins the examination of the contemporary literature by looking at the best-known and best-established relationships, those describing metabolic rates as a function of body size. Metabolic rate is a major component of the balanced growth equation that forms an organizing theme for this and subsequent chapters. Unfortunately, the task of assembling a complete picture of the allometry of energy flux is confounded by a plethora of units and activity levels. This chapter, therefore, begins by introducing the balanced energy equation, standardizing the choice of units, and discussing the major types of activity levels used in studies of metabolic rate. The best-known equations describing the influence of size on basal and standard metabolism are then examined in detail both because of their intrinsic interest and because they serve as models for later discussion of other equations. Next, the effect of size on other more active metabolic rates is presented and finally some of the implications of the standard metabolic relations are examined.

The balanced growth equation

The balanced growth equation applies the first law of thermodynamics to living organisms. Since neither energy nor mass can be destroyed in biological processes, what goes into an animal must come out and

$$\text{ingestion} = \text{somatic or individual growth}$$
$$+ \text{reproductive growth}$$
$$+ \text{respiration} + \text{egestion} + \text{excretion}$$

The units of this equation are usually given as rates of carbon or energy flux, but with little modification all interactions that involve the transfer of energy or material between an organism and its environment can be represented.

Ingestion represents the demands an organism places on its prey. This might be expressed, for example, as the amount of food derived from various classes of potential prey or as a proportion of the prey population or even as the amount of a specific nutrient removed from the prey. The two growth terms reflect the amount of resource that an animal will eventually supply to its predators; they also set the ability of a population to withstand exploitation and to recover from depredation. Egestion and excretion are important in the efficiency of utilization of ingested matter, in the retention and accumulation of toxic materials, and in the regeneration of nutrients. Finally, the rate of respiration is the summed power expenditure for all metabolic processes. It is an inclusive estimate of the energetic demands that an organism places on its own internal reserves and ultimately on its environment. Because of these broad implications, the balanced growth equation provides a basic framework for this book, which largely explores and expands the terms of the equation through allometric relations.

Respiration

The energy of respiration is a unique element in the balanced growth equation, because it is entirely a loss term. Egested matter and energy are loss terms for the initial consumer, but they are utilizable by other organisms. Waste products like urine, sweat, and sloughed skin also pass back to the environment. Even the energy and matter associated with growth are returned at death. The matter associated with respiration is recycled as CO_2 and water, but the energy of respiration is released as heat, which is eventually lost to space. The energy of respiration is a one-way outlet for organic energy reserves. An estimate of an animal's respiration rate, therefore, provides a measure of the energetic cost to the ecosystem of supporting that animal.

For the individual, respiration loses this unique character, since any outward flux of energy or material represents a loss. Nevertheless, the energy of respiration is of great interest in bioenergetic autecology, because it represents the energy an animal must expend to meet all the exigencies of life. With a few small exceptions, like hair, eggshells, and excreta, respiration represents the rate at which an animal uses its resources to meet the demands placed on it by the environment. Conversely, the demands that the environment places on an organism can be measured as the rate of respiration.

On a practical level, respiration is also a key term because it has been measured more thoroughly and with greater precision than any other process. Consequently, we can build better theories describing the effect of size, and other factors, on respiration than on ingestion, growth, or defecation.

The choice of units

In constructing and comparing allometric relations for metabolic rate, all relations must be in the same units. This is largely a matter of bookkeeping, but the plethora of units in use (mg, ml, or litres for O_2; J, cal, or kcal for energy; μg, mg, g, or kg in fresh or dry mass for size; and s, min, h, or d for time) has rendered such comparisons a daunting prospect. In this book, size is almost always expressed in kilograms of live mass and respiration rate, as a measure of power (i.e., energy flux rate), in Watts ($1\,\text{Watt} = 1\,\text{J s}^{-1} = 0.24\,\text{cal s}^{-1} = 0.05\,\text{ml }O_2\,\text{s}^{-1}$). A table of conversions among the most common units is given in Appendix I.

Metabolic typology: Basal, standard, maximum, and average realized metabolic rates

An organism's power demands in nature are usually estimated by some minimal metabolic rate that may be thought of as the power required simply to maintain the machinery of life. This value is then increased by estimates of the power demands associated with each additional activity, such as locomotion, growth, or food processing (Kendeigh, Dol'nik, & Govrilov 1977; Moen 1973; Wunder 1975). This approach, rather than a direct estimate of energy consumption in nature, is followed, because it is far easier to measure the metabolic rates of animals under standard laboratory conditions and most of the literature dealing with metabolism was developed by physiologists who wished to compare rates for different animals under standard conditions. Regressions built from such data are less desirable for ecological applications, but no alternative is available.

One effect of this physiological approach is the availability of a series of equations relating respiration at defined levels of animal activity to animal size. Since any discussion of the allometry of respiration involves these equations, such discussions almost inevitably begin with a discussion of the types of metabolic rate and the terms used to identify them.

The most highly defined conditions for the measurement of any metabolic type are those considered necessary to measure the *basal metabolic rates* of mammals. These are near-minimal estimates of the respiration rate of awake adult individuals. Most mammals maintain a nearly constant internal temperature. When the ambient temperature is low, the rate of heat loss from a warm animal increases, and body temperatures would fall unless metabolic heating were increased. Similarly, when the environment becomes very warm, animals expend energy to cool themselves and the metabolic rate rises. In consequence below some *lower critical temperature,* respiration rate rises with declining temperature, and above some *upper critical temperature,* it increases with temperature. A region of minimal metabolic rate, the *thermal neutral zone,* lies between these two critical temperatures. Basal metabolic rate must be measured within this zone. The physical and chemical processing of food also increases metabolic rate. To minimize this effect, basal metabolic rate is measured in fasting or "postabsorbtive" animals. Metabolic rate is also increased in actively exercising animals, juveniles, pregnant or nursing females, and diseased or nervous individuals. Consequently basal metabolic rate is the rate of respiration of awake, inactive, unexcited, healthy, nonreproductive adult animals that are fasting and within their thermal neutral zone.

The conditions for the estimation of minimal power demands for other taxa have not been so closely defined, and the term *standard metabolic rate* usually replaces basal metabolic rate to indicate that the values are not necessarily minimal but that they were obtained under standardized conditions normally to yield low values. The degree of standardization varies considerably. Requirements for the measurement of avian rates are nearly as stringent as those for mammalian basal rates. For vertebrate poikilotherms, standard rates imply that the animal was awake but inactive and at a constant temperature. For unicells and many invertebrate poikilotherms, "standard conditions" imply little more than constant external temperature.

Maximum metabolic rates are obtained by first training animals to run on a treadmill, fly in a wind tunnel, or swim in a flume; the metabolic rate of these trained animals are then measured as they move at maximum speed. Two types of maximum metabolic rate can be distinguished. The sustained metabolic rate is that which can be maintained for some extended period of time, usually minutes or tens of minutes. In contrast, *burst metabolism* can only be maintained for seconds. The former term would be applied to the respiration rate of a middle- or long-distance runner and the latter term to that of a sprinter. The method can only be applied to animals that perform appropriately in the laboratory and is better suited to animals that readily work at high activity levels for extended periods. It is, therefore, easier to measure maximum metabolic rates of animals that habitually travel long distances in nature at moderate to high speeds, like salmon or African wild dogs, than those that concentrate their maximum effort in brief spurts, like cheetahs or darters. Most estimates of maximum metabolic rate, therefore, apply to the sustained rate of respiration.

An estimate of the minimal power demands sets a floor to the cost of existence. Animals may require more energy than this but not less. Maximum metabolic rate sets a ceiling to energy expenditure. The difference between this floor and ceiling is the *scope for activity* (Fry 1947) and can be used to estimate how much, or how little, additional power an animal has available to meet any additional costs. A free-living animal, therefore, expends energy at some rate between the basal and maximum rates. This is usually called the *daily energy expenditure* or the *average daily metabolic rate*. However, it would be confusing to refer to a rate expressed in Watts as a daily rate, and so, in this account, respiration rates of animals in nature will be called *the average realized rate of metabolism* or simply the *realized rate*.

Basal, standard, sustained maximum, burst, and realized rates are necessary terms in this account of the allometry of metabolism. A large number of other variously qualified metabolic rates have been identified in the literature. Some of these will be discussed in passing, but the five rates introduced here are already too many. Instead, we should seek continuous variables besides body mass that would allow prediction of an unqualified metabolic rate from some complex of independent variables, such as size, ambient temperature, velocity, reproductive rate, and nutritional status.

Standard metabolic rate

Standard metabolic rates provide an ideal introduction to scaling in the balanced growth equation. These equations have extremely strong empirical bases and are, therefore, among the best confirmed of body size relations. They are important as descriptors of minimal rates of energy flow for higher animals and as estimates of realized rates for others. In addition, these equations establish patterns and exemplify calculations used throughout this book. Only the most general relations can be treated within the text but manipulations presented in this context can easily be transposed to more specific relations for metabolism (Appendix III) and often to other body size relations. All calculations developed in this book should be seen as models for some specific treatments of

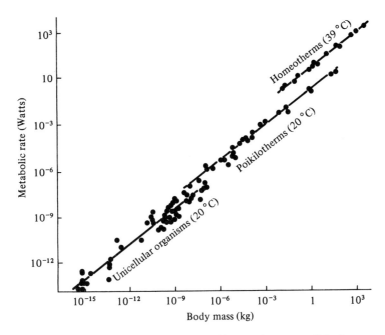

Figure 3.1. Standard metabolic rates of homeotherms, poikilotherms, and unicells presented as Equations 3.1–3.3. Modified from Hemmingsen (1960).

particular problems. Definitive solutions for all questions are not offered and should not be expected.

Hemmingsen (1960) found that the standard metabolic rate of homeotherms ($R_{s(h)}$, in Watts) could be predicted from fresh body mass (W, in kg) as

$$R_{s(h)} = 4.1\,W^{0.751} \tag{3.1}$$

(Appendix IIIa). The corresponding equations for the standard metabolic rates of poikilotherms ($R_{s(p)}$, in Watts) and unicellular organisms ($R_{s(u)}$, in Watts) are (Appendix IIIb)

$$R_{s(p)} = 0.14\,W^{0.751} \tag{3.2}$$

and

$$R_{s(u)} = 0.018\,W^{0.751} \tag{3.3}$$

where body size, W, is again in kilograms. These relations are presented graphically in Figure 3.1.

A number of points should be made with respect to Equations 3.1–3.3. First, the general form of the relations between metabolism and body mass is similar in all groups. This does not prove that only this formula may be fitted to the data, but it does show that the simple power relationship has been repeatedly successful. Differences among the elevations (a in Equation 1.1) indicate that, for a 1-kg organism, the metabolic rate of homeotherms is typically ($4.1/0.14 =$) twenty-nine times higher than that of a similarly sized poikilotherm, which would in turn be ($0.14/0.018 =$) eight times greater than that of a hypothetical 1-kg unicell. The constant slope (b in Equation 1.1) shows that these proportionalities are unaffected by size so a 1-ton homeotherm respires twenty-nine times faster than a 1-ton poikilotherm and 10-μg protozoan at one-eighth the rate of a 10-μg metazoan. Since the slope is positive, larger organisms within each metabolic class respire at a higher rate than do smaller members. Since the slope is less than unity, metabolic rate rises more slowly than body size. Finally the data are scattered around each regression; the equations should not be interpreted as deterministic laws giving the exact value of standard metabolism for any given animal but as very good statements of trend in the data and probabilistic estimates of metabolism for any given animal.

The equations show that homeotherms at rest draw more power than poikilotherms of similar size, which in turn place higher demands on their bodies and environment than do unicells. This may suggest that individual homeotherms must be more effective in resource utilization than other groups since they must extract far more energy from their environment to fuel respiration. Pough (1980) argues that these same differences show that poikilotherms and, by extension, unicells are better suited to low resource levels than homeotherms. Warm-blooded animals require high resource levels; if these were not available, homeotherms would rapidly exhaust their internal reserves just to supply their own standard metabolism. Poikilotherms and unicells require lower rates of resource supply, and, if required, they could resist relatively long periods of starvation. The large numbers of all three groups suggest that each strategy is successful.

Equations 3.1–3.3 are based on measurements taken under standardized conditions, and, strictly speaking, implications of these equations should be similarly restricted. On the other hand, allometric relations describing more active metabolic states are, qualitatively, not very different. Similar size effects and the same division of organisms into three metabolic classes are common characteristics of almost all descrip-

Table 3.1. *Calculations of "mass specific" or, simply, "specific" rates of respiration*

From Equation 3.1,

$$R_{s(h)} = 4.1W^{0.751}$$
$$\therefore \quad {}_wR_{s(h)} = R_{s(h)}/W$$
$$= 4.1W^{0.751}/W$$
$$= 4.1W^{0.751-1}$$
$$= 4.1W^{-0.249}$$

Specific metabolic rate declines with body size.

Note: These calculations are based on Hemmingsen's (1960) homeotherm relations; calculations for poikilotherms and unicells are homologous. $R_{s(h)}$, standard metabolic rate of homeotherms, in Watts; ${}_wR_{s(h)}$, specific metabolic rate of homeotherms in Watts kg^{-1}.

tions of metabolic rates. Some quantitative differences should be expected if different levels of metabolic activity are considered but even these differences are usually rather small. Calculations and conclusions based on standard rates may, therefore, serve as good models or paradigms for similar cases involving realized rates.

Equations 3.1–3.3 should further be considered in terms of specific power production. This term ($_wR_s$, in Watts kg^{-1}) is easily calculated by dividing both sides of each of the previous equations by body mass, W (Table 3.1):

$${}_wR_{s(h)} = 4.1 \ W^{-0.249} \tag{3.4}$$

$${}_wR_{s(p)} = 0.14 \ W^{-0.249} \tag{3.5}$$

$${}_wR_{s(u)} = 0.018 \ W^{-0.249} \tag{3.6}$$

These equations show more clearly (Figure 3.2) that the rate of energy expenditure per unit mass declines with increasing body size for the slope is negative. Within each metabolic group, the cost of maintaining a given biomass (i.e., the specific power production) of large animals is less than that required to maintain the same biomass of small ones. This argument leads to quantitative comparisons. For example, a half-ton moose degrades at least 440 J of chemical energy to heat every second; 1 half-ton of

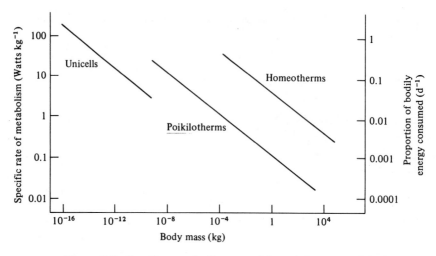

Figure 3.2. Specific metabolic rates of homeotherms, poikilotherms, and unicells (Equations 3.4–3.6). The proportion of the energy reserves used per day (right axis) was calculated as specific metabolic rate $(J s^{-1})$ times $86,400 s d^{-1}$ divided by $7 \times 10^6 J kg^{-1}$ fresh mass.

20-g mice would degrade at least 5,400 J in the same time (Table 3.2). Similar size effects occur in each of the other two groups.

A further manipulation of Hemmingsen's three equations allows a direct prediction of the maximum amount of living tissue supported per unit of energy flow ($_wR_s^{-1}$, in kg Watt^{-1}; W, body size, in kg). This is simply the inverse of Equations 3.4 to 3.6:

$$_wR_{s(h)}^{-1} = 0.24\,W^{0.249} \tag{3.7}$$

$$_wR_{s(p)}^{-1} = 7.1\,W^{0.249} \tag{3.8}$$

$$_wR_{s(u)}^{-1} = 56\,W^{0.249} \tag{3.9}$$

If an individual (or a population or an ecosystem) is to maintain a constant biomass, energy degradation through respiration must be balanced by the rate of energy supply. Thus, those equations that describe biomass per unit of heat production must also describe biomass per unit of energy supply. The same rate of energy supply could support about thirty times more poikilotherms than homeotherms because of the difference in metabolic rates. Unicells have a somewhat smaller advantage over poikilotherms. Within each metabolic class, a greater biomass of large organisms would be supported per unit of energy supply; the energy

Table 3.2. *Some calculations to exemplify the effect of size on metabolic rate*

I. Energy expenditure of half-ton moose ($W = 500\,\text{kg}$) (in Watts)

$$= 4.1W^{0.751}$$
$$= 4.1(500)^{0.751}$$
$$= 436$$

II. Energy expenditure of 500 kg of mice each weighing 0.02 kg assuming that Equation 3.4 holds

$$= \text{mass} \times {}_wR_{s(h)}$$
$$= 500 \times 4.1(0.02)^{-0.249}$$
$$= 500 \times 10.86$$
$$= 5{,}430$$

Small mammals, like mice, degrade ten times as much chemical energy to heat per unit time as does an equal mass of large mammals, like moose.

III. Mass of 0.02-kg mice, which produces 436 Watts (in kg)

$$= \text{power output/power per unit mass}$$
$$= 436/4.1(0.02)^{-0.249}$$
$$= 436/10.86$$
$$= 40.2$$

A given energy supply will support a much smaller biomass of mice than of moose.

Note: All calculations are based on Equation 3.1 for the standard metabolic rate of homeotherms (Hemmingsen 1960).

flow necessary to support one half-ton moose would only maintain about 40 kg of 20-g mice (Table 3.2).

If an estimate of the amount of stored energy in an animal's body were available, Equations 3.7 to 3.9 could be used to calculate turnover time of these stores. The energy content of animal tissue is approximately $22 \times 10^6\,\text{J}\,\text{kg}^{-1}$ dry mass (Cummins & Wuychek 1971) or about $7 \times 10^6\,\text{J}\,\text{kg}^{-1}$ fresh mass. When Equations 3.7–3.9 are multiplied by $7 \times 10^6\,\text{J}\,\text{kg}^{-1}$, their units become units of time ($\text{kg}\,\text{Watt}^{-1} \times \text{J}\,\text{kg}^{-1} = \text{s}$) and the equations give the time required to metabolize an amount of energy equal to the energetic content of the tissues: turnover time. For example, a 20-g homeotherm will release an amount equal to 100% of its body's energy in 7.3 d (Table 3.3). The turnover time of a 500-kg moose is 91 d. Because the energetic reserves of larger species within each metabolic group last longer, big animals are more resistant to food shortages than are small. Among animals of similar size, the shorter turnover times associated with homeothermy (Figure 3.3) lead to a dependence on stable food supply. Because the energy in poikilothermic tissues is mobilized

Table 3.3. *Calculated relations between metabolic power outputs and internal energy reserves in homeotherms, poikilotherms, and unicells*

I. Time required for a 0.02-kg mouse to release an amount of energy equal to its internal reserves

$$= E \times {}_wR^{-1}_{s(h)}$$
$$= 7 \times 10^6 \times 0.24(0.02)^{0.249}$$
$$= 6.3 \times 10^5\,s$$
$$= 7.3\,d$$

II. Time required for a 500-kg moose to metabolize the same amount of energy *per unit body mass*

$$= E \times {}_wR^{-1}_{s(h)}$$
$$= 7 \times 10^6 \times 0.24(500)^{0.249}$$
$$= 7.9 \times 10^6\,s$$
$$= 91\,d$$

Large mammals can survive much longer on their energy stores than can small ones.

III. Proportion of energy reserves used by a 0.020-kg poikilotherm in 7.3 d or $6.3 \times 10^5\,s$.

$$= time \times {}_wR_{s(p)}/E$$
$$= (6.3 \times 10^5) \times 0.14(0.02)^{-0.249}/7 \times 10^6$$
$$= 0.0334\,kg\,kg^{-1}$$
$$= 3.3\%$$

Poikilotherms use a smaller proportion of their energy stores per unit time than do homeotherms.

Table 3.3 (*cont.*)

IV. Mass of a unicell, which expends 100% of its energy reserves in 7.3 d, i.e., as fast as a 20-g homeotherm

$$_wR_{s(u)} = \text{specific energy expenditure/time}$$
$$= E/6.3 \times 10^5$$
$$= 11.1 \,\text{Watts kg}^{-1}$$
$$= 0.018W^{-0.249}$$
$$\therefore \quad W = (11.1/0.018)^{-1/0.249}$$
$$= 6.2 \times 10^{-12} \,\text{kg}$$
$$= 6.2 \,\text{ng}$$

Low metabolic rates permit unicells to exploit smaller sizes.

V. Energy expenditure of a hypothetical 6.2-ng mammal as a proportion of energy content

$$= _wR_{s(h)}/E$$
$$= 4.1(6.2 \times 10^{-12})^{-0.249}/7 \times 10^6$$
$$= 3.6 \times 10^{-4} \,\text{s}^{-1}$$
$$= 31 \,\text{d}^{-1}$$

Organisms of unicellular size that do not have low power demands would rapidly exhaust their energy reserves.

===

Note: $_wR_s^{-1}{}_{(h)}$, maximum mass supported per unit of metabolic power in kg Watt^{-1} (from Equation 3.7); $_wR_s$, specific metabolic rate in Watts kg^{-1} (from Equations 3.4 to 3.6); W, fresh body mass in kg; E, energy content of living tissue $= 7 \times 10^6 \,\text{J kg}^{-1}$.

much more slowly, cold-blooded animals resist starvation far better. In 7.3 d, a 20-g poikilotherm would expend less than 3.5% of its energy reserves (Table 3.3). This conservatism is increased in unicells and certainly contributes to their ability to be very small. According to Hemmingsen's unicell equation, a unicell that expends energy as fast as a 20-g homeotherm (in Watts kg^{-1})would weigh only 6.2 ng (Table 3.3). This is a relatively big unicell, but if it followed the homeotherm relation, instead of that for unicells, it would require thirty-one times its energy content per day (Table 3.3) just to fuel its metabolism. So active an algal cell would never survive the night.

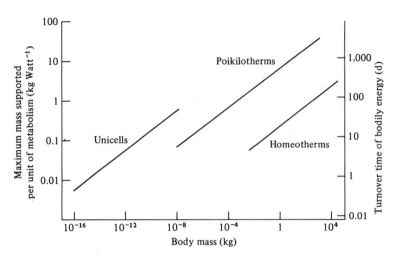

Figure 3.3. Biomass supported per unit of energy flow at standard metabolic rate for homeotherms, poikilotherms and unicells as calculated in Table 3.2 and shown as equations 3.7–3.9. Turnover time of the body's energy reserves (right axis) was calculated as biomass per unit power times 7×10^6 $J\,kg^{-1}$.

These estimates are, of course, very crude. Since the energy content of animal tissue varies (Cummins & Wuychek 1971) among species and individuals, so will the length of time that a given mass can fuel metabolism. Alternate relations yield somewhat different estimates of standard metabolism; although several authors (Brody 1945; Kleiber 1961; Robinson, Peters, & Zimmerman 1983; Appendix IIIa) have confirmed Hemmingsen's relation for homeotherms, significantly different elevations have been obtained for both poikilotherms and unicells (Robinson, Peters, & Zimmerman 1983; Appendix IIIb). In any case, general curves are only approximations; more precise predictions may be possible if one applies more taxonomically restricted relations. Because these calculations are based on standard rates, they will underestimate the metabolic rates achieved by homeotherms and some poikilotherms in the field and overestimate turnover times. Finally, body size explains only part of the variation in metabolic rate of different organisms; the residual variation around all allometric relations shows that more precise predictions will require the use of other independent variables.

One would suspect that some of the residual variation around these allometric equations might be explained by considering the ecology of the animals as well as their taxonomy. Although such attempts are surprisingly rare, some trends have been identified. Among mammals, fossorial rodents seem to follow a relation with a slope less than 0.75 (McNab 1979) and the metabolic rates of rodents from arid habitats are 25 to 50% lower than those of other mammals of similar size (McNab 1974). Insectivorous bats and foliovores have lower metabolic rates than would be expected (McNab 1978a; 1980). A number of authors have noted seasonal changes in metabolic rate at a given temperature. Usually (Casey, Withers, & Casey 1979; Kendeigh, Dol'nik, & Govrilov 1977; Appendix IIIa); but not always (Weathers & Calcomise 1978), winter rates are higher than summer rates under otherwise identical conditions. Homeotherms also show a diel pattern in standard metabolic rate. Normally diurnal animals have higher rates in the day and normally nocturnal animals at night (Aschoff & Pohl 1970). Among poikilotherms, Mautz (1979) reported that the metabolic rate in a family of reclusive lizards was 30 to 50% lower than that of other lizards of similar size. Naturally, factors controlled in the measurement of basal metabolic rate (temperature, nutrition, reproduction, activity, health, and age) have ecological implications, but since they are standardized in experiments, they cannot explain the residual variation in the body size relations. The search for other ecological variables to explain this variation has not been pursued as vigorously as it might be. However, the successes described in this paragraph are quite limited, and the awkward mathematical treatments often employed do not present compelling examples.

Maximum metabolic rate

Hemmingsen (1960) compared maximum rates to standard rates for homeotherms and concluded that the maximum rate was rate about ten times standard. Other estimates (Appendix IIId) have confirmed this order-of-magnitude difference. The allometric slope (0.845) describing the relation of maximum metabolic rate to body mass for 55 species of wild and domestic mammals is significantly greater than 0.75 (Taylor et al. 1980; Appendix IIId). This suggests that the maximum rate may rise faster with size than basal rate and the scope for activity, therefore, is disproportionately large in bigger animals.

A similar situation may exist among nonflying poikilotherms (Appendix IIId). Hemmingsen's (1960) collection again suggested that the maximum rate of these organisms was about an order of magnitude above their standard rate, but the data were quite variable. The regression equation of Bennett and Dawson (1976) shows that the maximum metabolism of reptiles at 30°C is about five times the standard rate (Appendix IIIb) regardless of size. Wilson (1974) reports similar results. In contrast, Brett's (1965) intensive study of sockeye salmon showed an increase in scope with size: The ratio of maximum to standard metabolic rate increased from 4 among the smallest fish (3.4 g) to 16 among the largest ones (1.43 kg; Appendix IIId). The literature, therefore, suggests that, in general, the maximum metabolic rate is an order of magnitude greater than standard or basal rates, but there are conflicting reports as to whether this ratio increases or is unchanged as size increases.

Flying insects are a dramatic exception to this rule. Hemmingsen (1960) found that the maximum metabolic rate of insects in flight was similar to that which would be expected from a similarly sized homeotherm exercising at its maximum rate – an increase of 300 times above the standard metabolic rate of insects. Bartholomew and Casey (1978a) found a similar increase (150 times) in the flying metabolism of moths relative to their resting rate. The tenfold increase may be useful rule of thumb, but it does not apply in all cases.

A possible rationalization of these differences is offered in the work of Prothero (1979). He suggested that burst metabolism is less affected by metabolic grouping and one equation could fit estimates of maximum power output in mammals, birds, insects, fish, and plants. Regression analysis of his data (Appendix IIId) shows that burst metabolism is about ten times the basal metabolic rate of homeotherms. Because burst metabolism is a nearly instantaneous phenomenon, it reflects the intrinsic capacity of tissue to produce power; the broad base for Prothero's relation suggests that this capacity is unaffected by metabolic or taxonomic groupings.

Sustained metabolic rate depends on this intrinsic capacity but also on the ability of the respiratory and circulatory systems to supply substrate and oxygen for continuous, high rates of respiration and to remove the waste products of metabolism. The physiologies of homeotherms and flying insects allow maintenance of extremely high rates of respiration. Other organisms can maintain such rates only briefly – for fish, less than 15 s (Webb 1975). The sustainable maximum rate for such poikilotherms is, therefore, considerably less than burst metabolism.

In summary, the sustained metabolic rate of most animals is about ten times standard; flying insects, however, can maintain a rate 150 to 300 times the resting rate. Larger animals seem to sustain a disproportionately higher metabolic rate than smaller animals. Burst metabolism is a similar function of body size for all organisms, and these high rates can be maintained for an appreciable length of time by birds, mammals, and flying insects. Burst and sustained maximum metabolic rates are similar in those three groups.

Both sustained and burst metabolism can be used to define the scope for activity. If the activity is of very short duration, the upper limit to scope is set by burst metabolism. This large scope is reflected in the alacrity of a lizard pouncing on a fly or an aquarium fish dodging the net. Alternatively, if we wish an estimate of longer-term power output, such as the energy available for migration, or to provide a ceiling for the average realized rate of metabolism of free-living animals, we would use sustained rate as the appropriate estimate of maximum metabolic rate.

The average realized metabolic rate

The average realized rate of free-living homeotherms is moderately higher than the basal or standard rate (Appendix IIIe). King (1974) summarized published information on average realized rate for rodents and birds in two equations (Appendix IIIe). These show that average realized metabolic rate increases with body size and realized rate is between 2.6 and 3.5 times basal. Wunder (1975) calculated the average realized rate of mammals to be about 1.7 times basal by considering the added costs of locomotion and thermoregulation. Kendeigh, Dol'nik, and Govrilov (1977) used an analogous but more complete calculation to arrive at figures between 1.6 and two times basal rate for birds in the thermal neutral zone.

These estimates concur in suggesting that the average realized metabolic rate is usually far less than the maximum metabolic rate. Except for small animals at low temperatures, two to three times basal seems a reasonable approximation for homeotherms. A similar figure is often used for poikilotherms (Bennett & Nagy 1977; Ware 1975; Winberg 1960). This chapter uses standard rates in most calculations; analogous arguments based on realized metabolic rates estimated as twice basal or standard would yield essentially similar results.

Interpretations and implications

The earlier development of Equations 3.4 to 3.9 was largely a matter of rephrasing Equations 3.1, 3.2, and 3.3. Tests of these hypotheses involve little more than comparisons with similar data relating size to metabolism.

More ambitious extensions of these equations require additional information to make predictions in new areas. In one example, described below, the scaling of weight loss during fasting is calculated from standard metabolism and the average energy content of animal tissue. Such calculations are subject to many sources of error. Poor predictions may indicate that the energy source during fasting has an energetic content that differs from average, metabolism during fasting differs from standard, or weight loss is not a direct reflection of the consumed energy stores. Consequently, testing of these predictions tests several hypotheses simultaneously. As the potential for testing increases, the proposals should become more fertile and interesting to other scientists (Kuhn 1970). At present, some of these extensions can be provisionally tested against available information; others serve only heuristically by suggesting probable allometries for which data must still be collected.

Scaling other variables

Because metabolism is the summed energetic cost of an organism's biological processes, relations between size and respiration should be reflected in other biological functions. We might, therefore, expect that power functions are frequently successful in describing scaling, that the slopes of relations for rates would approximate $\frac{3}{4}$, those for specific rates would be $-\frac{1}{4}$, and those for periods or physiological times would be $\frac{1}{4}$. We would also expect that the elevations of many of these relations would depend on metabolic grouping: homeotherm, poikilotherm, or unicell. There is no necessity to these speculations. They would hold only if the many components of metabolism contribute a similar proportion of total metabolic work in animals of very different taxon and size. In other words, if organisms are fundamentally similar, size-related changes in most biochemical and physiological processes should parallel the scaling of metabolism.

This presumed parallelism is sometimes termed the *principle of similitude* (Thompson 1961). Much of this book is a confirmation of that principle, and I will repeatedly refer to expected slopes of $\frac{3}{4}$, $\frac{1}{4}$, and $-\frac{1}{4}$ (as opposed to measured values, which will be presented as decimal fractions), when I wish to compare observation to expectation. Like all rules, the principle of similitude has exceptions but the impatient reader may assure himself of the generality beyond the exceptions by scanning the allometric relations listed at the end of this book.

Fasting

The simplest extension of equations for metabolic rate is in the calculation of weight loss and survival time in unfed animals. Since the energy of
animal tissue is relatively constant, power outputs in fasting animals
correspond to rates of weight loss. According to *Chossat's rule* (Kleiber
1961), animals die when total weight loss approaches half of the initial
body mass. Hence, survival times can be calculated from metabolic rate
and body size. Two aspects of fasting will be considered here: overnight
fat loss in birds and survival times during starvation.

Overnight weight loss. Sleeping birds fuel their metabolism by mobilizing
a portion of their fat reserves. The amount of fat burned can be estimated
from the nighttime resting rate of metabolism (Aschoff & Pohl 1970),
which is slightly lower than the daytime rate. Since fat contains about
4×10^7 J kg^{-1} dry mass (Kleiber 1961), the hourly rate of fat loss for a
22-g bird can be calculated as about 0.14% (Table 3.4). Doln'ick (cited in
Calder 1974) measured the overnight fat loss from six species of finch
($\bar{W} = 22$ g) to be 0.15% (range, 0.05 to 0.21%).

Starvation. Starving birds cannot depend solely on stored fat but must
consume other tissues as well. Assuming that the tissues so utilized
contain 7×10^6 J kg^{-1} wet mass, that starving birds respire at the rate
predicted by Zar's (1969; Appendix IIIa) equation for standard metabolism, and that animals die of starvation when their body mass is reduced
by half, the survival time of birds in the thermal neutral zone can be
calculated as the ratio of one-half the body's energy content to the rate of
energy consumption (Table 3.4). Similar calculations using the rate of
energy consumption at 0°C (Kendeigh, Dol'nik, & Govrilov 1977;
Appendix IIIc) estimate the survival time below the thermal neutral zone
(Table 3.4). Since small birds are disproportionately stressed at low
temperatures, their survival is clearly shortened by lower temperatures
(Figure 3.4). The calculated relations compare favorably both quantitatively and qualitatively with two regression lines (Calder 1974; Appendix
IVa) relating survival time to body mass for birds exposed at different
temperatures (Figure 3.4).
    The lower metabolic rate of poikilotherms should allow those animals
to survive much longer without food than homeotherms of similar size.

Table 3.4. *Calculations of the effects of fasting on birds and poikilotherms*

I. Fat loss in a 0.022-kg bird as a proportion of body mass per unit time: Resting metabolic rate, $R_{r(b)} = 5.56W^{0.726}$ (Aschoff & Pohl 1970; Appendix IIIa) in Watts; W, body mass in kg; and E_f, energy content of fat $= 4 \times 10^7 \, J \, kg^{-1}$

$$
\begin{aligned}
\text{Fat loss} &= R_{r(b)}/(E_f \times W) \\
&= 5.56(0.022)^{0.726}/(4 \times 10^7 \times 0.022) \\
&= 3.96 \times 10^{-7} \, \text{Watt} \, J^{-1} \\
&= 1.432 \times 10^{-3} \, J \, J^{-1} h^{-1} \\
&= 0.14 \, \% \, h^{-1}
\end{aligned}
$$

Overnight, birds burn 0.14% of their body mass hourly.

II. Survival time of birds in the thermal neutral zone assuming Chossat's rule that a 50% loss of initial mass is fatal (Kleiber 1961), that the metabolic rate $R_{s(b)} = 3.76W^{0.739}$ in Watts (Zar 1969; Appendix IIIa), and that the energy content of tissue (E) is $7 \times 10^6 \, J \, kg^{-1}$. Survival time at thermal neutrality

$$
\begin{aligned}
&= 0.5 \times E \times W/R_{s(b)} \\
&= 0.5 \times 7 \times 10^6 \times W/3.76W^{0.739} \\
&= 931,000W^{0.261} \, s \\
&= 10.6W^{0.261} \, d
\end{aligned}
$$

Larger birds take longer to starve than small birds.

III. Survival time of passerines at 0°C given that $R_{s(b)} = 5.88W^{0.531}$ (Kendeigh, Dol'nik, & Govrilov 1977; Appendix IIIc)

$$
\begin{aligned}
&= 0.5 \times E \times W/R_{s(b)} \\
&= 0.5 \times 7 \times 10^6 \times W/5.88W^{0.531} \\
&= 595,000W^{0.47} \, s \\
&= 6.9W^{0.47} \, d
\end{aligned}
$$

Smaller birds are more adversely affected by low temperatures than large.

IV. Survival time of poikilotherms at 20°C given that $R_{s(p)} = 0.14W^{0.751}$ (Hemmingsen 1960; Appendix IIIb)

$$
\begin{aligned}
&= 0.5 \times E \times W/0.14W^{0.751} \\
&= 2.5 \times 10^7 W^{0.249} \, s \\
&= 289W^{0.249} \, d
\end{aligned}
$$

Poikilotherms are more resistant to starvation than homeotherms.

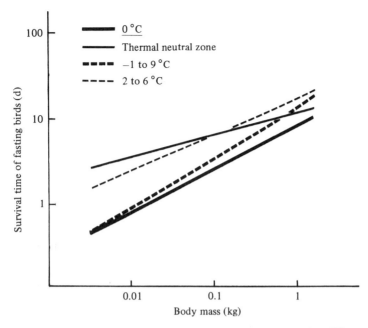

Figure 3.4. Survival times (Y, in d) for fasting birds under different ambient temperature regimes as a function of body mass (W, in kg). Dashed lines indicate regressions to observed data (Calder 1974; Appendix Vc) and solid lines represent relations calculated in Table 3.4. The equations are, observed: -1 to $-9°C$, $Y = 12.4W^{0.58}$ and 2 to 4°C, $Y = 15W^{0.58}$; calculated: thermal neutral zone, $Y = 10.6W^{0.26}$ and 0°C, $Y = 6.9W^{0.47}$. Both sets of relations are quite comparable.

Their calculated survival time (Table 3.4), based on Hemmingsen's (1960) relation for poikilotherms (Equation 3.2; Appendix IIIb), approximates that observed experimentally (Threlkeld 1976) in small planktonic crustaceans (Figure 3.5). The increased resistance of larger animals in this and similar experiments may have important consequences. In experimental studies of zooplankton competition, low food availability typically leads to the disappearance of smaller species and juvenile animals (Goulden & Hornig 1980). It is entirely plausible that similar effects apply in natural populations of zooplankton and of other animals.

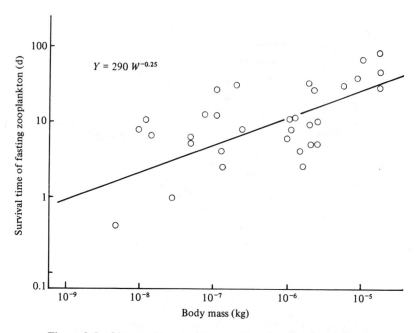

Figure 3.5. Observed survival times of starving planktonic crustaceans compared to predicted survival times predicted on the basis of allometric relations for metabolic rate (Table 3.4). Data from Threlkeld (1976).

4

Physiological correlates of size

Introduction

The most general allometric equations describe metabolic rate as a linear function of body mass raised to an exponent of approximately $\frac{3}{4}$ (Hemmingsen 1960; Appendix III). This regularity holds a special fascination for biologists, because its wide applicability suggests that this may be a rare example of a general biological law (Wilkie 1977). Before this claim is accepted, it should be examined as closely as possible.

One could seek confirmation by examining those allometric equations that relate metabolic rate to body size in particular taxa. The equations assembled in Appendix III suggest that the generality holds at these more restricted levels. The mode of the frequency distribution of the slopes (Figure 4.1) lies between 0.725 and 0.750; the median of the distribution is 0.735 and the mean is 0.738 ($SD = 0.11$; $N = 146$). There is no great advantage in promoting 0.74 over 0.75, for the two values are not significantly different and the latter is widely accepted and slightly easier to compute (Kleiber 1961). Figure 4.1, therefore, confirms the $\frac{3}{4}$ law or *Kleiber's rule* as a valid statistical generalization. This does not imply that $\frac{3}{4}$ is the "true" value of the slope for all equations, only that $\frac{3}{4}$ is a reasonable approximation.

Figure 4.1 may overestimate the dispersion in the slopes, because the slope in body size relations is not completely independent of the intercept. This is most easily demonstrated when the intercept lies to one side of the data. For example, since all homeotherms weigh more than 1 g, the intercept always lies below the data swarm when mass is expressed in grams. As Figure 4.2 shows, this results in an inverse relationship between slope and intercept: High slopes are associated with low intercepts and vice versa. If the data swarm instead lay to the right of the intercept (for example, if mass were expressed in tons), steep slopes would be associated with high intercepts, and low slopes with low ones. When mass is expressed in kilograms, no trend is seen. Because slopes and intercepts covary, they should not be considered in

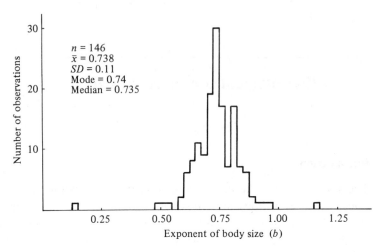

Figure 4.1. Frequency distribution of the exponents of body mass in allometric relations for metabolic rate (Appendix III). Values derived from homeotherms outside of their thermal neutral zones have been omitted.

isolation. Comparisons among allometric regressions are best made by graphing the curves.

Further confirmation of the $\frac{3}{4}$ law could be sought by examining the physiological components of metabolism, especially gas exchange. Since metabolic heat production is closely tied to oxygen consumption, one would expect to find that the total effect of the processes underlying oxygen transport would vary with metabolism and, therefore, as $W^{3/4}$. Moreover, from the principle of similitude, one might suppose that other physiological processes may vary as $W^{3/4}$. This may seem a digression from our primary goal of applying allometric relations to ecology, but no review of body size would be complete if it ignored the morphological, physiological, and biochemical correlates of size. In addition, examination of these relations provides strong support for the $\frac{3}{4}$ law and, therefore, circumstantial evidence for application of the principle of similitude to other, more ecological, relations.

In overview, this chapter presents a mammalian model for the scaling of physiological and morphological traits to body size. The model suggests that gross physiological rates of mass flow, like air, blood, and urine fluxes, vary linearly with metabolic rate and, therefore, as $W^{3/4}$. Physiological volumes, like lung or blood volume, and the masses of many organs vary directly with body mass. Physiological times, like the time between heartbeats or breaths, vary as $W^{1/4}$ and their inverses,

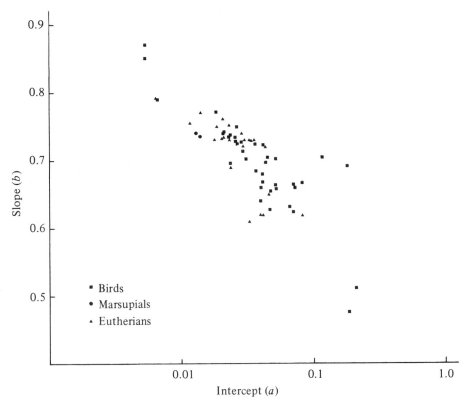

Figure 4.2. The inverse relationship between slope (*b*) and elevation (*a*) in allometric relations between basal or standard metabolic rate and body mass (in g) for birds and mammals. So strong a trend is only apparent when body mass is expressed in grams since the intercept then lies far from the mean body size.

physiological frequencies, as $W^{-1/4}$. The same exponents, $\frac{1}{4}$ and $-\frac{1}{4}$, apply to turnover times and their inverses, turnover rates, of various processes. These are, therefore, close analogs to the metabolic relations (Equations 3.1, 3.4, and 3.7) discussed in the preceding chapter. Finer-scale physiological and biochemical processes also vary with size, but the exponents of these relations frequently depart from the values expected on the basis of similitude (1, $\frac{3}{4}$, $\frac{1}{4}$, and $-\frac{1}{4}$). These exceptions might suggest that adherence to the $\frac{3}{4}$ law is imposed by pressures external to the organism and that similarities in whole organism function can be achieved by somewhat different processes at the physiological and biochemical levels.

Bias toward mammalian relations was dictated by the available relationships, for mammals are certainly the most studied group. Presumably, analogous relationships exist for other taxa and these will be formalized with further research. Regressions for other aspects of the morphology and physiology of mammals and other taxa are given in Appendix IV.

Mammalian models of respiratory and circulatory physiology

In very coarse terms, the role of the respiratory and circulatory systems in gas exchange is the reduction of the diffusion distance between the sites of oxidative metabolism and an oxygen-rich, carbon-dioxide-poor environment. Each breath flushes fresh air over the alveolar surfaces in the lung. Oxygen in this air diffuses into the blood, where it combines with hemoglobin, while carbon dioxide in the blood diffuses into the lung whence it is expelled on exhalation. The right ventricle of the heart drives the blood through the lung and back to the heart; the left ventricle then pumps the blood to the body. There, carbon dioxide from the respiring tissues enters the capillaries, while oxygen dissociates from the blood pigment and diffuses into the cells. Oxygen then passes into the mitochondria, the sites of oxidative metabolism, where it receives electrons from the cytochromes of the electron transport system. This reduced oxygen eventually combines with hydrogen ions to form water.

Respiratory gas exchange

The first step in this process is the exchange of air in the lungs. The rate of this exchange depends on the volume of gas moved with each breath and on the number of breaths taken per unit time. Both *tidal volume,* the volume of air exchanged at each breath, and *vital capacity,* the maximum volume that can be inhaled in one breath, increase almost directly with body mass (Adolph 1949; Guyton 1947a; Stahl 1967; Appendix IVa). The exponents of allometric relations describing these volumes range from 1.0 to 1.03. Since lung volumes also bear a nearly constant proportionality to body mass – reported exponents (Stahl 1962; Tenney and Remmers 1963; Weibel 1973; Appendix IVa) vary from 0.99 to 1.05 – the proportion of the lungs' contents exchanged at each breath is virtually independent of size (Table 4.1). In contrast, respiratory frequency declines as $W^{-0.25}$ to $W^{-0.28}$ (Adolph 1949; Guyton 1947b; Stahl 1967; Appendix IVa) and its inverse, breath time, rises as

Table 4.1. *Aspects of the scaling of pulmonary air flow to body size*

Given

$$\text{Lung volume} = 5.35 \times 10^{-5} W^{1.06}\,\text{m}^3$$
$$\text{Tidal volume} = 7.69 \times 10^{-6} W^{1.04}\,\text{m}^3$$
$$\text{Respiratory frequency} = 0.89\, W^{-0.26}\,\text{s}^{-1}$$

then

I. Rate of air flow

  = tidal volume × respiratory frequency
  $= 7.69 \times 10^{-6} W^{1.04} \times 0.89 W^{-0.26}$
  $= 6.86 \times 10^{-6} W^{0.78}$

  Air flow is approximately proportional to metabolic rate.

II. Breath time

  $= (\text{respiratory frequency})^{-1}$
  $= (0.89 W^{-0.26})^{-1}$
  $= 1.12 W^{0.26}$

III. Turnover time of pulmonary air

  = lung volume/air flow
  $= (5.35 \times 10^{-5} W^{1.06})/(6.86 \times 10^{-6} W^{0.78})$
  $= 7.8 W^{0.28}$

  Biological times increase as the fourth root of body mass.

IV. Turnover rate

  $= (\text{turnover time})^{-1}$
  $= 1/(7.8 W^{0.28})$
  $= 0.128 W^{-0.28}$

  Specific rates decline as the inverse of the fourth root of body mass.

V. Turnover per breath

  = tidal volume/lung volume
  $= (7.69 \times 10^{-6} W^{1.04})/(5.35 \times 10^{-5} W^{1.06})$
  $= 0.15 W^{-0.02}$
  $= 0.15$ fraction/breath (no units)

  Dimensionless ratios are often biological constants.

Note: Body size (W) is in kg.
Source: Empirical relations are from Stahl (1967).

$W^{0.25}$ to $W^{0.28}$ (Table 4.1). If respiratory frequency falls as about $W^{-1/4}$ and tidal volume rises as $W^{1.0}$, the rate of air flow should increase as $W^{3/4}$ (Table 4.1). Empirical estimates of the slopes of air flow–body size relations support this calculation for they range from 0.74 to 0.80 (Adolph 1949; Guyton 1947b; Stahl 1967; Appendix IVa). Turnover rate, the fraction of the lungs' volume cleared per unit time, therefore, declines approximately as $W^{-1/4}$ and turnover time, the period required before a volume equal to lung volume has been exhaled, rises as $W^{1/4}$ (Table 4.1). These components of air flow in the lungs parallel metabolic rates, energy turnover time, and specific metabolic rates and, therefore, support the $\frac{3}{4}$ law.

From the lung, oxygen diffuses into the blood. Diffusion follows a modification of Fick's law, which states that the net rate of diffusion (F) depends on the area through which diffusion occurs (A), the thickness of the diffusion barrier through which diffusion must occur (Δz), the difference in partial pressure of the diffusing substance across the barrier (Δp), and on some constant (K), which depends on the nature of the diffusing material and that of the diffusion barrier (Dejours 1975). This law can be stated as

$$F = AK \, \Delta p / \Delta z \qquad\qquad (4.1)$$

Since respiration involves diffusion from lung to blood, the total effect of these terms on pulmonary gas exchange must be to vary F as $W^{3/4}$. For biological systems, the value of K, *Krogh's constant*, should be independent of body size and, therefore, K ought not scale as $W^{3/4}$. In mammals, all oxygen diffusion occurs through the alveolar surface of the lung; since the area available for diffusion varies nearly directly with body mass (Gehr et al. 1981; Tenney & Remmers 1963; Weibel 1973; Appendix IVa), this term too cannot lead to the appropriate scaling of gas flow. The thickness of the respiratory epithelium, the blood–air barrier, increases slightly (i.e., as $W^{0.05}$) with size (Weibel 1973; Appendix IVa). If the change in gas partial pressure across this barrier were a constant, the increase in barrier thickness with size should result in a slight decline in oxygen flux into the blood of larger mammals per unit area of alveolar surface or per unit mass. The measured *diffusion capacity* of the mammalian lung, the flux of oxygen across the lung when the change in oxygen partial pressure is 1 Pa, has been reported (Gehr et al. 1981; Weibel 1973; Appendix IVa) to rise almost directly with W. Stahl (1967) gives a higher value (1.14) for the exponent in this relation but no reported value would, by itself, generate an oxygen flux

Table 4.2. *Calculation to show the probable change in oxygen partial pressure* (P_{O_2}, *in Pa) between the lung and blood of mammals with body size* W *(in kg)*

From equation 4.1,

$$P_{O_2} = F_{O_2} z_1 A_a^{-1} K^{-1}$$

F_{O_2} = basal metabolic rate/oxycalorific coefficient

$= 4.1 W^{0.751}/20.1 \times 10^6$

$= 0.204 \times 10^{-6} W^{0.751}$

From Appendix IVa,

$$A_a = 3.31 W^{0.98}$$

$$z_1 = 1.41 \times 10^{-6} W^{0.05}$$

$$\therefore \quad P_{O_2} = (0.204 \times 10^{-6} W^{0.751})(1.41 \times 10^{-6} W^{0.05})/$$
$$(3.31 W^{0.98})(2.4 \times 10^{-16})$$

$$= 360 W^{-0.179}$$

The difference in oxygen partial pressure between the lung and the blood is reduced in larger mammals.

Note: This assumes that net oxygen diffusion (F_{O_2}, in $m^3 s^{-1}$) follows Equation 4.1 and equals basal metabolic rate (Hemmingsen 1960). A_a is the alveolar surface area in m^2, z_1 is the thickness of the blood–air barrier in m (Weibel 1973), K is Krogh's constant here set to $2.4 \times 10^{-16} m^2 s^{-1} Pa^{-1}$ (Dejours 1975), and the oxycalorific coefficient to convert power expenditure to oxygen consumption is $20.1 \times 10^6 J m^{-3} O_2$.

proportional to $W^{3/4}$. This proportionality could only be maintained if the air–blood difference in oxygen partial pressure declines in larger mammals. The oxygen carrying capacity of the blood may be independent of mammalian size (Prothero 1980), although opinions differ on this point (Porer & Metcalfe 1967). If the blood O_2 concentration were independent of size, any allometric change in the concentration gradient would reflect changes in the effective oxygen concentration of the lung. Although I know of no empirical support for this, a probable relation can easily be calculated (Table 4.2). Gehr et al. (1981) give a similar argument to suggest that the partial pressure of oxygen in the lung must decline with size. Perhaps this putative decline in partial pressure is related to the longer residence time of each breath in the lungs of larger species.

Circulation

Allometric relations governing the flow of blood (Appendix IV) are often analogous to those describing air flow in the lung. The volumes involved, total blood volume (Brody 1945; Prothero 1980; Stahl 1962, 1967), heart volume (Stahl 1962), and volume expressed per stroke (Günther 1975; Holt, Rhode, & Kines 1968), are directly proportional to body mass, since the exponents in such relations range from 0.98 to 1.05. Hemoglobin mass (Adolph 1949), red cell volume, and plasma volume (Prothero 1980) also vary directly with size. The time between each stroke rises approximately as $W^{1/4}$ and cardiac frequency consequently declines as $W^{-1/4}$ (Adolph 1943; Brody 1945; Holt, Rhode, & Kines 1968; Stahl 1967). The combination of increasing stroke volume and decreasing cardiac frequency should result in the scaling of blood flow and cardiac output to $W^{3/4}$. Observation supports this conclusion (Holt, Rhode, & Kines 1968; Stahl 1962, 1967).

Ultimately oxygen is discharged from the blood and diffuses to the tissue, where it is used to fire metabolism. The reduction of oxygen in metabolism is confined to the mitochondria and the concentration of these organelles in isolated tissues falls as $W^{-0.06}$ to $W^{-0.3}$ (Mathieu et al. 1981; Munro 1969) and, so, roughly parallels the decline in specific metabolic rate with size. Oxygen reduction depends on the presence of cytochromes and cytochrome oxidase to facilitate the passage of electrons from the fuel to the oxidant. The concentration of these biochemicals is also nearly proportional to metabolic rate (Adolph 1949; Munro 1969; Appendix IVb).

Minor allometric trends

Allometries that are associated with subcomponents of physiological and biochemical function are sometimes not directly proportional to either size or metabolic rate. Many fine-scale mechanisms – diffusion distance, oxygen affinity of hemoglobin, certain enzyme activities, and blood sugar concentrations – change with size (Appendix IVb). The magnitude of these changes is usually smaller than changes in blood flow or respiration rate, but the changes are statistically significant and usually promote higher levels of activity in smaller mammals.

For example, there is good evidence that diffusion distances increase with size. Gehr et al. (1981) and Weibel (1973) have reported that the blood–air barrier is slightly thicker in larger mammals. Two lines of evidence suggest that the diffusion distance between blood and the sites of oxygen reduction also increases with size. The capillary density in the musculature of large mammals is less than that of small species (Mathieu et al. 1981; Schmidt–Neilsen & Pennycuick 1961; Appendix IVb). An increase in capillary density implies a reduction in the average distance between the blood and mitochondria in the respiring tissues. This effect may be enhanced by a reduction in average cell diameter among smaller animals (Maldonado et al. 1974; Rensch 1960; Sealander 1964; Appendix IVb). Reduced cell diameter may decrease diffusion distance from the cell surface to the respiration sites and increase the ratio of cell surface area to cell volume. Both effects should increase rates of diffusion in the tissues of smaller mammals.

The overall significance of such minor processes is both statistically and physiologically uncertain. When exponents of mass are calculated, they are usually very small, between -0.1 and 0.1, and variation among the data is frequently high. It is possible that the effects of different processes tend to cancel one another, and, in any case, the net effect of the entire complex of minor factors cannot be estimated. Nevertheless, they are quite consistent in suggesting that respiration is faster in smaller species.

There seems little need for such additional processes at basal metabolic rate, for small and large mammals probably perform equally well at that level. At higher levels of activity, rather subtle differences may become important (Schmidt–Nielsen 1979). Since small animals habitually respire at rates well above basal (French et al. 1976), they may have adopted structural and physiological mechanisms that support these higher metabolisms. Larger mammals may depend instead on "behavioral" adjustments, like increased respiratory and cardiac flows, to meet their rarer needs for increased metabolic power. Baudinette (1978) found disproportionate increases in maximum cardiac frequency among larger mammals.

In summary, the mammalian model of gas exchange suggests that flux rates are proportional to metabolic rate. This is achieved by increasing physiological volumes in direct proportion to body mass and by decreasing turnover rates and physiological frequencies as $W^{-1/4}$. Superimposed on this rather strict pattern is a tendency for smaller species to develop ancillary mechanisms to promote gas exchange, which may be important in the maintenance of elevated metabolic rates over extended periods.

5

Temperature and metabolic rate

For most animals, body size is that universal characteristic that is most easily measured. As a result, many researchers include estimates of size in descriptions of their experimental animal, and empirical theories can be built from published data. Other variables, for example, protein content or metabolic rate, may be equally universal and, perhaps, even better predictors. However, because their determination is more difficult, they are far less frequently reported. Body size is also attractive because, as a continuous variable, it may be more easily treated mathematically and the power curve's repeated success makes the researcher's choice of statistical models clear. In part, body size is a good independent variable, because it is practical and convenient.

Only body temperature is as practical and convenient an independent variable as body size. Every organism has some body temperature, and, as a rule, this is easily measured. Long experience has shown that body temperature influences physiological rates, and temperature is, therefore, reported frequently. Like body mass, temperature is a continuous variable and is, therefore, amenable to regression analysis. Although no one mathematical function dominates the description of thermal response, the choice of statistical models is small (Bottrell 1975; McLaren 1963), and so statistical effort is reduced. Unlike body mass, there is a strong physicochemical rationale for the effect of temperature on chemical and biochemical rates. Consequently, we have perhaps more reason to suppose that body temperature will have a significant effect on biological rates than we have in the case of body size.

This chapter examines the joint effect of temperature and body size on metabolic rate. I hope it will serve as a model for the inclusion of temperature, and other variables, in multiple-regression models to explain the substantial residual variation in all allometric plots. At the very least, this chapter will show that our predictions may be significantly improved if both temperature and size are considered when treating rates of respiration. The chapter will also show that more factors besides these two are important, since the residual variation is still considerable.

54

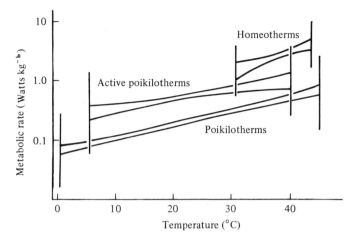

Figure 5.1. The effect of temperature on metabolic rate of homeotherms, poikilotherms, and active reptiles and fish. The two solid curves for each class represent 95% confidence limits around the regression lines (Appendix IIIa,b); vertical lines indicate the 95% confidence limits for individual data points. Metabolic rate has been normalized for the effects of size through division by W^b, where W is body size in kilograms, and b is the exponent of mass in the multiple regression of respiration rate on body mass and temperature.

Temperature, size, and metabolism: A regression model

For over a decade, modern statistical packages have rendered multiple regression as simple as univariate analysis. Nevertheless, few authors have attempted general multiple regressions of metabolic rate on both body mass and temperature. Robinson, Peters, and Zimmerman (1983) analyzed the metabolic rate of poikilotherms, homeotherms, and unicells in an attempt to refine Hemmingsen's equations. This regression model assumed a simple exponential rise in rate with temperature

$$\log R = \log a + b \log W + ct \quad \text{or} \quad R = aW^b e^{cT} \tag{5.1}$$

The results of these analyses were somewhat different from expected (Figure 5.1; Appendixes IIIa,b). Both fish and reptiles moving at an unstipulated level and inactive poikilotherms showed temperature responses ($c = 0.036$ and 0.051 which correspond to $Q_{10} = 1.4$ and 1.7, respectively) that were less pronounced than expected, and there was no significant effect of temperature on the metabolic rate of unicells. The flat temperature responses in these analyses may reflect a general acclimation of the organisms in the data set to the experimental temperatures. The more commonly cited, steeper responses represented by Q_{10}'s of 2 to 3 may represent acute responses (Ivlea 1980) or experimental artifacts (Holeton 1974) of less ecological relevance. Flat thermal response curves also imply that temperature does not remove a large amount of the residual variation, and, in fact, the increases in explained variance in our analyses were not large. For poikilotherms, only a further 25% of the scatter remaining after regression on size alone was explained by reference to body temperature.

In contrast to our analyses of unicells and poikilotherms, the effect of body temperature on the metabolic rate of homeotherms was as high as would be expected ($c = 0.087$; $Q_{10} = 2.4$). Kayser and Heusner (1964) reported a similar value for the Q_{10} of hibernating mammals. Other authors have reported similarly steep responses for both crustaceans (Ivlea 1980) and reptiles (Bennett & Dawson 1976).

In the absence of experimental data, generalized temperature relations like those in Figure 5.1 can be used to estimate the responses of individual organisms and species. These relationships can be substituted wherever the more familiar equations describing metabolic rate as a function of size have been applied.

The estimation of body temperature

The multiple regressions developed by Robinson, Peters, and Zimmerman (1983; Appendix IIIa,b) can only be applied if the body temperature is known. For most animals, this can be approximated as ambient temperature. Most animals are small: They heat and cool rapidly so that their body temperature differs little from that of their immediate surroundings. This is also true of quite large aquatic and soil poikilotherms, because the high conductivities of these media quickly dissipate any temperature difference between animals and their environment (Stevens & Fry 1974; Spigarelli, Thommes, & Beitinger 1977). Body temperatures of terrestrial animals weighing less than 0.1 g and most aquatic poikilotherms differ little from their immediate environments.

Larger terrestrial poikilotherms often maintain some temperature differential between their bodies and surroundings. The size of this differential depends on size (see "Poikilothermic heating and cooling" below) but also on a combination of behavioral and physiological traits and appropriate environmental opportunities for exogenous heating or

Table 5.1. *Normal and upper lethal body
temperatures* (T_b), *in nontorpid homeotherms*

Taxon	Normal T_b (°C)	Lethal T_b (°C)
Passerines	39–44	47
Nonpasserines	39–41	46
Eutherian mammals	36–39	42–44
Marsupials	35–36	40–41
Monotremes	30–31	37

Sources: Schmidt–Nielsen (1979), Hensel, Brück, and Raths (1973).

cooling (Jankowsky 1973). The body temperatures of *behavioral thermo-regulators* should be measured whenever a general relation describing the effect of temperature is applied.

Among homeotherms, body temperature is relatively constant and, if measurement is not possible, an estimate may be had by reference to taxonomic averages (Table 5.1). Rodbard (1950) proposed that the body temperatures of birds decline with size, whereas those of mammals rise to a maximum at about 1 kg and then decline with further increase in size. Neither Morrison and Ryser (1952) nor McNab (1970) were able to confirm this mammalian result. McNab (1966b) also analyzed the relation between avian body mass and temperature, confirming and quantifying Rodbard's earlier conclusions (Appendix Vb).

The effects of ambient temperature

All animals lose heat to a cooler environment and gain heat from a warmer one. This exchange can have important metabolic consequences. If a poikilotherm loses or gains heat, its body temperature will change and its metabolic rate will follow. Equations describing metabolism as a function of body temperature still apply but body temperature must be corrected for the effects of any change in ambient temperature. If a homeotherm loses or gains heat, it must increase its metabolic rate to heat or to cool its body. Outside the thermal neutral zone, relations describing homeothermic metabolism as a function of body size and temperature are inappropriate unless corrections for the increased cost of thermoregulation are applied.

The expression of heat flux

The rate of heat gain or loss is usually expressed as the *rate constant of temperature change* (the fractional change in the temperature differential between body and environment, per unit time) or as the *half-time of temperature change* (the time required to halve the initial temperature difference) or as *conductance* (the rate of heat flux per unit of temperature difference).

The determination of a rate constant of temperature change (λ) assumes that body temperature (T_b) of a poikilotherm is a simple function of time (t), initial body temperature (T_0), and ambient temperature (T_a):

$$T_b - T_a = (T_0 - T_a) \, e^{-\lambda t} \qquad (5.2)$$

This equation describes a simple exponential decline in temperature differential ($T_b - T_a$) from a high initial value ($T_0 - T_a$) toward zero. It implies that the instantaneous change in body temperature is a constant fraction of the temperature differential

$$dT_b / (T_b - T_a) dt = \lambda \qquad (5.3)$$

Since λ is a constant, the rate of temperature change is great when $T_b - T_a$ is large and less when the temperature differential declines. Some authors prefer to deal with a half-time that varies as the inverse of the rate constant (Table 5.2) and is calculated under the same assumptions of exponential decline in temperature differential. Heat loss may also be presented as conductance, a measure of heat flow, rather than temperature change. Since temperature changes because of heat flux, heat flux is also an exponentially declining function of time and so conductance can easily be determined from a rate constant or half-time of temperature change (Table 5.2).

Strictly speaking, the term *conductance* is reserved for heat flux per unit area per degree of temperature change and is expressed in the units Watts per square meter per degree centigrade. Although animal heat flux is sometimes expressed per unit area, these units are inconvenient because animal surface area can only be measured approximately and biologists are more interested in heat flux per unit mass or per animal. A more biologically useful concept is *specific conductance,* which refers to heat flux per unit of animal mass per unit of temperature gradient and has the units Watts per kilogram per degree centigrade. This is the most common expression of biological conductance. In this book, however, biological conductance is used primarily to correct total metabolic rate of homeotherms for the effect of low temperatures. It is, therefore, more convenient to express conductance as heat flux per animal per unit of temperature differential. This expression of conductance also removes the statistical danger of autocorrelation in plots of mass-specific rates against mass. Readers familiar with the expression of biological conductance as a specific rate should bear in mind that, in this account, the units of biological conductance are Watts per degree centigrade.

For poikilotherms, the expression of heat flux and temperature change is largely a matter of convenience since all terms are readily interconverted (Table 5.2). Half-time is the least useful of the alternatives, because it can only be used for comparisons, whereas the other two, rate constant and conductance, are as effective in comparison and more effective in calculation. Rate constant is most convenient when calculating changes in body temperature of poikilotherms and conductance when calculating increased metabolism (heat output) from homeotherms. Homeothermic heat loss should not be converted to a rate constant of temperature change, because their body temperature does not change. In this account, rate constant is used to treat poikilotherms and conductance in all calculations and comparisons involving homeotherms.

Poikilothermic heating and cooling

As one would expect, small animals cool and heat faster than large animals exposed to a similar temperature gradient (Appendix Va). Available relations describe the rate constant of cooling for three quite different groups of poikilotherms: fish (Spigarelli et al. 1977; Stevens & Fry 1970, 1974), reptiles (Bartholomew & Lasiewski 1965; Bartholomew & Tucker 1964; Ellis & Ross 1978; Spray & May 1977), and some larger insects (Bartholomew & Epting 1975; May 1976; Pyke 1980). In all cases, the rate constants of cooling are described as power functions of mass (Figure 5.2) with similar negative slopes but different elevations. At a given size, fish cool faster than reptiles and reptiles faster than insects.

Thermal equilibration. One way of visualizing these relations is to determine the time required for equilibration of body temperature with new ambient conditions. Figure 5.3 and Table 5.3 show that equilibration time increases with animal size and initial temperature differential. Although the trends with size are similar, there are marked absolute differences

Table 5.2. *Interconversion of the rate constant of change of body temperature (λ, in s^{-1}), the half-time of temperature change ($t_{1/2}$, in s), and conductance (c, in Watts $°C^{-1}$)*

I. Half-time and rate constant
   Given (Equation 5.2)

$$T_b - T_a = (T_0 - T_a)e^{-\lambda t}$$

By definition, when $t = t_{1/2}$,

$$(T_b - T_a)/(T_0 - T_a) = 0.5$$

From Equation 5.2,

$$e^{-\lambda t} = (T_b - T_a)/(T_0 - T_a) = 0.5$$
$$\therefore \quad e^{-\lambda t_{1/2}} = 0.5$$
$$\therefore \quad t_{1/2} = -\ln 0.5/\lambda$$
$$= 0.69/\lambda$$
$$\therefore \quad \lambda = 0.69/t_{1/2} \qquad \text{or} \qquad t_{1/2} = 0.69/\lambda$$

II. Conductance and rate constant
   By definition,

$$\text{Conductance } (c) = dQ/dt(T_b - T_a)$$
$$\text{Total heat capacity of body} = S_q W$$

By definition,

$$dT_b = dQ/(S_q W)$$
$$\therefore \quad c/(S_q W) = dQ/(S_q W)\, dt\, (T_b - T_a)$$
$$= dT_b/dt\, (T_b - T_a)$$

but from Equation 5.3,

$$\lambda = dT_b/dt\, (T_b - T_a)$$
$$\therefore \quad \lambda = c/(S_q W)$$
$$= c/(3{,}550\ W)$$
$$= 0.000282\ c/W$$

or

$$c = 3{,}550\lambda\ W$$

Note: T_0, T_b, and T_a are, respectively, initial body temperature, body temperature at time t, and ambient temperature. Q is the amount of heat in an animal's body (in J), W is body mass (in kg), and S_q is the specific heat capacity of animal tissue ($3{,}350\,\text{J kg}^{-1}\ °C^{-1}$; if an animal weighing 1 kg lost 3,350 J its body temperature would fall 1°C).

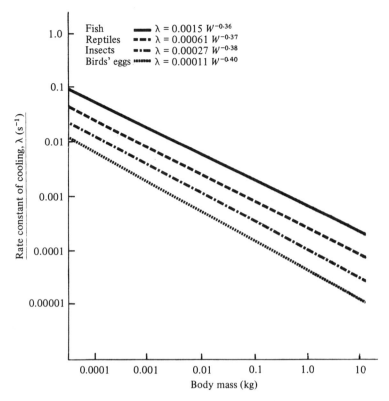

Figure 5.2. The effect of body mass on the rate constants of cooling of fish (Spigarelli, Thommes, & Beitinger 1977), reptiles (Bartholomew & Tucker 1964), insects based on cooling of temperate-zone dragonflies at 25°C (May 1976), and birds' eggs (Kendeigh, Dol'nik, & Govrilov 1977).

among taxa. For example, a 10-g insect equilibrates as slowly as a 100-g reptile or a 1-kg fish, but within each class, a 100-fold increase in body size results in about a 10-fold increase in equilibration time.

This increase in thermal inertia with size has several ecological implications. Larger insects and lizards may take advantage of their long equilibration times by heating their bodies in a warm microhabitat and then foraging in a cool one. Provided that the length of each foraging bout was less than equilibration time, the warmer predator would have some advantage in performance over its cooler prey. The converse also holds; species that forage in excessively hot environments will heat up and must return to cool burrows or shade to lose their excess heat (Schmidt–Nielsen 1964). Small species will heat and cool rapidly so the foraging and

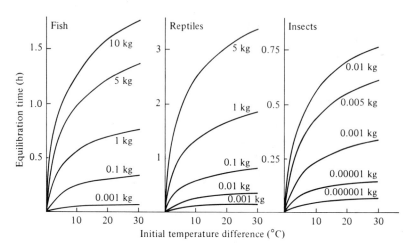

Figure 5.3. The effect of body size and initial temperature difference between body and environment on the equilibration time of lizards, fishes, and insects. These figures are solutions to the equations for equilibration time in Table 5.3 given the equations for the rate constant of cooling in Figure 5.2.

Table 5.3. *Calculation of equilibration time* (t_e) *as a function of body mass* (W) *and the initial temperature difference between the body and environment* ($T_0 - T_a = \Delta T_0$)

From equation 5.2,

$$\Delta T_t = \Delta T_0 \, e^{-\lambda t}$$
$$\therefore \quad e^{-\lambda t} = \Delta T_t / \Delta T_0$$
$$\therefore \quad -\lambda t = \ln(\Delta T_t / \Delta T_0)$$
$$\therefore \quad t = -[\ln(\Delta T_t / \Delta T_0)]/\lambda$$

When $t = t_e$, $\Delta T_t = 0.5$

$$\therefore \quad t_e = -[\ln(0.5/\Delta T_0)]/\lambda$$
$$= [-\ln 0.5 - \ln(1/\Delta T_0)]/\lambda$$
$$= [0.69 + \ln \Delta T_0]/\lambda$$
$$\because \quad \lambda = aW^{-b}$$
$$\therefore \quad t_e = (0.69 + \ln \Delta T_0)/(aW^{-b})$$
$$= W^b \, (0.69 + \ln \Delta T_0)/a$$

Equilibration time rises with body size and with initial temperature difference.

Note: In this calculation, the time to thermal equilibrium is approximated as the time required to reduce the temperature gradient between the body and environment (ΔT_t) to 0.5°C. The rate constant of cooling, λ, is expressed as s^{-1}.

Table 5.4. *The calculation of body temperature* (T_b) *of a poikilotherm of body mass* W *as a function of time and size following a change in ambient temperature from* T_0 *to* T_a

From Equation 5.2,

$$(T_b - T_a) = (T_0 - T_a) \, e^{-\lambda t}$$
$$\therefore \quad T_b = T_a + (T_0 - T_a) \, e^{-\lambda t}$$

but

$$\lambda = aW^{-b}$$
$$\therefore \quad T_b = T_a + (T_0 - T_a)\exp(-aW^{-b}t)$$

The body temperature of large animals changes more slowly than that of small animals experiencing the same thermal gradient.

Note: λ is the rate constant of heating or cooling (Appendix Va).

cooling bouts will be brief; larger species require more time, and very large species may be able to wait until cooler nights. Very large terrestrial poikilotherms cool so slowly that their body temperatures may never equilibrate with a fluctuating external temperature. Spotila et al. (1973) developed this idea to show that dinosaurs living in a mild temperate climate may have achieved functional homeothermy for a large part of the year. Short equilibration times for smaller species suggests that they must experience large and rapid changes in body temperature in association with temporal and spatial variation in ambient temperature. In a given environment, large species should enjoy greater thermal independence, and small species should be more thermally tolerant. A case in point is the rise in minimum flight temperature with body size in dragonflies (May 1976). Small species have less opportunity to heat to high body temperatures and are able to perform over a greater range of temperature.

The calculation of equilibration time gives only a partial picture of poikilothermic cooling, because body temperature changes over the equilibration period and large temperature differences occur only at the beginning of equilibration. The effects of size and taxon on the rate constant can also be demonstrated by calculating body temperature (T_b) as a function of time (t) and size (W). This can be calculated (Table 5.4) by simply rearranging Equation 5.2:

$$T_b = T_a + (T_0 - T_a) \exp(- taW^{-b}) \tag{5.4}$$

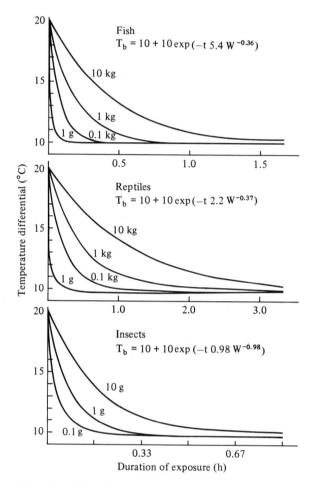

Figure 5.4. The effect of exposure time (t in s) and body size (W, in kg) on the body temperatures (T_b) of fish, reptiles, and insects that experience a change in ambient temperature from 20 to 10°C. Body temperature was calculated according to specific solutions of the relation determined in Table 5.4 in which $T_0 = 20$°C; $T_a = 10$°C; and the constants a and b taken from the appropriate equations in Figure 5.2.

Solutions to this equation are shown in Figure 5.4. In each case, large animals cool more slowly than small ones, and insects more slowly than reptiles or fishes of similar size.

Endothermic heating in poikilotherms. The values in Figures 5.3 and 5.4 are uncorrected for any gain of heat from metabolism. For most poikilotherms, such a correction is unnecessary because metabolic heating is only a small fraction of heat loss under any substantial temperature gradient. An estimate of this fraction can be made easily by comparing allometric relations for metabolic rate with that for total heat loss calculated from conductance (Table 5.5). Solution to these equations shows that metabolic heating would offset an increased proportion of heat loss in large animals but that, under a temperature gradient of 5°C or more, metabolism is unlikely to replace more than 5% of heat loss.

This is not so for flying insects. If conductance is unchanged in flight, their metabolic rates should often exceed heat loss (Figure 5.5), leading to a rise in body temperature. This will increase until the thermal gradient between the body and the environment is so steep that metabolic rate is balanced by heat conductance. In flying insects, the thermal gradient necessary for temperature equilibrium may be calculated (Table 5.5) from allometric relations for insect conductance and an estimate of their metabolic rate in flight. This calculation shows that flying insects maintain a temperature 8 to 20°C above ambient. In warm weather, such animals would be in danger of overheating, and conductance is probably increased to avoid this problem.

Poikilothermic heating. This discussion of poikilothermic heat flux has centered on cooling because more information is available about that process. Heat gain is an equally important aspect of an animal's existence and can be treated in a manner analogous to heat loss.

When the environmental temperature is constant, the heating of a cooler animal is an exponential function of time and is again described by Equation 5.2. The rate constant of heating, like that of cooling, is a power function of size. The slopes of these relations are similar to those describing the rate constant of cooling and therefore indicate, as expected, that large animals warm more slowly than small (Appendix Va). For fish and aquatic reptiles, the rate constant of heating is often greater than that of cooling; these poikilotherms appear to heat faster than they cool, and this difference in rate constant is accentuated with size (Spray & May 1977; Spigarelli, Thommes, & Beitinger 1977). Comparisons involving terrestrial reptiles have shown no such trends, and the rate constants of heating and cooling are similar functions of body mass (Spray & May 1977).

Heat flux in homeotherms

Homeotherms, like poikilotherms, lose heat to a cold environment and gain heat from a warm one. Homeotherms differ from poikilotherms

Table 5.5. *The expression of metabolic rate as a proportion (P) of total heat flux from different poikilotherms as a function of body size and temperature gradient*

I. For inactive poikilotherms at 20°C, standard metabolism, $R_{s(p)}$, in Watts, can be estimated (Robinson, Peters, & Zimmerman, 1983) as

$$R_{s(p)} = 0.19W^{0.76}$$

From Figure 5.2 (Spigarelli, Thommes, & Beitinger 1977),
$$\lambda_f = 0.0015W^{-0.36}$$

From Table 5.2,
$$c_f = 3{,}350\ \lambda_f W$$
$$= 5.33W^{0.64}$$

$$\because \quad \text{heat loss} = \Delta T c_f$$

$$\therefore \quad P_f = \text{metabolic rate/heat loss}$$
$$= R_{s(p)}/c_f \Delta T$$
$$= 0.19W^{0.76}/\Delta T (5.33W^{0.64})$$
$$= 0.036W^{0.12}/\Delta T$$

Similarly for reptiles,
$$c_r = 2.2W^{0.63}$$
$$\therefore \quad P_r = 0.086W^{0.13}/\Delta T$$

and for insects,
$$c_i = 0.95W^{0.62}$$
$$\therefore \quad P_i = 0.20W^{0.14}/\Delta T$$

Metabolism can replace a significant proportion of heat loss from poikilotherms only if heat loss is small. This occurs only if the temperature differential is small or if the animals are enormous.

II. For actively flying insects metabolic rate may be approximated as the maximum metabolic rate, R_{max} (Watts), reported by Prothero (1979)

$$R_{if} = R_{max} = 38.7W^{0.763}$$

assuming c_i does not change with level of activity

$$c_{if} = c_i = 0.95\ W^{0.62}$$
$$P_{if} = R_{if}/\Delta T c_{if}$$
$$= 40.4W^{0.14}/\Delta T$$

$$\therefore \quad \text{when } P_{if} = 1, \qquad \Delta T = 40.4W^{0.14}$$

Metabolic heating in flying insects is sufficiently great that quite large temperature differentials can be maintained. Insects are "warm-blooded" when in flight.

Note: In these equations, λ represents the rate constant of cooling (s^{-1}), c the conductance (Watts °C^{-1}), W is animal size (kg), and ΔT is the thermal gradient between body and environment. Subscripts f, r, i, and if represent fishes, reptiles, insects, and insects in flight, respectively.

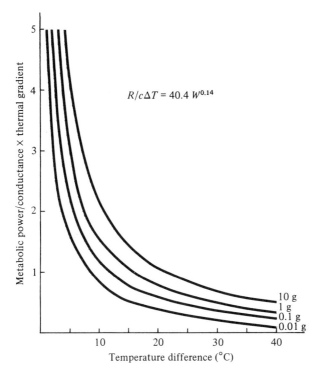

Figure 5.5. The metabolic rate of flying insects, expressed as a proportion of heat flux from the body, as a function of body mass and the air–body gradient in temperature (Table 5.5). If metabolic rate exceeds heat loss, body temperature will rise until the temperature gradient is so large that the ratio of metabolic rate to heat loss is unity. At this body temperature, flying insects are in thermal equilibrium.

because they use their higher metabolic capacities to offset this heat flux and stabilize internal temperature. A large part of the description of a homeotherm's response to external temperature is, therefore, a description of changes in its metabolic rate. Although this response appears more complex than that of poikilotherms, homeotherms have been studied more, and we can provide a nearly complete, if basic, description of the combined effects of size and ambient temperature on the metabolic rate of homeotherms.

The ten components of a homeotherm's response to changes in ambient temperature are

    1. The lower lethal temperature below which metabolic heating is insufficient to replace heat loss and body temperature falls

2. The lower critical temperature below which metabolic heat production rises to offset heat loss and stabilize body temperature
3. The upper critical temperature above which metabolic rate rises as the animal uses metabolic power for cooling, the thermal neutral zone lies between the lower and upper critical temperatures
4. The body temperature
5. The upper lethal temperature at which the animal dies
6. Summit metabolism, the maximum rate of metabolic heat production that can be induced by chilling the animal
7. The basal or standard metabolic rate, the rate of heat production in the thermal neutral zone
8. The maximum rate of metabolism that can be induced by high temperatures
9. Conductance, the rate of change in metabolic rate with change in ambient temperature between the lower critical and lower lethal temperatures
10. The coefficient of heat stress, the rate of change in metabolic rate with change in external temperatures lying between the upper critical and upper lethal temperatures

These parameters are shown graphically in Figure 5.6, and solutions for those aspects that have been related to size are given in Appendix V or, for metabolic rates, Appendixes IIIa and c.

Conductance. After basal and standard rates of metabolism (Chapter 3), the most studied of these components is conductance (c). In practice, this is determined as the slope of the line relating metabolic rate to ambient temperature below the thermal neutral zone. It, therefore, has the units Watts per degree centigrade and is a measure of heat loss per unit of thermal gradient. In animals, the thermal gradient is the difference between the temperature of the body (T_b) and that of the environment (T_a). The total heat flux (R, in Watts) can then be calculated as

$$R = c(T_b - T_a) \qquad \text{if} \quad T_a < T_{lc} \tag{5.5}$$

where T_{lc} is the lower critical temperature. Homeothermic heat production exactly balances heat loss and so Equation 5.5 gives both the rate of heat loss and the metabolic rate between the lower lethal and lower critical temperatures.

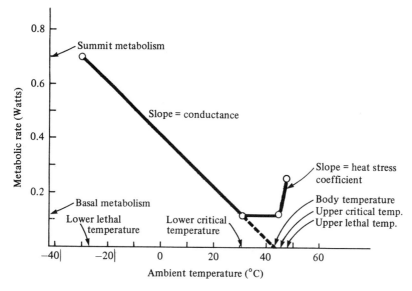

Figure 5.6. A schematic description of the effects of external temperature on homeothermic metabolic rate. The various components that must be predicted to describe this relationship are shown in the figure and defined in the text. This figure should describe the response of a nonpasserine bird weighing 0.01 kg.

Although conductance is measured from changes in body temperature among poikilotherms and from changes in metabolic rate among homeotherms, it is a measure of heat flux in both cases. For all studied groups, conductance is a power function of body size (Figure 5.7; Appendix Va). The slopes of these relations lie between 0.45 and 0.65, which suggests that under similar temperature gradients the rate of heat loss rises more slowly with body size than does basal metabolic rate. Thus, the increase in metabolic rate required to offset heat loss induced by a given gradient is proportionately less for large birds and mammals than for small ones (Table 5.6). As a result, when environmental temperature is low, homeothermic metabolism rises more slowly with size than it does within the thermal neutral zone (Appendix IIIc). Low temperatures place greater relative demands on small homeotherms than on large ones.

Lower critical temperature. Lower critical temperatures (T_{lc}) can be calculated (Table 5.7) from conductance (c), basal rate (R_b), and body temperature (T_b).

$$T_{lc} = T_b - R_b/c \tag{5.6}$$

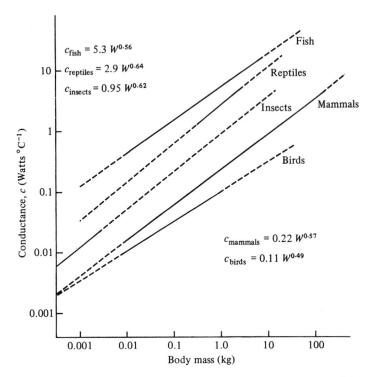

Figure 5.7. Conductance for fish (Spigarelli, Thommes, & Beitinger 1977), reptiles (Spray & May 1977), temperate dragonfiles (May 1976), birds (Calder & King 1974), and mammals (Bradley & Deavers 1980). Solid lines indicate the ranges of body size used to generate these relations.

Since body temperature is little affected by body mass, this implies that lower critical temperatures decline as size increases. Thus, large mammals and birds resist low temperatures better than small species not only because their increase in heat loss is small relative to basal rates but also because their metabolic rates do not rise until lower temperatures are experienced.

Equation 5.6 shows that body temperature, conductance, and basal metabolic rate are all interrelated components of thermoregulation. Animals that have high basal metabolic rates will, if conductance and body temperature are constant, have lower critical temperatures. If conductance or body temperature is decreased, lower critical temperature will fall. Since birds have higher body temperatures (Table 5.1), higher metabolic rates (Appendix IIIa), and lower conductances (Figure 5.7), their lower critical temperatures are reduced relative to

Table 5.6. *Low temperatures induce a greater proportional rise in metabolic rate* (R) *relative to basal metabolic rate* (R$_b$) *in smaller birds and mammals than in larger ones because conductance* (c) *and heat loss for a given thermal gradient* (ΔT) *between body* (T$_b$) *and ambient temperatures* (T$_a$) *rise more slowly with size than does basal metabolic rate*

I. For mammals

From Kleiber (1961),

$$R_{b(\text{mammals})} = 3.28W^{0.756}$$

From Bradley & Deavers (1980),

$$c_{\text{mammals}} = 0.224W^{0.574}$$

From Equation 5.5, at low temperature,

$$R = c\Delta T$$
$$\therefore \quad R/R_{b(\text{mammals})} = (0.224W^{0.574})\ \Delta T/(3.28W^{0.756})$$
$$= 0.068\Delta T W^{-0.182}$$

Low temperatures induce a greater increase in metabolism, relative to basal metabolism, among small mammals.

II. For birds

From Zar (1969)

$$R_{b(\text{birds})} = 3.76W^{0.739}$$

From Calder & King (1974),

$$c_{\text{birds}} = 0.11W^{0.49}$$
$$\therefore \quad R/R_b = c_{\text{birds}}\Delta T/R_{b(\text{birds})}$$
$$= 0.029\Delta T W^{-0.249}$$

Among both birds and mammals, metabolic costs increase relative to basal metabolism at lower temperatures but less so at greater body sizes.

Note: R and *R*$_b$ in Watts, *W* in kg.

those of mammals over most their common range in body mass (Figure 5.8). An increase in metabolic rate, through exercise, digestion, reproduction, and so forth, should also reduce an individual's lower critical temperature (Kleiber 1961).

Table 5.7. *Calculation of the lower critical (T_{lc}, °C) and lower lethal (T_{ll}, °C) temperatures of homeotherms from conductance (c, in Watts °C^{-1}), body temperature (T_b, in °C), and basal (R_b, Watts) or summit (R_{summit}, Watts) metabolism*

From Equation 5.5,

$$\text{heat loss} = \text{metabolic rate } (R) = c(T_b - T_a)$$

I. Lower critical temperature

$$R_b = c\,(T_b - T_a) \quad \text{when } T_a = T_{lc}$$
$$\therefore\ T_b - T_{lc} = R_b/c$$
$$\therefore\ T_{lc} = T_b - R_b/c$$

For mammals, from Table 5.6,

$$R_{b(\text{mammals})} = 3.28W^{0.756}$$
$$c_{\text{mammals}} = 0.224W^{0.574}$$
$$\therefore\ T_{lc} = T_b - (3.28W^{0.756})/(0.224W^{0.574})$$
$$= T_b - 14.6W^{0.182}$$

From Table 5.1, T_b is independent of size and, consequently, T_{lc} declines with size. Small homeotherms increase metabolism to offset heat loss at higher temperatures than do large ones.

II. Lower lethal temperatures

At $T_a = T_{ll}$, $R = R_{summit}$
$$= c(T_b - T_{ll})$$
$$\therefore\ T_{ll} = T_b - R_{summit}/c$$

In mammals, from Hensel et al. (1973),

$$R_{summit} = 5\,R_b$$
$$= 16.4W^{0.756}$$
$$\therefore\ T_{ll} = T_b - (16.4W^{0.756})/(0.224W^{0.574})$$
$$= T_b - 73.2W^{0.18}$$

Lower lethal temperatures decline with size.

Note: W is body mass (kg), and T_a is ambient temperature.

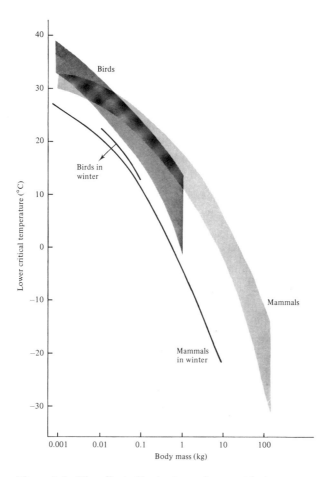

Figure 5.8. The effect of body size on lower critical temperature of birds and mammals. The two envelopes include all values calculated from equations in Appendix Vb assuming an average body temperature of 41°C for passerines, 39°C for nonpasserines, and 38°C for mammals. Lower critical temperature is reduced in winter in both birds and mammals.

Both seasonal and diurnal cycles affect heat loss from homeotherms. In winter, birds (Kendeigh, Dol'nik, & Govrilov 1977) and mammals (Casey, Withers, & Casey 1979) reduce both conductance and lower critical temperature (Figure 5.8), presumably by increasing insulation. During periods of rest, the conductance of homeotherms typically declines (Aschoff 1981), but since the standard metabolic rates (Aschoff & Pohl 1970) and body temperatures also show diel fluctuations, the net effect on lower critical temperature is not clear.

Summit metabolism and lower lethal temperature. Summit metabolism is usually measured by chilling the animals in an ice water bath or by some equally stressing method. This may put valuable experimental animals at risk, and so summit metabolism is normally reported only for the more common, and usually smaller, domestic species. In general, summit metabolism is about five times basal metabolic rate (Dawson & Dawson 1982; Hensel, Brück, & Raths 1973) and, hence, rises as $W^{0.75}$. Summit metabolism can be used to calculate lower lethal temperatures (Table 5.7), and such a calculation suggests that lower lethal temperatures decline in parallel to the lower critical temperature as body size increases.

Body temperature. If the high metabolic rates of homeotherms were primarily an adaptation to reduce temperature change in cooler environments, then heat production would continue to fall as temperatures rose above the lower critical temperature. When ambient temperature equaled body temperature, heat flux would cease, and metabolism would be zero. Of course, this does not occur, because homeotherms do not reduce their metabolic rates below basal, indicating that their high basal rates have functions other than warming the body. Mammals and birds must instead increase conductance in the thermal neutral zone to avoid overheating

If heat production and conductance are known at some ambient temperature below the thermal neutral zone, Equation 5.5 can be further rearranged to determine body temperature; this is shown graphically (Figure 5.6) as the extrapolation of the lower limb of the thermal response curve to intersect the *x* axis at the point where ambient temperature equals body temperature and any metabolic heating requirement is zero. This corresponds closely to measured body temperatures in mammals but is less effective in treating birds (Schmidt-Nielsen 1979).

Coefficient of heat stress. The upper limb of the thermal response curve has been far less extensively studied. It is a narrow region, and accurate measurements of the coefficient of heat stress are difficult to make. All temperatures in the upper limb are close to lethal, and, in nature, few animals will voluntarily remain above their upper critical temperatures for long periods. Thus, studies above the thermal neutral zone have less applicability and have been less actively pursued.

The slope of the upper limb of the thermal response curve has been termed the *coefficient of heat stress* (Weathers 1981). Some confusion may arise because, although this term is calculated analogously to conductance and has the same units (Watts $°C^{-1}$), it is not a measure of heat flux per unit of temperature gradient and is, therefore, not a measure of conductance. The increase in metabolic rate above the upper critical temperature is only an incidental reflection of increased heat gain from a warmer environment. Most of the increased heat load for homeotherms at high temperatures derives not from the environment but from their own bodies. Thus, the increased power demands at high temperatures represent the cost of increased blood flow and evaporation necessary to transport deep body heat to the body surface for dissipation into the environment. At high temperatures, such dissipation becomes increasingly difficult and more metabolic work is required. The metabolic rate (R) at temperatures between the upper critical (T_{uc}) and upper lethal (T_{ul}) can be calculated from the coefficient of heat stress (h_s) and basal metabolic rate (R_b) as

$$R = R_b + h_s(T_a - T_{uc}) \qquad \text{if} \quad T_{uc} < T_a < T_{ul} \qquad (5.7)$$

The only available relation (Weathers 1981) that describes the heat stress coefficient as a function of size applies to birds. This shows (Appendix Va) that the additional power demands associated with cooling at high temperatures rise with size but more slowly than does basal metabolic rate. Calder and King (1974) earlier suggested a similar trend, and, of course, this mimics the slow rise in conductance with size. Larger animals are, therefore, less stressed by extreme temperatures, both high and low, than small animals.

Upper critical temperature. The lower critical temperature can be calculated from basal metabolic rate, conductance, and body temperature (Equation 5.6), but no analog exists for upper critical temperature. Weathers (1981) collected data from 6 species of birds that suggest a slight tendancy for upper critical temperature to increase with body mass; S. Chin (unpublished) found data for 10 species that suggested a decline in upper critical temperature with size (Appendix Vb). Neither study is based on enough points to provide a definitive statement of any trend between size and upper critical temperature.

Upper lethal temperature. Finally, a complete description of thermal response requires an estimate of the upper lethal temperature. Lethal body temperatures are about 6°C higher than normal body temperature (Schmidt–Nielsen 1979), but most homeotherms cannot survive for long at such extremes. A more likely approximation of the upper lethal temperature of the environment may be about 3°C above body temperature (S. Pyne unpublished).

With this last approximation, we can estimate the metabolic response for any nontorpid homeotherm. Figure 5.9 compares the observed thermal response of two animals with the predicted responses. The comparisons are generally close, but some departures are apparent and not unexpected.

The set of equations in Appendix V suggest that larger animals will be less affected by extreme temperatures than small species. Big animals have larger thermal neutral zones and may meet the demands of heating or cooling with relatively moderate increases in metabolic output. Small species are far less able to withstand thermal extremes and must seek shelter from both cold and heat. Fortunately, small size permits the utilization of a number of milder microclimates – under the snow, in burrows, in shade, and under rocks – that are not open to larger species.

Other factors and other processes

The ecological importance of almost all relations presented in this chapter might be questioned because they are necessarily determined under artificial laboratory conditions. In nature, a number of other physiological, behavioral, physical, and climatic factors will also affect thermal relations. The relevance of such factors could be determined by examining the applicability of this generation of relations derived in the laboratory to organisms in the field. Shortcomings might then indicate which further variables should be included in multiple regressions to produce better predictions of heat exchange in nature. Moen (1973) provides one example of such extensions in his work with white-tailed deer.

One of the dangers of any thematic investigation in biology, like this book, is that the theme so restricts one's viewpoint that other factors are forgotten. At least mild tunnel vision is needed to finish any book, and it is not within my scope to review the nonallometric intricacies of heat exchange (or any other characteristic that forms a topic for discussion in this book). A number of excellent texts (Gates 1980; Gates & Schmerl 1975; Precht, et al. 1973; Prosser 1973; Schmidt–Nielsen 1979) are already available for that purpose. Such omissions and simplifications,

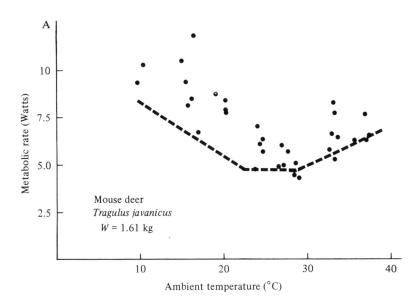

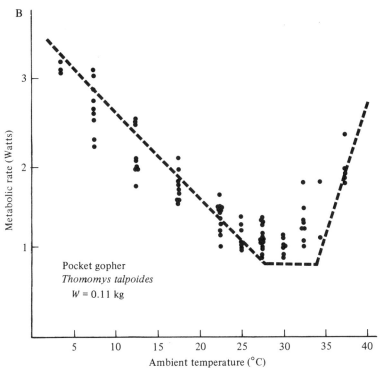

Figure 5.9. A comparison of the predicted and observed thermal response curves for two homeotherms. The solid lines represent the responses calculated from Appendixes IIIa and Va (Bradley & Deavers 1980; Hensel, Brück & Raths 1973; Kleiber 1961) and Equations 5.5, 5.6, and 5.7.

both conscious and not, will gall some readers. Hopefully, these short-comings will stimulate the development of better descriptions. The present predictions are, however, the best available.

This chapter has dwelt on the interaction between ambient temperature and metabolism, because this has been better studied than other temperature responses. Temperature also influences ingestion, production (Ames 1980; Hensel, Brück, & Raths 1973; Kendeigh, Dol'nik, & Govrilov 1977), locomotion (Bennett 1980), excretion (Peters & Rigler 1973), and so forth but general relations describing these effects are developed only infrequently. Temperature will, therefore, assume a minor role in the remainder of this book, not because it is unimportant but because it is less studied. One must hope for future development of more comprehensive relationships.

6

Locomotion

Locomotion is the most obvious and most characteristic of animal activities. Like any activity, movement requires energy and, therefore, increases an animal's metabolic rate. From our own experience with walking, running, or swimming, we know that the power demands for movement can be very large. The demands of locomotion may, therefore, be a large component of respiration in the balanced energy equation and a significant energetic cost for the moving animal. This raises a quantitative question: How important are locomotive costs?

This chapter examines the interrelations between the metabolic costs of movement, body size, velocity, distance traveled, time spent traveling, and mode of locomotion – flying, swimming, or running. It provides the basic information required to estimate the metabolic rate of a moving animal. This is achieved by first considering the results of empirical studies relating metabolism to body mass and velocity for terrestrial, aquatic, and aerial locomotion. Because such data are largely derived in the laboratory, equations describing these data have limited ecological relevance when considered in isolation. To use them, we also require some estimate of an organism's speed in nature. Available allometric descriptions of average velocity are, therefore, introduced and then used to calculate the metabolic rates of moving animals. Whenever possible, the ecological importance of these relations is considered by comparing speeds and costs with the demands of existence.

Descriptions of the metabolic costs of locomotion

The costs of locomotion are usually determined by training an animal to exercise on a treadmill or in a wind or water tunnel. The subject is then forced to move at a given velocity, determined by the rate of movement of the substrate or medium, for a given time during which its metabolic rate is measured. Such experiments, therefore, yield the moving metabolic rate ($J s^{-1}$ = Watts) at a series of velocities ($m s^{-1}$). These values are often recalculated to give *transport cost*, the energy required to move an animal over a unit distance:

$$\text{transport cost} = \text{moving metabolic rate/velocity} \qquad (6.1)$$

Calculated in this way, transport cost would have units Joules per meter. The comparative literature usually presents both moving metabolic rate and transport cost as mass specific values with units Watts per kilogram and Joules per kilogram per meter, respectively. I will follow this convention for only the first part of this chapter, because individual rates of energy expenditure seem more ecologically relevant. However, the choice of unit has no effect on the validity of any argument presented here.

Walking and running

Experimental evidence shows that the moving metabolic rate of terrestrial animals rises in direct proportion to velocity from a velocity of almost zero to nearly maximum. For an animal of a given body mass, this relation can be approximately described by a linear relation:

$$_{w}R_{run} = {_{w}T_{c(net)}}V + {_{w}R_0} \qquad (6.2)$$

where $_{w}R_{run}$ is the specific metabolic rate while running or walking (Watts kg^{-1}), $_{w}T_{c(net)}$ is the slope (J m^{-1} kg^{-1}), V is velocity (m s^{-1}), and $_{w}R_0$ is the y intercept of the line.

The intercept in Equation 6.2, $_{w}R_0$, is somewhat higher than basal metabolic rate. This is attributed (Schmidt–Nielsen 1979) to the postural costs of holding the body in an erect running position. For homeotherms, $_{w}R_0$ is usually approximated as 1.7 times basal rate (Taylor, Schmidt–Nielsen, & Raab 1970). Bakker (1972, 1975) suggests that a similar figure may apply to poikilotherms. Paladino and King (1979) argue that 1.2 times basal rate is a better average value. I have adopted the latter value here because it is based on more species.

The slope in Equation 6.2 is usually termed *specific net transport cost*, because it represents the additional energy needed to move 1 kg of animal over a distance of 1 m. Since the slope is unaffected by velocity, the relation implies that the greater expense of running faster is offset by the shortened time needed to cover the distance. Surprisingly, net transport cost, at a given size, is the same for terrestrial birds, mammals, lizards (Paladino & King 1979), hoppers, and runners (Fedak & Seeherman 1979). All these groups can be described with a single curve that gives the

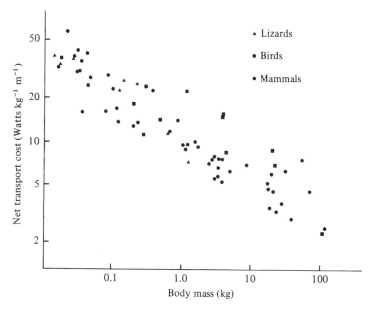

Figure 6.1. Specific net transport costs ($_wT_{c(net)}$, in J kg^{-1} m^{-1}) as a function of body mass (W, in kg) for running birds, mammals, and lizards. The curve is described as $_wT_{c(net)} = 11.3\ W^{-0.28}$. Modified from Fedak and Seeherman (1979).

specific metabolic rate for any running animal:

$$_wR_{run} = 1.2\ _wR_b + V\ _wT_{c(net)}$$
$$= 4.24W^{-0.25} + V11.3W^{-0.28} \tag{6.3}$$

where specific basal metabolism ($_wR_b$) is determined from Kleiber (1961), the postural correction for all groups (1.2), from Paladino and King (1979), and net transport costs ($_wT_{c(net)}$) from Fedak and Seeherman (1979; see Figure 6.1).

Equation 6.3 is a more complex allometric relation than those treated previously. The relation suggests that the total power production will be somewhat higher in running birds than in mammals, because of the higher avian basal rate of metabolism, and considerably less in running poikilotherms, because of their low standard rates. These differences diminish at high speeds, because the second term ($V_wT_{c(net)}$), which is independent of metabolic grouping, increases with velocity, thereby overpowering the effect of differences in specific basal metabolism.

Because the exponents of the two body mass terms in Equation 6.3 are essentially similar (-0.25 and -0.28), the metabolic rate of an animal moving at a given velocity is very close to a constant multiple of basal or standard metabolic rate. Other relationships (Paladino & King 1979; Taylor 1977; Taylor, Schmidt–Nielsen, & Raab 1970; Tucker 1970, 1973b) suggest that the cost, relative to basal metabolism, of moving at a given speed should fall with increasing body size. This follows because, in most published relations (Appendix VIa), net transport costs appear to fall more quickly than basal metabolism. The shallower relation in Equation 6.3 was selected because it is based on more species.

Total transport costs. Division of both sides of Equation 6.3 by velocity yields specific total transport costs $_wT_{c(tot)}$ in Joules per meter per kilogram:

$$_wT_{c(tot)} = 11.3W^{-0.28} + 1.2_wR_b/V \tag{6.4}$$

which shows that the total cost of transport is not independent of velocity but declines as velocity increases. The decline is more important in homeotherms because postural costs $(1.2_wR_b/V)$ represent a greater portion of the total. Equation 6.4 therefore, shows that the most energy-efficient speed for running is the highest an animal can maintain. Further discussion of transport costs will be deferred until we have examined both the metabolic rates associated with flight and swimming and the relation of speed to size.

This discussion is limited to the cost of movement across a flat surface. Several interesting extensions are possible if one considers the additional costs of climbing (Reichman & Aitchison 1981; Taylor 1973) and of carrying a load (Taylor, Heglund, McMahon, & Looney 1980). Basically, climbing is more onerous for large animals and carrying a load more so for smaller organisms. Unfortunately, these topics cannot be examined more extensively here.

Swimming

The metabolic rate of aquatic animals, like that of terrestrial animals, increases with speed, but the rate of increase is not linear (Figure 6.2). The specific metabolic rate of swimming organisms ($_wR_{swim}$, in Watts kg^{-1}) has been described (Ware 1978) as a power function of velocity (V, in $m s^{-1}$):

$$_wR_{swim} = _wR_s + _wR_p V^c \tag{6.5}$$

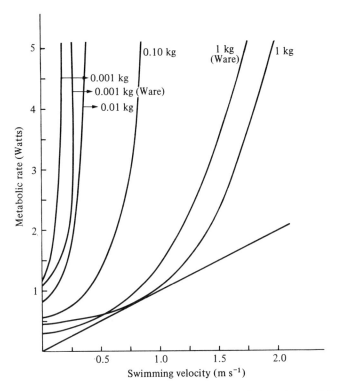

Figure 6.2. The effect of swimming velocity (V) and size (W) on the specific metabolic rate ($_wR_{swim}$) of fish. Shown are solutions to the equation of Beamish (1978), $_wR_{swim} = {_wR_s} + 0.116\exp(1.884W^{-0.36}V)$, and to Ware's (1978) equation $_wR_{swim} = {_wR_s} + 1.17W^{0.44}V^{2.42}$. For these calculations, $_wR_s$ was taken from Winberg (1960): $_wR_s = 0.285W^{-0.19}$. The straight line through the origin touches Beamish's curve for 1–kg fish at the speed where total transport costs are minimal.

and as an exponential function of velocity (Beamish 1978):

$$_wR_{swim} = {_wR_s} + {_wR_p}e^{cV} \qquad (6.6)$$

where $_wR_s$ is the specific rate of standard metabolism, and $_wR_p$ is an additional cost, which, like the coefficient of velocity c, is a constant derived by statistically fitting a curve to the data. Since $_wR_p$ is the difference between standard rate and the y intercept ($V = 0$) of the curve describing metabolic rate as a function of velocity, it is anaiogous to postural costs in running animals.

As both curves show (Figure 6.2), above a certain size-dependent velocity, metabolic rate rises sharply. This dramatic rise puts a distinct upper limit on swimming speed and confines swimming animals to a relatively narrow range of velocities. Larger fish exploit a much broader range of speeds than small ones.

Within these limits, the costs of swimming can be expressed as total transport costs instead of metabolic rate by dividing both sides of Equation 6.5 or 6.6 by velocity. Transport costs of swimming are more strongly affected by velocity than are those of running and are usually compared at some standard velocity. Most frequently, comparisons are made at that velocity at which transport costs are minimal. This is also the speed at which fishes normally swim (Glebe & Leggett 1981). The speed at minimum transport costs can be calculated as the speed at which the first differential of metabolic rate with respect to speed is minimized and the second differential is zero, or it can be determined graphically as the speed at which a straight line drawn through the origin of a plot of metabolic rate against speed is tangent to the curve (Figure 6.2). The division of the corresponding metabolic rate by this velocity yields minimum transport costs of swimming. The specific minimum cost of transport for swimmers is a declining function of size (Beamish 1978; Tucker 1973a; Appendix VI) as is the specific transport cost of running. Unlike the case for runners, transport cost of swimming and flying animals is usually expressed as total transport costs. Net cost may be approximated, if desired, by subtracting standard metabolic rate divided by velocity from total cost.

Flying

The relationship between velocity and metabolic rate of flying animals is still more complex. When running or swimming animals stand still, their metabolic rates approach standard, and any increase in motion increases respiration. For flying animals, hovering is extremely expensive. A hovering 1-kg animal respires at a rate double that of normal flight (Calder 1974; Appendix IIId). As forward motion increases, metabolic rate falls to some minimum and then rises again (Figure 6.3). The breadth and height of this response are affected by both body size and wingspan. Larger animals find hovering more expensive but have reduced costs at higher speeds. Birds with shorter wings have higher metabolic rates at the same body size and velocity (Tucker 1973b). Soaring, which requires relatively long wings, also reduces the cost of flight (Hails 1979).

The complex shapes of the curves in Figure 6.3 imply that transport costs change markedly with speed. In consequence, the transport costs

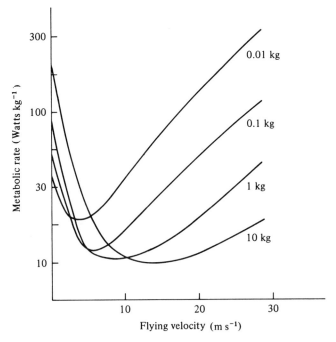

Figure 6.3 The effect of size (W, in kg) and velocity (V, in m s^{-1}) on the specific metabolic rates ($_wR_{fly}$, in Watts kg^{-1}) of flying animals. Curves in this figure are solution's to Tucker's (1973b) semi-empirical equation:

$$_wR_{fly} = 32W^{-0.34}V^{-1} + 0.0033W^{-0.34}V^{2.5} + 0.0058W^{-0.51}V^{2.5} + 1.1_wR_b,$$

where $_wR_b$ is specific basal metabolic rate.

are normally compared at the "optimum speed" at which transport costs are minimal. This is determined in a manner analogous to the minimum transport costs of swimming. The minimum transport costs of flight, like those of swimming and running, decline with body size (Berger & Hart 1974; Greenwalt 1975; Schmidt–Nielsen 1972; Tucker 1973a,b).

The appropriate equation describing avian transport costs is a point of some debate. Greenewalt (1975) suspects that the high costs observed by Tucker (1970, 1973a,b) reflect experimental artifacts associated with the use of wind tunnels; he prefers to use estimates from free-flying birds. Greenewalt (1975) shows that, given Tucker's estimates of transport cost, some birds would be incapable of the nonstop flights they are thought to perform. On the other hand, Greenewalt's (1975) relations have a less powerful empirical base. The twofold difference in elevation of the two curves is rather small in allometric relations and could easily be introduced by slight changes in constants used to calculate transport costs. Pennycuick (1972) suggests that size may have little effect on avian transport costs. Although my account follows Tucker's relations, this is a choice of convenience. I do not endorse one author over another in this field.

Speeds of locomotion

The previous section presents equations that relate metabolic rate to size and velocity. If these equations are to be used to determine the additional costs of movement, we must measure or estimate an animal's likely speed. For fish and birds, a typical value can be calculated as the optimum speed; for terrestrial animals, the typical speed must be established by observation. This section, therefore, introduces and compares relations that scale maximum and average velocity to size for each locomotory mode.

Calculation of metabolic rates and transport costs at these speeds will be deferred until later in the chapter. After summarizing the available relations between velocity and body mass, this section instead examines some of the direct implications of the scaling of speed to size.

The allometry of speed

Very few relations describe animal speed as a function of body size. This paucity no doubt reflects wise caution in treating so plastic a phenomenon as animal movement. Nevertheless, few will doubt that, on average, large animals move faster than small. One of the premises of allometry is that estimates can be improved if size effects are considered, even though considerable residual variation remains. Naturally, one hopes that further improvements will be found, but this hope should not deter attempts to make use of our current knowledge.

The literature dealing with animal velocities frequently presents animal size as length and animal velocity as lengths covered per unit time (Bonner 1965; Buddenbrock 1934; Hempel 1954; Webb 1975). When necessary, I have converted animal sizes to mass (using the mass–length regressions in Appendix IIa) and relative velocities to absolute velocities. This combination of regressions introduces a further approximation into the relationships between size and speed. Although this "chaining" is statistically undesirable (Smith 1980), it is made necessary by the absence

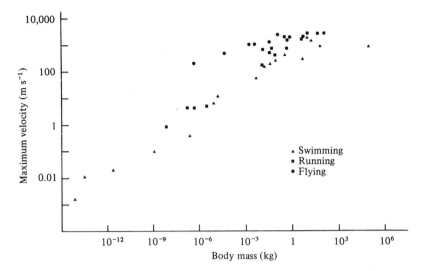

Figure 6.4. The effect of size (W. in kg) on the maximum velocity (V_{max}, in m s^{-1}) of flying ($V_{max(fly)} = 22.2W^{0.144}$), swimming ($V_{max(swim)} = 2.37W^{0.35}$), and running ($V_{max(run)} = 10.4W^{0.38}$). Modified from Bonner (1965).

of alternatives; relations built directly from measured masses and absolute speeds would be preferable.

Maximum velocity. Bonner (1965) relates the maximum (nonsustained) speeds of flying, running, and swimming to animal size. His estimates serve to set an upper bound on velocity, for he purposely selected faster animals within each locomotory mode and size class. As expected, maximum speed increases with size, and larger animals can usually outrun small ones. For animals of similar size, swimming is slower than running, and flying is fastest (Figure 6.4) The data also suggest that maximum flight velocities are much less sensitive to size than are running or swimming speeds and the maximum speed of large animals is nearly independent of locomotory mode (Bonner 1965). Very roughly, the maximum speed of flight is about 22 m s^{-1} (80 km h^{-1}), which is reached at a body size of about 1 kg (Bonner 1965), and maximum speed of mammals is 19 m s^{-1} (70 km h^{-1}), reached at a body size of approximately 10 kg (Schaller 1972). Bonner's data suggest that the maximum speed of swimmers is perhaps 11 m s^{-1} (40 km h^{-1}) for animals weighing more than 80 kg, but Webb (1975) estimates maximum burst swimming speeds to be about twice as great.

Other velocities. Animals do not habitually move at their maximum speeds but at lower, less easily defined rates. Taylor et al. (1981; Appendix VIb) give a relation describing the running velocity of mammals at maximum sustained metabolic rate in 2- to 10-min trials. This is less affected by size than maximum speed in Figure 6.4 (Bonner 1965) and markedly underestimates the maximum speed of larger animals. For salmonid fish, Brett (1965; Brett & Glass 1973) uses the maximum sustained swimming speed in 1-h trials as a standard comparative velocity. These speeds are only one-fourth (Webb 1975) to one-half (Bonner 1965) burst and maximum speeds and are less affected by size. I know of no standard flight speed other than maximum and optimum velocities.

Normal speeds. Relations built from Buddenbrock's (1934) data (Appendix VIb) describing the normal walking speeds of mammals suggest that an animal weighing 1 kg usually walks at only 3% of maximum running speed. Since the slope of the relation for normal walking speed is less than that for maximum speed (Figure 6.5), this proportion declines with increasing size. The relation describing the walking speed of beetles (Buddenbrock 1934) is remarkably similar to that for mammals, however, flies and cursive insect predators, like carabid beetles, normally run much faster (Hempel 1954).

Ware (1978) calculated the optimal swimming speeds (the velocity at minimal transport costs) for salmonids from the data of Brett (1965; Brett & Glass 1973). Over a large range of sizes, the optimal swimming speeds of fish are quite close to, but slightly above, the normal walking speeds of mammals and beetles (Figure 6.5).

All estimates of flight velocity at minimum power requirements (Berger & Hart 1974; Greenewalt 1975; Tucker 1973b) show that flight is much faster than walking or swimming and lies close to the maximum flight velocity.

Some implications

Average speeds permit several simple calculations of interest. For example, many animals, like seals, bees, nesting birds, and burrowing rodents, must move from a resting place or nest to forage elsewhere. If we assume or, better, measure the maximum time an animal will devote to this travel, we can estimate the maximum commuting distance or *foraging radius* (Pennycuick 1979) as the product of velocity and time. Figure 6.5 offers a qualitative indication of how foraging radius might change with

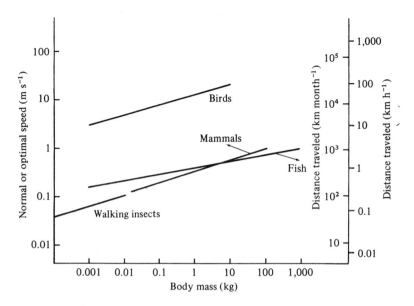

Figure 6.5. The effect of size (W, in kg) on optimal speed of fishes ($V_{n(fish)} = 0.39W^{0.136}$, in m s^{-1}; Ware 1978), birds ($V_{n(bird)} = 13.3W^{0.21}$; Tucker 1973b), mammals ($V_{n(mammals)} = 0.33W^{0.21}$; Buddenbrock 1934), and beetles ($V_{n(beetles)} = 0.30W^{0.29}$; Buddenbrock 1934). On the right-hand vertical axis, foraging radius is the distance an animal can cover in 1 h at average speed: this suggests that if all animals can devote only 1 h d^{-1} to commuting between nest and foraging area, the commuting distance will increase with size. Migration distance is the distance animals will cover at average speeds moving for 30 d at 12 h d^{-1}. The times permitted for commuting and migration were arbitrarily selected to show qualitative trends.

size. If all animals were to limit themselves to commuting only 1 h between nest and foraging area, flying animals would have the largest area followed at a considerable distance by swimmers and then runners. Within each locomotive mode, larger animals have the greater foraging radii. An upper limit of 1 h is, of course, an entirely arbitrary figure. On principle, I would expect that such a maximum, if it exists at all, would rise with body size, probably as $W^{1/4}$, since most physiological times bear such a relation to size (Lindstedt & Calder 1981). This would emphasize the advantage of larger animals over smaller ones as commuters.

A parallel calculation can give the effect of size on migration distance. If the time available for migration is limited to a given period, say 1 month, and it is assumed that animals move for only 12 h d^{-1}, then migration distance can be calculated (Figure 6.5) as the product of average velocity and available time. Under these assumptions, flying animals weighing more than 1 mg should be capable of migrations of 1,000 km or more, whereas mammals and fishes would need to weigh about 100 kg to complete so long a trip in so short a time. This does not prohibit terrestrial or aquatic migrations. The migrations of anadromous fishes cover thousands of kilometers but are achieved over the course of 1 or more years (Leggett 1977) rather than months. In other words, slow organisms can undertake long migrations by relaxing temporal constraints. Migration may become attractive even to very small organisms, like zooplankton, if a sufficiently steep environmental gradient is available. Average speeds set a lower limit to the size of migrators only if the necessary distance is large and set an upper limit to migration distance only if time ⸲ limited.

Transport costs

Calculations

Given the foregoing estimates of average speed, we can now compare transport costs for the three modes of locomotion. To this point, discussion has followed convention by treating specific transport costs. Although the choice of unit is incidental, individual rates, rather than specific rates, would seem of more direct ecological relevance, because an entire organism is a more likely ecological unit than is a kilogram of tissue. For this reason, further treatment centers on the total transport cost per individual.

Figure 6.6 and the equations in Table 6.1 show clear differences in the total transport costs among metabolic groups and locomotory modes. Poikilothermic swimmers spend the least energy in locomotion followed by flying animals, which compensate for high power demand with high velocity. The calculated total transport cost of swimming homeotherms is considerably higher than that of swimming poikilotherms, because, at optimal speed, basal metabolic rate accounts for most energy expenditure. For homeotherms, flight is a more efficient mode of transport, because high flight velocities reduce the relative importance of high basal metabolism, essentially spreading this irreducible cost over greater distances. In other words, when the ratio of basal metabolism to speed is high, as with swimming homeotherms, total transport costs are increased. This same effect is seen among running animals. Terrestrial homeotherms have higher total costs than terrestrial poikilotherms even though net transport costs are the same (Fedak & Seeherman 1979;

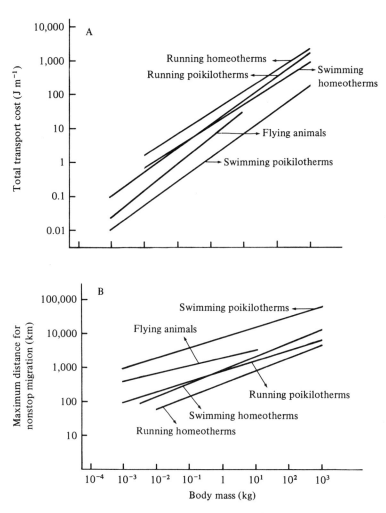

Figure 6.6. (A) Transport costs ($J\,m^{-1}$) associated with swimming, running, and flying homeotherms and poikilotherms as calculated in Table 6.1. (B) The effect of size on the maximum distance for nonstop migration if the animals' energy reserves as fat amount to 25% of body mass (Table 6.2).

Paladino & King 1979) because of the high basal rate of homeotherms. In fact, running homeotherms have the highest transport costs of any group.

Implications

Transport costs are usually considered in relation to migration and particularly to nonstop migration. If an animal is to undertake a relatively

Table 6.1. *A comparison of transport costs (T_c, in J m^{-1}) for flying, swimming, and running poikilotherms and homeotherms*

I. For flying animals

From Tucker (1973b),

$$_wT_c = 5.2W^{-0.227}$$
$$\therefore \quad T_c = W_wT_c$$
$$= 5.2W^{0.773}$$

II. For swimming poikilotherms

From Tucker (1973a),

$$_wT_c = 1.24W^{-0.295}$$
$$\therefore \quad T_c = 1.24W^{0.705}$$

III. For swimming homeotherms[a]

From Ware (1978),

$$R_{swim} = R_b + 1.17W^{0.44}V^{2.42}$$
$$V_n = 0.39W^{0.136}$$
$$\therefore \quad R_{swim} = R_b + 1.17W^{0.44}(0.39W^{0.136})^{2.42}$$
$$= R_b + 1.17W^{0.44}(0.10W^{0.33})$$
$$= R_b + 0.117W^{0.77}$$

From Hemmingsen (1960),

$$R_b = 4.1W^{0.75}$$
$$\therefore \quad R_{swim} = 4.1W^{0.75} + 0.117W^{0.77}$$

From Equation 6.1,

$$T_c = R_{swim}/V_n$$
$$= (4.1W^{0.75} + 0.117W^{0.77})/(0.39W^{0.136})$$
$$= 10.51W^{0.61} + 0.44W^{0.63}$$
$$\sim 11W^{0.61}$$

IV. For running homeotherms

From Equation 6.3,

$$R_{run} = 1.2R_b + V11.3W^{0.72}$$

From Hemmingsen (1960),

$$R_b = 4.1W^{0.75}$$

Table 6.1. (*cont.*)

From Equation 6.1,

$$T_c = R_{run}/V$$
$$= (4.92W^{0.75}/V) + 11.3W^{0.72}.$$

From Buddenbrock (1934),

$$V_n = 0.33W^{0.21}$$
$$\therefore \quad T_c = 4.92W^{0.75}/(0.33W^{0.21}) + 11.3W^{0.72}$$
$$= 14.92W^{0.54} + 1.3W^{0.72}$$

V. Similarly for running poikilotherms[a]

From Hemmingsen (1960),

$$R_s = 0.14W^{0.75}$$

From Buddenbrock (1934),

$$V_n = 0.30W^{0.29}$$

From Equations 6.1 and 6.3

$$T_c = 1.2(0.14W^{0.75})/(0.30W^{0.29}) + 11.3W^{0.72}$$
$$= 0.56W^{0.46} + 11.3W^{0.72}$$
$$\sim 12W^{0.72}$$

In all cases, transport costs rise with body mass.

===

Note: In this table, W is body mass (kg); V is velocity and V_n is normal or optimal speed (m s^{-1}); R_b is basal metabolic rate and R_s is standard metabolic rate (both in Watts), $_wT_c$ is specific transport cost (J m^{-1} kg^{-1}). R_{swim} and R_{run} refer to the moving metabolic rates (Watts) associated with each locomotory mode.
[a]Neither the equation for running poikilotherms nor that for swimming homeotherms has, as yet, empirical support.

long trip without feeding, it must meet the energy expenditure of travel by drawing on its own stored reserves. Prior to migration, animals may accumulate a fat store of 25 or even 50% of body mass (Pond 1978). The maximum migration distance can be calculated from the energetic cost of transport and the amount of stored energy (Table 6.2). These calculations show that a fish weighing 1 g is capable of migrating 1,000 km without feeding, whereas a 10-kg mammal cannot walk so far although it is four orders of magnitude larger. Flying animals occupy an intermediate

Table 6.2. *Maximum distance for nonstop migration assuming that fat content = 0.25W and that velocity is the normal or optimal speed for each group*

l_m = energy reserves/transport costs

$$= 40 \times 10^6 \times 0.25W/T_c$$

$$= 1 \times 10^7 W/T_c$$

From Table 6.1,

I. For flying animals

$$T_c = 5.2W^{0.77}$$

$$\therefore \quad l_m = 1 \times 10^7 W/5.2W^{0.77}$$

$$= 1.9 \times 10^6 W^{0.23}$$

II. For swimming poikilotherms

$$T_c = 1.24W^{0.705}$$

$$\therefore \quad l_m = 8.1 \times 10^6 W^{0.295}$$

III. For swimming homeotherms

$$T_c = 11W^{0.61}$$

$$\therefore \quad l_m = 0.9 \times 10^6 W^{0.39}$$

IV. For running homeotherms

$$T_c = 14.9W^{0.54} + 11.3W^{0.72}$$

$$\therefore \quad l_m = 1 \times 10^7 W/(14.9W^{0.54} + 11.3W^{0.72})$$

$$= 0.67 \times 10^6 W^{0.46}/(1 + 0.76W^{0.18})$$

V. For running poikilotherms

$$T_c = 0.56W^{0.46} + 11.3W^{0.72}$$

$$\sim 12W^{0.72}$$

$$\therefore \quad l_m = 0.8 \times 10^6 W^{0.28}$$

Migration distances rise with size.

Note: Transport cost is T_c in $J\,m^{-1}$; W is body mass in kg; l_m is migration distance in m; and 1 kg of fat is considered to contain $40 \times 10^6\,J$.

place: A bird need weigh only 60 g to migrate approximately 1,000 km without food.

Such figures are only approximations; most animals can and do stop and feed during migrations. Moreover, variation in the initial fat reserves (Table 6.2; Pond 1978) will greatly influence the calculation. Nevertheless, the differences among the three empirically based relations (i.e., those for poikilothermic swimmers, homeothermic runners, and all fliers) are quite substantial, and the curves in Figure 6.6 are probably representative. These show that fish have the greatest migratory capability and terrestrial homeotherms have the least. Within locomotory modes, larger animals are capable of longer migrations than small ones. Long-range migration is a viable strategy for a wide range of swimming animals, the bigger fliers, and the largest walkers.

Several observations lend anecdotal support to these interpretations. The lower size limit for migrating mammals appears to be about 20 kg, the mass of smaller African gazelles. Much smaller flying animals, like the monarch butterfly, which weighs only 1 g, undertake substantial migrations, as do many relatively small fish. There is also some evidence that larger fish are better able to undertake longer and more demanding migrations. Within restricted geographic regions, Atlantic salmon caught in larger rivers tend to weigh more than those from shorter streams (Schaffer & Elson 1975; Thorpe & Mitchell 1981). Among birds, Greenewalt (1975) has compared the distance known to be covered in nonstop migration by three species with the distance such birds could cover based on their fat reserves. Observed and calculated values are close but the calculated value is about 33% larger than observed weight loss. This may represent a safety factor to allow straying from the most direct path or the effects of headwinds or suboptimal speeds, and so forth.

Moving metabolic rates

We now return to the question that opened this chapter and attempt a comparison of metabolic rates of moving animals with their basal or standard rates. Unfortunately, only a partial answer can be given. We can calculate the metabolic rate of moving animals, but, unless we know what proportion of their time such animals move, we cannot include the costs of locomotion in the balanced growth equation. This book provides no solution to the second half of the problem, the proportion of an animal's day spent in active foraging. Instead, this section limits itself to comparisons between the respiration rates of resting and moving animals.

An allometric relationship describing typical metabolic rates during locomotion can be calculated as the product of total transport costs and the average velocity of movement, or it may be obtained directly by regression of metabolism at a given speed against size (Appendix IIId).

Table 6.3. *Moving metabolic rates* (R_m, *in Watts*) *of animals moving at their normal or optimal velocities* (V_n, *in* ms^{-1})

I. For flying animals

From Tucker (1973a),

$$R_m = 84.7W$$

II. For swimming poikilotherms

From Ware (1978),

$$R_m = 0.14W^{0.75} + 1.17W^{0.44}V^{2.42}$$
$$V_n = 0.39W^{0.136}$$
$$\therefore \quad R_m = 0.14W^{0.75} + (0.39)^{2.42}(W^{0.136})^{2.42}$$
$$= 0.14W^{0.75} + 0.117W^{0.77}$$
$$\sim 0.26W^{0.76}$$

where, from Hemmingsen (1960), standard metabolic rate $= 0.14W^{0.75}$

III. Similarly for swimming homeotherms

From Hemmingsen (1960),

$$R_b = 4.1W^{0.75}$$
$$\therefore \quad R_m = 4.1W^{0.75} + 0.117W^{0.77}$$
$$\sim 4.2W^{0.75}$$

IV. For running homeotherms

From Equation 6.3,

$$R_m = 1.2R_b + T_{c(net)}V_n$$

From Fedak & Seeherman (1979),

$$T_{c(net)} = 11.3W^{0.72}$$

From Buddenbrock (1934),

$$V_n = 0.33W^{0.21}$$
$$\therefore \quad R_m = 1.2(4.1W^{0.75}) + (11.3W^{0.72})(0.33W^{0.21})$$
$$= 4.9W^{0.75} + 3.73W^{0.93}$$

V. Similarly for running poikilotherms

$$R_m = 1.2(R_b) + 11.3W^{0.72}V_n$$

From Buddenbrock (1934),

$$V_n = 0.30W^{0.29}$$
$$\therefore \quad R_m = 1.2(0.14W^{0.75}) + (11.3W^{0.72})(0.30W^{0.29})$$
$$= 0.17W^{0.75} + 3.4W^{1.0}$$

Note: R_b is basal metabolic rate (Watts), W is body mass (kg), and $T_{c(net)}$ is net transport cost (Jm^{-1}). Empirical relations for metabolic rate are described in Appendix III, and those for speed and transport cost in Appendix VI.

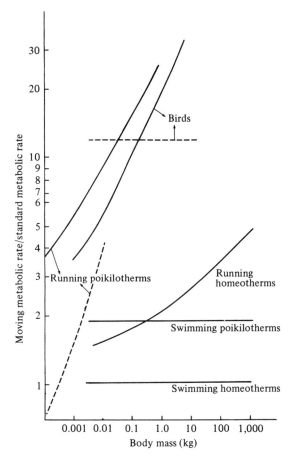

Figure 6.7. Moving metabolic rate relative to basal or standard metabolic rates based on the equations for moving rates in Table 6.3 and Hemmingsen's equations for basal or standard rate. Dotted lines for walking beetles (Bartholomew & Casey 1978b) and flying birds (Berger & Hart 1974) were derived from measured values and provide a comparison with some of the calculated curves.

Most of the equations listed in Table 6.3 are calculated from other equations we have met in this chapter. When equations are divided by allometric relations for standard metabolic rate, they yield moving metabolism relative to basal or standard metabolic rates (Figure 6.7). This ratio of moving to standard metabolism indicates the proportionate increase in the cost of existence induced by movement.

At average velocities, both the absolute and relative costs of movement are greater in large than small animals. This trend is probably negligible in swimming animals, for the increase is very small for both homeothermic and poikilothermic swimmers.

The cost of walking is markedly increased among larger terrestrial homeotherms, and one might expect to see this cost offset by a reduction in the amount of movement in big mammals. Perhaps this contributes to the lazy behavior of lions (Schaller 1972), which move only $2\frac{1}{2}$ h d^{-1}. The power expenditures of the remaining two classes, walking poikilotherms and flying animals, are more controversial. The calculated relation for walking insects is two to three times higher than that observed for a few beetles by Bartholomew and Casey (1978b), but the measured and calculated trends agree that the cost of locomotion rises precipitously in larger insects. The same trend should hold for other terrestrial poikilotherms as well. The metabolic rate of flying birds has been reported to rise in parallel with basal rate (Berger & Hart 1974; Hails 1979; King 1974), to rise with body mass (Greenewalt 1975; Tucker 1973b), or to rise slightly more slowly than basal metabolism (Kendeigh, Dol'nik, & Govrilov 1977). Of these relations, those that propose that moving metabolic rate should rise as $W^{3/4}$, like basal metabolism, are usually based on regression analyses, whereas others arrive at their conclusions by a combination of conceptual analysis with empiricism. The range of solutions suggests that no one equation can currently be selected as preferable.

In terms of the balanced growth equation, the values in Figure 6.7 suggest that locomotion is a negligible cost only for aquatic homeotherms. For fish and walking homeotherms, movement approximately doubles basal metabolic rate, and, for walking insects, movement increases metabolism approximately fivefold. Since the costs of locomotion for walking vertebrate poikilotherms should demand a similar increase in metabolic rate, it seems likely that terrestrial poikilothermic vertebrates move more slowly or less frequently than other animals. Flying birds seem to be able to increase their power output by an order of magnitude, although this demand can be reduced by soaring (Hails 1979) or increased wingspans (Tucker 1973b). With the exception of swimming homeotherms, these comparisons suggest that most animals should minimize their activity levels. This should be particularly pronounced in walking poikilotherms, birds, and larger walking homeotherms. The low cost of swimming for aquatic homeotherms may contribute to the playfulness so apparent in captive whales, seals, and otters. For these animals, motion is virtually no more tiring than standing still. In walking animals and perhaps in fliers, the ratio of moving metabolic rate to basal metabolism rises with size. Larger animals may compensate for these higher costs by moving less. Two studies among birds suggest that larger birds may be less active than smaller ones (Pearson 1968; Gibb 1954).

Locomotion is an essential, but expensive, part of animal existence. A full representation of an animal's energetic relations must include some term for the cost of movement. The seemingly more difficult and less engaging determinations of moving metabolic rates and transport costs have now been made. A more precise statement of the role of locomotion in power output now awaits the development of behavioral studies that report the amount of time that wild animals spend moving and their velocity. In the absence of such information, the physiological literature permits comparisons of transport costs, metabolism, and the speed of moving animals. These show that, in general, swimming is a slow but inexpensive means of transportation. Flight is expensive per unit time, but, because it is fast, it imposes only an intermediate transport cost per unit of distance. Running allows a greater range of speeds, but it is very expensive per unit of distance and of intermediate expense per unit of time. Within each locomotory mode, large species move somewhat more quickly and consume more energy per unit time and per distance. Specific costs both per unit distance and per unit time fall with increasing size. Finally, costs relative to basal power expenditure appear to rise with size among walking animals and, perhaps, birds.

7

Ingestion

Ingestion rate is the largest term in the balanced growth equation and sets an upper limit to all other variables. Allometric relations describing the components of the energy budget can seem reasonable and coherent only if ingestion is greater than any other single rate and equal to the sum of all other rates. Such checks on coherence are particularly important, since all allometric relations have a large residual variation and always contain unevaluated sources of error.

Ingestion rate is also the basis for any calculation of the efficiency with which an animal converts food to new tissue or metabolic power. Such efficiencies, calculated as the ratios of the energy used in growth or respiration to that eaten, normalize differences in absolute rates among animals of different size or metabolic grouping. In this and in subsequent chapters, *energetic efficiencies* highlight important differences and similarities within the animal kingdom. For example, such calculations will show if homeothermy has resulted in more efficient use of food for growth or if this strategy has simply increased heat loss through respiration.

Ingestion is far more than a tool for ecological bookkeepers. It also represents predation, certainly the most apparent and probably the most significant interaction between an animal and its community. As a measure of a predator's demands, ingestion rate estimates the impact of a given animal on its ecosystem. If the recipients of this demand, the prey, could also be identified, one would be well toward a description of both the magnitude and direction of the predator's influence. The reciprocal of this relationship is, of course, the influence of a given prey population on the ecology and behavior of its predators.

This chapter consists of three sections. The first presents basic empirical models that describe ingestion rate as a function of body size and discusses the significance of these descriptions in relation to metabolic demands. Ingestion rate is influenced by more variables than size alone and so the second section discusses other factors that influence ingestion

100

rate. The third section provides some simple predictions and comments about predation, community structure, and competition.

Some basic properties

Ingestion rate

By now, most readers will surely suspect that ingestion rate will rise as $W^{3/4}$ and homeotherms will eat more than poikilotherms. These expectations will not be disappointed, but it is important to realize that allometric trends need not be so regular. Neither logic nor statistics nor more general empirical laws, such as the laws of thermodynamics, impose similitude. For example, more complete digestion associated with the longer guts of large species or with the higher body temperatures of homeotherms could lead to higher assimilation efficiencies. (Assimilation efficiency is the ratio of assimilation to ingestion when both are expressed in the same units). If this were the case, body size could have less effect on ingestion than on respiration. The marked similarity of most known allometric relations too easily leads to the suspicion that all body size relations must be similar. In fact, each new allometric relation should be treated as a new and independent hypothesis to be tested as thoroughly as possible, like any scientific theory.

The scaling of ingestion rate to size. The equations of Farlow (1976) relating ingestion rate and body mass (Appendix VIIa) have the largest empirical base. He treated homeothermic carnivores and herbivores with separate regressions, but the two curves fit to these data are virtually identical (Appendix VIIa), and reanalysis of the data (Figure 7.1) showed no significant effect of diet on ingestion rate in homeotherms (R. H. Peters unpublished). A single regression suffices. Peters (1978b) also suggested that there is no substantial difference between the ingestion rate of herbivores and carnivores. Several authors have provided similar regressions for mammals (Evans & Miller 1968; Harestad & Bunnell 1979) and birds of prey (Calder 1974; Schoener 1968). In all cases, the relations are power curves with slopes ranging from 0.63 to 0.75; these values ($\bar{X} = 0.70$, $SD = 0.04$, $N = 9$) do not differ greatly from the $\frac{3}{4}$ slope describing the effect of size on respiration. The elevations of these feeding rate relations are also remarkably similar, ranging from 7.1 to 15.7 Watts for a 1-kg animal. The intercept calculated from Farlow's data (10.7 Watts) lies close to the midpoint of this range.

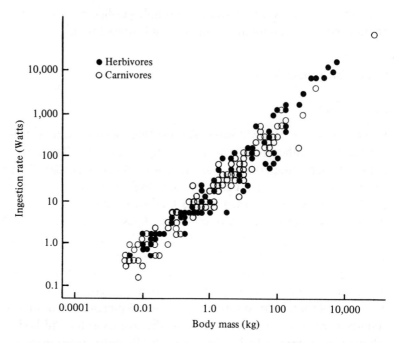

Figure 7.1. Ingestion rate of herbivorous and carnivorous endotherms as a function of animal body mass. Data from Farlow (1976). The regression line through these data is ingestion rate = $10.7W^{0.70}$.

Farlow (1976) also collected published ingestion rates for carnivorous, poikilothermic tetrapods. These data (Figure 7.2) also fit a power function of body mass with a similar slope (0.82) but a much lower intercept (0.78 Watt). Other equations for poikilotherms (Cammen 1980; Ikeda 1977; Nival & Nival 1976; R. H. Peters & J. A. Downing unpublished) confirm this slope ($\bar{X} = 0.80$; $SD = 0.14$; $N = 9$) and share a similar range of elevations – from 0.12 to 2 Watts for a 1-kg animal. Since comparison of elevations requires extrapolations of several orders of magnitude beyond the sizes of animals used to construct the regression lines, this degree of similarity is quite striking.

Comparisons of ingestion and other rates. Realistic estimates of ingestion rate must exceed basal or standard rates by a substantial margin. In homeotherms, we would expect that ingestion rate must be at least two to three times higher than basal, just to meet the average metabolic expenditure (French et al. 1976; Kendeigh, Dol'nik, & Govrilov 1977; King 1974). For homeotherms, Farlow's ingestion rates are about 2.6 times

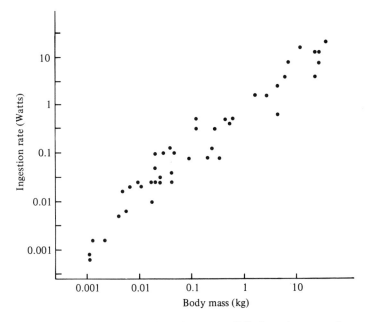

Figure 7.2. Ingestion rate of carnivorous poikilothermic tetrapods as a function of body mass. The regression line through these data is ingestion rate $= 0.78W^{0.82}$. Modified from Farlow (1976).

basal metabolic rate, and this ratio may decrease slightly with size. The ratio in nature may be somewhat higher than that calculated here, because Farlow's equations incorporate data for animals in captivity that may reduce activity and food demands. For poikilotherms, the ratio is somewhat higher and may increase with body size (Table 7.1).

Such ratios are important for the generation of new predictions and evaluations of the corpus of allometric relations. The ratios offer comparisons to check on the internal consistency of allometric theory; reasonable values for such ratios, together with the growing number and similarity of other allometric relations, provide us with more assurance in any single regression. In addition, new predictions are produced by the now familiar process of combining two empirical relations. The resultant constant or near constant suggests that animals that respire at greater than average rates will also ingest food at a similarly high rate. Thus, metabolically active animals, like weasels, should eat more, whereas animals with low metabolic rates, like sloths, should eat less (McNab 1978a, 1980). This can be rephrased in the claim that basal metabolic rate may be a better predictor of ingestion rate and other physiological

Table 7.1. *The ratio of ingestion rate (I, in Watts) to the basal metabolic rate of homeotherms (*R_b*, in Watts) and to the standard metabolic rate of poikilotherms (*R_s*, in Watts)*

I. For homeotherms:

From Hemmingsen (1960),

$$R_b = 4.10W^{0.751}$$

From Farlow (1976),

$$I = 10.7W^{0.703}$$
$$\therefore \quad I/R_b = 10.7W^{0.703}/4.10W^{0.751}$$
$$= 2.61W^{-0.048}$$

II. For poikilotherms

From Robinson, Peters, & Zimmerman (1983) at 20°C,

$$R_s = 0.19W^{0.76}$$

From Farlow (1976),

$$I = 0.78W^{0.82}$$
$$\therefore \quad I/R_s = 0.78W^{0.82}/0.19W^{0.76}$$
$$= 4.1W^{0.06}$$

The residual exponents of mass for both groups are so small (-0.048, 0:06) as to be negligible. It seems that all animals eat two to four times their basal or standard metabolic rate.

Note: Body mass (W) is in kg.

processes than body size. Unfortunately, this potential has been little investigated (Boddington 1978; McNab 1980).

Specific dynamic action

Ingestion and the subsequent processing of food is as much an activity as running or thermal regulation. The metabolic cost of this processing, which is usually called *specific dynamic action*, is apparently fixed for various foodstuffs (Table 7.2). Thus, the handling of 1,000 J of protein requires 310 J of additional metabolic power regardless of whether the protein is consumed by fish or fowl or by mouse or moose. Given the composition of the diet and the ingestion rate, one can determine the

Table 7.2. *The specific dynamic action associated with various foods*

	Protein	Fat	Carbohydrates	Specific dynamic action[a]
Energy content $(J g^{-1})$	24,000	40,000	18,000	
Specific dynamic action (%)	31	13	6	
Food composition (% fresh mass)				
Tropical fruit	1.4	7.5	12.3	11.3
Temperate fruit	1.02	0.48	13.2	8.6
Grains	10.9	2.4	73.2	23.7[b]
Insects	17.7	3.4	2.3	25.2[b]
Marine invertebrates	16.5	1.1	1.33	28
Fish	19.4	6.2	0	24.7
Reptiles and amphibians	18.1	0.4	0	30.4
Birds and mammals	22.9	5.7	0	25.7

Note: For each food, specific dynamic action (%) was calculated from the composition of the food by mass (Ricklefs 1974), the energy of these components (Kleiber 1961), and their specific dynamic action (Ware 1975). Listed values of specific dynamic action represent the proportion of the food's energy that an animal must expend to process that food: This is an inescapable cost associated with ingestion and appears unaffected by meal size, animal mass, or metabolic group.
[a]For example, specific dynamic action of birds and mammals was calculated as $[(24,000 \times 31 \times 22.9) + (40,000 \times 13 \times 5.7)]/(24,000 \times 22.9 + 40,000 \times 5.7)$.
[b]The specific dynamic action of starch (23%) was used for carbohydrates in this calculation.
Source: Modified from Ricklefs (1974).

increase in metabolic rate that is the inevitable consequence of eating food. Table 7.2 suggests that ingestion increases the metabolic rate of herbivores by about 30% over basal; this calculation assumes that ingestion rate is three times the basal metabolic rate and the specific dynamic action is about 10% of the energy content of the ingesta. The higher specific dynamic action of proteins would increase carnivore metabolism from 75 to 100% over basal.

Differences in the specific dynamic action of different diets must affect patterns of energy utilization of herbivores and carnivores. Plant eaters assimilate a smaller proportion of their food (Table 7.3) and must compensate for lower assimilation efficiencies by reduced costs of food processing (lower specific dynamic action), reduced levels of activity, or increased food intake. Farlow (1976) has shown that ingestion rates are not increased among herbivores; however, Table 7.2 indicates that specific dynamic action is lower for most plant foods. Seeds are exceptions to this rule; they are highly assimilable, but the high specific

Table 7.3 *Assimilation efficiencies (percentage) for various food*

In general		birds	
Meat and fish	<90	Fish	81
Insects	70–90	Insects	67
Seeds	<80	Seeds	75–80
Aquatic plants	50–90	Berries	31
Young terrestrial foliage	60–70	Willow buds	32
Mature foliage	30–40	Pine needles	30
Wood	15		

Note: Listed values are the percentages of all energy in the food that can be metabolized by the predator. All figures are approximate.
Source: Modified from Ricklefs (1974).

dynamic action of seeds may partly offset the apparent advantages offered by such a diet. For example, insectivorous or granivorous birds assimilate about 75% of their ingested food but the specific dynamic action associated with these foods is so high (about 25%) that the net gain in energy is only 50% of the food's energy content. An herbivorous bird can assimilate only 31% of its food, but if the specific dynamic action is only 6%, the net gain from the food is 25%. Differences in specific dynamic action thus modulate, but do not completely offset, differences in the assimilability of plant and animal food. A full explanation of the similarity in ingestion rates probably requires analysis of other forms of activity. Carnivores may be more active than herbivores and so burn off more of their assimilated energy. Parenthetically, it should be noted that the distinction between herbivores and carnivores ought not be drawn too rigidly. Most, if not all, herbivores will eat meat when available and often seem to prefer it (Peters 1977).

Other factors

As one would expect, many of the variables that affect metabolic rate have qualitatively similar effects on ingestion rate. Unfortunately, few of these effects have been sufficiently analyzed that we may use allometric relations to compare their quantitative effects. The lowered ingestion rates of zoo animals (Evans & Miller 1968), relative to other estimates (Farlow 1976; Harestad & Bunnell 1979), may reflect lower levels of activity, although all curves incorporate some data from zoological parks.

Growth, reproduction, and the accumulation of fat stores probably require periodic increases in ingestion rate in both homeotherms (Kendeigh, Dol'nik, & Govrilov 1977; Pond 1977) and poikilotherms (Geller 1975; Hayward & Gallup 1976; Nauwerck 1963; Zankai & Ponyi 1976). High temperatures increase ingestion rate in poikilotherms (R. H. Peters & J. A. Downing unpublished) but depress ingestion rate in homeotherms. Low temperatures have the opposite effects (Ames 1980; Hensel, Brück, & Raths 1973).

Temperature. Kendeigh, Dol'nik, and Govrilov (1977) have provided one of the most thorough analyses of temperature effects in their studies of avian *existence metabolism.* Existence metabolism is measured as the amount of food energy consumed by birds held in small cages in a given environment less the amount of energy egested. Under the reasonable assumption that assimilation efficiency is not affected by external temperature, existence metabolism should be proportional to ingestion rate. The effects of temperature on existence metabolism (Appendixes IIIa, Va) should, therefore, reflect the effects of temperature on ingestion rate.

Existence metabolism parallels resting metabolic rate both within the thermal neutral zone and at $0°$ C (Figure 7.3), and, at both temperatures, existence metabolism is about 1.4 times the resting rate. Low temperatures lead to a greater increase in both existence and resting metabolism in small birds.

Qualitatively similar changes are suggested by equations describing the *temperature coefficient of existence metabolism* as a function of body size (Appendix Va). This parameter has the units Watts per degree centigrade and is an analog to conductance, for it measures an increase in power demand per degree of temperature differential between the body and a cool environment. Like conductance, this temperature coefficient increases with body size, indicating that a given temperature gradient requires a greater increase in total ingestion rate in larger animals. Also like conductance, the exponent of mass in these relations is less than $\frac{3}{4}$, showing that the increase in ingestion of large animals, relative to that in the thermal neutral zone, is less than the relative increase of small animals (see Table 5.6). The net effects of both size and low temperatures on rates of ingestion and respiration are similar (Figure 7.3). However, the allometric relations for the temperature coefficient of existence metabolism and for conductance have markedly different slopes (Figure 7.4). If real, these differences must be offset by differences in the lower

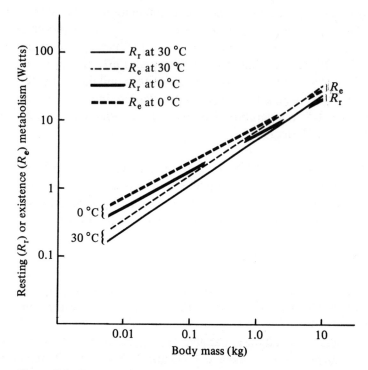

Figure 7.3. Temperature has a similar effect on both resting respiration rate (R_r, in Watts) and existence metabolism (R_e, in Watts), an index of ingestion rate. Both variables are similar functions of body mass and both undergo similar upward displacements at low ambient temperatures. The equations here describe the response of passerine birds kept in a winter light regime of about 10 h light : 14 h dark. Similar equations exist for nonpasserine birds and for summer light conditions (Appendix III).

critical temperatures of ingestion and metabolism. The results in Figure 7.4 suggest that some differences do exist between the effect of low temperature on resting metabolism and ingestion (or existence metabolism) but the net effect (Figure 7.3) suggests that these differences are not very important. The point requires further research.

Prey size

Ingestion rate is only a part of the interaction between an animal and its food. This relation would be still better defined if we could also indicate

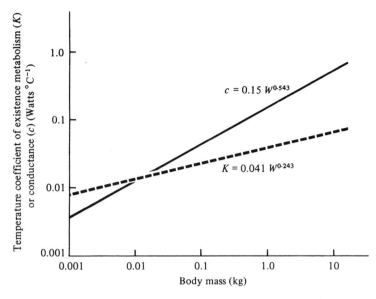

Figure 7.4. The effect of size on conductance (c, in Watts C^{-1}) and the temperature coefficient of existence metabolism (K, same units) of passerine birds in a winter light regime (10 h light : 14 h dark). Similar relations for other birds and light conditions are given in Appendix Va (Kendeigh, Dol'nik, & Govrilov 1977).

which prey will support a predator's demands. In the context of this book, an important step would identify the size relation of predator and prey.

Most of us intuitively expect some positive trend between predator and prey size. Supporting evidence is available for insectivorous birds (Hespenheide 1973), mammals (Rosenzweig 1966), raptors (Schoener 1968), and a variety of other animals (Wilson 1975). Unfortunately, these trends frequently appear weak; the relations have rarely been given quantitative treatment; and it is easy to think of probable exceptions to any general trend. For example, the prey of parasites are usually very large relative to predator size and those of baleen whales very small. Even predators of similar size and taxonomic group may show consistent differences in prey size (Hespenheide 1973). In other words, one can easily imagine why no general rule should apply. Perhaps we ought to expect a family of curves (Enders 1975) or a continuum among different foraging strategies.

The search for allometric trends always leads to this dilemma: As scientists, we want generality in our theory; but, as biologists, we recognize the existence of many exceptions. This problem may be partially resolved by an arbitrary decision to sacrifice detail for generality, the choice of this book, or to prefer the specific over the general, the choice of the scientific specialist. The decision depends on the goals and scientific questions of the individual researcher.

Basic relations

The most thorough analysis of the size relations of predator and prey treats published data for 103 species of terrestrial vertebrates (A. Vézina unpublished). For that study, prey were assigned to base 2 logarithmic size classes – in other words, the classes range from 0.25 to 0.5, from 0.5 to 1, from 1 to 2, from 2 to 4 kg, and so forth – in order to generate frequency distributions of prey number versus size that were approximately normally distributed about the mean. This normalization allows the use of parametric statistical analyses and produces a sufficient number of size classes for statistical treatment. Once these distributions were constructed, the mean prey size and the variance in prey size (actually the geometric mean and the log variance, since the data had been logarithmically transformed) were calculated for each predator.

Regression of mean prey size, calculated in this way, shows a strong effect of predator size. The data fell into two groups requiring separate statistical treatment (Figure 7.5). Mammals and birds of prey eat relatively big animals; on average, their prey weigh about 10% of their own weight. The average prey of lizards, amphibians, piscivorous, and insectivorous birds weigh only 0.2% of the predator's body mass. The exponents of mass in both power equations relating prey and predator size (Appendix VIIb; Figure 7.5) do not differ significantly from unity; mean prey size is a constant proportion of predator size. The two classes of predators were called *large-prey eaters* and *small-prey eaters* to avoid new terms until the phenomena to be described were firmly established. The division may represent one between animals that rend their prey (large-prey eaters) and those that swallow it whole (small-prey eaters). Further, preferably independent, analyses are required to show how constant such a separation might be.

In contrast to these obvious trends in mean size, the variances of the transformed frequency distributions of prey sizes showed no effect of size within each predator group. Nevertheless large- and small-prey eaters differ with respect to variance, and inferences can be drawn concerning the ranges of prey sizes taken by each predator class. Constant variance of prey size within each predator group implies that a log-normal

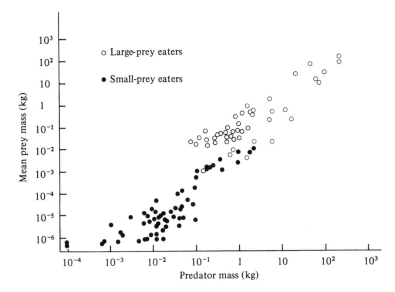

Figure 7.5. The relationship between mean prey size (W_{prey}) and predator size (W_{pred}), in kilograms, for mammals and birds of prey (large-prey eaters) and for lizards, amphibians, seabirds, and insectivorous birds (small-prey eaters). Equations describing these relations are $W_{prey} = 0.109W_{pred}^{1.16}$, for large-prey eaters, and $W_{prey} = 0.0018W_{pred}^{1.18}$, for small-prey eaters. Neither slope differs significantly from unity. Modified from A. Vézina (unpublished).

approximation of the prey size distribution for a given predator can be constructed from the appropriate equation in Figure 7.5 and the average variance for the foraging mode.

Large-prey eaters are more variable in their choice of prey sizes (mean variance = 4.0) than are small-prey eaters (mean variance = 2.9), and the variances themselves vary more among the large-prey eaters. The analysis was extended to estimate the maximum and minimum prey sizes. These were estimated for each predator as the upper and lower 99% limits of each prey size frequency distribution, and these limits were regressed against size. For small-prey eaters, this treatment resulted in three parallel lines corresponding to the maximum, mean, and minimum prey size. The lines for large-prey eaters appear homologous, but their statistical basis is far weaker because variation in prey size among predators is much greater for this foraging group. These statistics indicate that mammals and birds of prey enjoy a larger and less predictable range of food sizes than other terrestrial vertebrates.

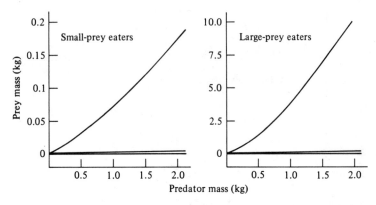

Figure 7.6. An untransformed representation of the relation between predator size and the maximum, minimum, and mean size of its prey. Once transformed to logarithms the data suggest a log-normal distribution about the mean; the untransformed data are heavily skewed toward smaller-sized animals for both large- and small-prey eaters. Equations describing these relations are given in Appendix VIIb. Modified from A. Vézina (unpublished).

Extensions and implications

Frequency distributions of prey size. The reader should recognize that a near-normal distribution of the logarithm of prey size implies that, without transformation, the distribution of prey size about the average is highly skewed toward small prey. This is shown graphically in Figure 7.6, which shows that the upper limit of prey size rises rapidly with predator body mass, whereas the mean and minimum size both remain small. This is one effect of logarithmic transformation. If the frequency distribution is log-normally distributed about the mean size (say about 10 g), then there will be as many prey in the interval between 5 and 10 g as there are in the 10- to 20-g class. This, of course, implies that there are fewer prey in the class between 10 and 15 g than there are in the logarithmic classes from 5 to 10 g. This effect is demonstrated in Figure 7.6.

The difference in the range of prey size reflects real differences between the two groups. Large-prey eaters frequently feed on large prey when it is available and small ones when it is not (Wilson 1976). This flexibility may result in a bimodal distribution of prey size for some large-prey eaters. Such prey size distributions are not normalized by logarithmic transformation, and this undoubtedly contributes to the greater variance associated with large-prey eaters. Small-prey eaters

Table 7.4. *Calculation of killing rate (Kr, number of individual prey killed d^{-1}) from ingestion rate (I, in kg animal^{-1} d^{-1}) and prey size (W_{prey}, in kg) for poikilothermic and homeothermic large- and small-prey-eaters (subscripts p, h, L, and S, respectively)*

From Farlow (1976),

$$I_h = 0.130W_{pred.}^{0.69}$$
$$I_p = 0.0096W_{pred.}^{0.82}$$
$$\therefore \quad Kr = IF_i/W_{prey\,(i)}$$

where F_i is the frequency of occurrence of size class i in the predator's diet, i is the position of class i with respect to the mean size class ($i = 0$), and $W_{prey\,(i)}$ is the midpoint of size class i. Since the frequency of occurrence follows a normal (actually lognormal) distribution, F_i can be described by the equation for a normal curve:

$$F_i = [1/(s\ \sqrt{2\pi})]\exp[-(i/s)^2/2]$$

where s is the standard deviation of the size distribution around the mean. For large-prey eaters,

$$s_L = 2.00$$

and for small-prey eaters,

$$s_S = 1.70$$

The mean prey size for each foraging group is given by the equations in Figure 7.5:

$$_LW_{prey\,(0)} = 0.109W_{pred.}^{1.16}$$
$$_SW_{prey\,(0)} = 0.00187W_{pred.}^{1.18}$$

I. The killing rate of large-prey-eating homeotherms

$$Kr_{Lh} = \sum I_h F_i/_L W_{prey\,(i)}$$

but since prey are grouped in $\log_2$ size classes $_LW_{prey\,(i)}$ is a function of mean prey size such that

$$_LW_{prey\,(i)} = 2^i \left(_LW_{prey\,(0)}\right)$$
$$\therefore \quad Kr_{Lh} = (I_h/_L W_{prey\,(0)}) \sum (F_i/2^i)$$
$$= (0.130W_{pred.}^{0.69}/0.109W_{pred.}^{1.16}) \sum (F_i 2^{-i})$$
$$= 1.19W_{pred.}^{-0.47} \sum (F_i\ 2^{-i})$$

If the summation term is considered over nine size classes as an approximation of the 95% limits of the prey distribution, then

$$\sum_{-4}^{4} F_i 2^{-i} = 2.52$$

and

$$Kr_{Lh} = 3.00W_{pred.}^{-0.47}$$

The killing rate of large-prey-eating homeotherms falls rapidly as predator body mass increases.

Table 7.4. (*cont.*)

II. Similarly the killing rate of small-prey-eating homeotherms can be calculated as

$$Kr_{Sh} = 137W_{pred.}^{-0.49}$$

III. The killing rate of small-prey-eating poikilotherms

$$Kr_{Sp} = 10.1W_{pred.}^{-0.36}$$

IV. For hypothetical large-prey-eating poikilotherms

$$Kr_{Lp} = 0.22W_{pred.}^{-0.34}$$

Killing rate is increased when ingestion rate is high (i.e., among homeotherms and among small predators) and decreased when prey size is large (among large-prey eaters and among bigger predators).

Note: Ingestion rates were transformed from those of Farlow (1976) assuming that 1 Watt = 0.012 kg fresh mass d^{-1}. These equations make no allowance for incomplete utilization of the prey among large-prey eaters. This would increase killing rates by about 20% (Farlow 1976) but is difficult to include in these calculations, since predators must eat a greater proportion of their prey's body as prey size decreases relative to predator size. $W_{pred.}$ is the body mass of the predator in kg.

cannot respond to seasonal declines in the abundance of their preferred prey with a facultative switch to larger food. Instead, they respond by hibernation or migration to regions where appropriate prey can be found.

Killing rates and their effects on foraging. Prey size can be combined with ingestion rate to yield estimates of the number of kills a predator must make to feed itself. Conceptually, this "killing" rate equals daily ingestion rate divided by prey size, but the calculation in Table 7.4 is slightly more complex, because it incorporates the observation that predators kill more prey that are smaller than average than prey that are larger. However, both calculations yield similar trends. Homeotherms that specialize in small prey must forage continuously to get enough food to supply their high metabolic rates. For example, a 10-g warbler must catch 1,300 prey d^{-1} (Figure 7.7) or almost two prey items per minute of an active day lasting 12 h. A 10-g poikilotherm requires so much less energy

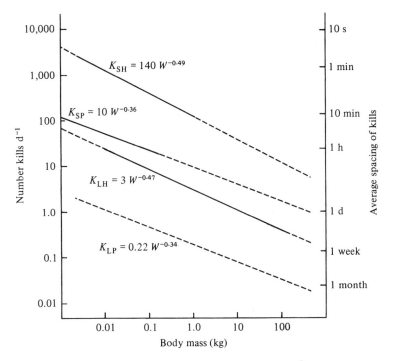

Figure 7.7. Killing rate (Kr, number of prey killed d^{-1}) for predators with different metabolic and prey-size strategies as calculated in Table 7.4. (The subscripts L, S, h, and p indicate large-prey eaters, small-prey eaters, homeotherms, and poikilotherms, respectively.) Dotted lines indicate extrapolations, and the right-hand axis indicates the average wait between kills if hunting is continuous. Periods of rest will, therefore, decrease this spacing during periods of activity.

that it could satisfy its daily needs with only 50 prey, one every quarter-hour. Large-prey eaters need kill far less frequently. For example, a jaguar weighing 100 kg might meet its needs with one prey every 3 d and a hypothetical large-prey-eating poikilotherm, perhaps a large constrictor or a dinosaur, weighing 100 kg would kill only once in 3 weeks (Table 7.4).

The prey of the largest carnivores must frequently exceed the predator's capacity. For example, the mean prey size for large-prey eaters weighing 4 kg or more exceeds daily food demands. Above 40 kg, predators kill less than once per day (Table 7.4). Large mammalian predators must frequently waste food or they must adopt strategies to use the excess. Both alternatives hold implications for predator behavior. For example, increase in the size of potential prey should lead to increase in gorging (Curio 1974), scavenging, pack hunting (Bowen 1981; Owens & Owens 1980), and food caching in large mammal communities.

Other organisms. The applicability of these predator categories and regression equations to other animals has not been demonstrated. Because fish eat food similar to seabirds and usually swallow their food whole, one would suppose that they may be described by the small-prey eater curve. This seems to be the case for cod (Ware 1980; Appendix VIIb) but A. Vézina (unpublished) found data (Lawler 1965) that suggest that pike eat larger prey than would be predicted. Enders (1975) suggested that predatory arthropods may lie on an extension of the large-prey eater relation. Lawton's (1970) data for a damsel fly nymph support this (A. Vézina unpublished). A. Vézina's data included no poikilotherm that could be considered a large-prey eater, but larger snakes (Godley 1980), crocodilians, or varanids may fall into this group. Obviously, far more analyses are required before the concept of two basic foraging strategies can be accepted or rejected.

Some size-related trends also exist in the food of herbivores and detritivores, which deserve mention here. Cammen (1980) has used multiple regression to show that aquatic detritivores ingest more substrate when the concentration of organic matter is low. Larger detritivores normally feed on less rich substrate and, therefore, pass more material. Cammen found (Appendix VIIa) that although total ingestion rate rises as W^1 the ingestion rate of organic material rises only as $W^{0.75}$. Apparently the concentration of organic material in the diet of benthic detritivores declines approximately as $W^{-1/4}$. Smaller animals are, therefore, more selective than large; Cammen's data suggest that this is achieved by the selection of richer sites rather than the selection of richer material from similar sites. Case (1979) suggests that an analogous situation exists among African ungulate herbivores. Larger animals tend to graze on food of lower nutritional value, and smaller species tend to browse more selectively. Wasserman and Mitter (1978) similarly suggest that large caterpillars are more catholic in their tastes than small species.

In overview, this chapter began by establishing basic descriptions of the effect of size on ingestion rate. As expected, ingestion rate rises as $W^{3/4}$ and homeotherms require much more food than poikilotherms. The effect of other factors, like temperature, was then considered briefly. The third section in this chapter describes the size relations between predator and prey. Two types of predator were identified, large- and small-prey eaters; the mean sizes of prey of the two types differ by two orders of magnitude. Finally, some of the implications of these relations were examined by comparing prey size and ingestion rate.

8

Production: Growth and reproduction

Production is the last component of the balanced growth equation that we can examine in depth and probably the most important. Production determines the amount of exploitation by man or natural predators that a population can withstand, the capacity of a population to recover from depredation, and its ability to resist control. Conversely, production defines the role of an animal population as a continuing resource for other members of its community, and, therefore, it determines much of the population's role in directing mass and energy to other parts of the community. Finally, the relationships of production to ingestion and respiration describe efficiencies of resource use around which animal communities must be organized and upon which human utilization of both wild and domestic stocks depends (Ames 1980).

The primary goal of this chapter is an equation or set of equations that predicts average rates of total production from animal size. These equations are essential for the allometric definition of the balanced growth equation, which is an underlying theme of this book, and for further analysis of the implications of body size in animal ecology. However, relations that predict average individual rates of production are rare and often imprecise. Were this chapter to limit itself to predicting individual production rates, it would seriously underestimate our knowledge of the scaling of the production process. Body size–production relations for use in the balanced growth equation will be treated at the end of the chapter. First, two related areas will be examined: the allometry of individual life history and that of population production.

The scaling of life history

Life history includes all the events that mark an organism's path from conception to death. The most important events are birth, sexual maturity, reproduction, and death, but the term also includes less dramatic events, like the attainment of half-adult mass or the dissolution of the family unit. Life history, therefore, includes most, if not all, of what was once termed *ontogeny* and *reproduction*. The scaling of life history parameters to body size is most thoroughly known for eutherian mammals. These mammalian relations are, therefore, examined first so they may serve as a standard in comparison with other, less well-known taxa.

A eutherian model

Equations describing much of the life history of eutherian mammals are listed in Appendixes VIIIa,b and shown graphically in Figure 8.1. These mammals begin life as a fertilized egg of relatively constant size (Brody 1945), and they begin to grow after implantation of the fertilized egg in the uterus. Taylor (1965) suggests that this usually occurs $3\frac{1}{2}$ days after conception, but, in weasels and other animals, implantation may be long delayed (Sadleir 1969). Once begun, fetal growth rate accelerates with time so that fetal size can be described either as an exponentially increasing function (Brody 1945; Taylor 1965) or as a modified power function (Huggett & Widdas 1951; Payne & Wheeler 1967a) of time. This period of accelerating growth ends near birth and is replaced by a phase in which growth rate is a decelerating function of time until adult mass is achieved. This body size is maintained until the animal approaches death. Mammalian growth is, therefore, described as a sigmoid curve like that in Figure 8.1a. The precise shape of the curve partly reflects the logarithmic transformation of the time axis. If growth is plotted on arithmetic axes, the accelerating phase of growth is confined to a narrow region near the vertical axis. If the axes are both logarithmic, the accelerating portion of the curve is more obvious and the decelerating phase less so.

The timing of ontogeny. This growth curve is punctuated by four major events: birth, weaning, sexual maturity, and death. Adult size has a very regular effect on spacing: Almost all exponents in equations scaling ontogenetic timing to size are close to the expected value of $\frac{1}{4}$ (Appendix VIIIa). This implies that, regardless of size, a given developmental phase requires a constant proportion of a mammal's life. About 2% of the maximum life span is passed between conception and birth, about 3% is over at weaning, and the average mammal still has 90% of its maximum life span ahead when sexual maturity is achieved; the average age at death is roughly half the maximum span. This proportionality also extends to arbitrary stages in animal development. Taylor (1965, 1968) has shown that most mammals take an equal fraction of their lives to reach a given fraction of their adult mass. Consequently, if time is

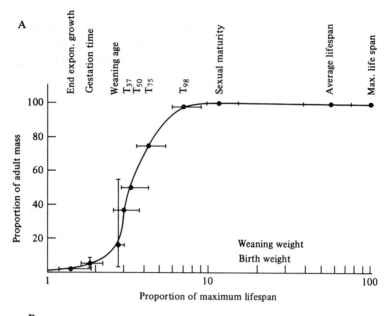

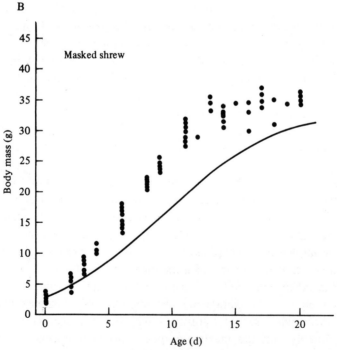

Figure 8.1. (A) Generalized plot illustrating the life history of a typical mammal. The independent variable, time, is scaled as a proportion of maximum life span and the dependent variable, body mass, as a proportion of adult mass. Because the timing of various life history events and the body masses associated with these events scale similarly with size, the plot should apply to all eutherian mammals. This model

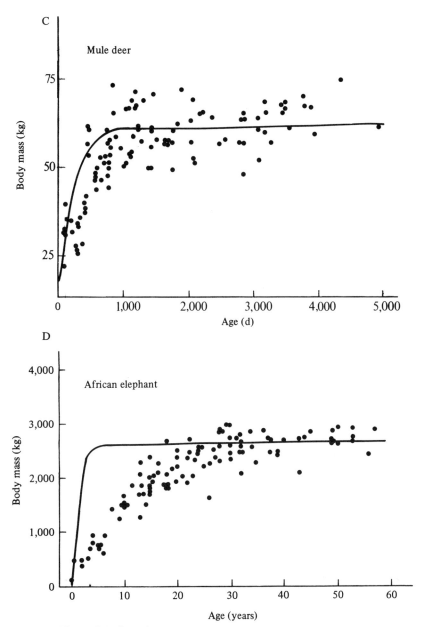

Figure 8.1. (*cont.*)
is compared with data describing the growth of (B) shrews (Forsyth
1976), (C) male deer (Walmo 1981), and (D) elephants (Laws, Parker,
& Johnstone 1975). The general model describes the growth of deer
quite well but underestimates the size of juvenile shrews and overestimates that of young elephants.

expressed as a proportion of maximum age and size as a proportion of adult mass, the resultant curve (Figure 8.1) should apply to all eutherian mammals.

Similar exponents are repeatedly encountered in dealing with physiological time (Lindstedt & Calder 1981). For example, the body mass exponents for allometric equations describing the time for a single breath or heartbeat (calculated as the inverse of pulmonary or cardiac frequency; Appendixes IVa,b) are also close to $\frac{1}{4}$. This has led to the gothic calculation that, on average, a full life is metered by each of 250 million breaths and 1.2 billion contractions of the heart.

Size in ontogeny. Body sizes achieved at each ontogenetic stage are similarly scaled (Figure 8.1). For 1-kg mammals, individual birth mass is about 6%, and litter mass is about 16% of maternal mass. This proportionality suggests that the average number of young per litter is approximately ($\frac{16}{6} = $)2.7. A slight (Millar 1981; Tuomi 1980), but usually statistically insignificant, trend (Blueweiss et al. 1978; Millar 1977; Sacher & Staffeldt 1974) to smaller litters in larger species is reflected by small differences in the exponents of the neonate and litter mass relations (Appendix VIIIb).

The slopes of relations between offspring mass and adult mass are less than unity. At birth and at weaning, the offspring of larger mammals are, therefore, farther from adult size than are the offspring of smaller species. If some disadvantage is associated with relatively small size of newborn or newly independent individuals, this disadvantage must fall more heavily on the young of larger species. This may be offset by greater parental care or by greater absolute size, but, as the evidence for differential juvenile mortality rates is equivocal, this problem can only be resolved with further study.

Somatic growth rates. Several mathematical models are routinely fitted to somatic growth curves like those in Figure 8.1. Three of the most popular are listed in Table 8.1 with two modifications to treat the earlier phase of accelerating growth. Underlying mechanisms have been suggested for each model, but these are irrelevant to the present purposes. Usually, the listed alternatives are equally good descriptors of the data (Ricklefs 1967), and their rationales have fallen into disuse. In all three equations, constants are determined by fitting the selected model to size-at-age data for each individual or population under study. In each case, the equation requires three constants: The minimum size to which the equation applies

Table 8.1. *Three equations that describe individual body mass* (W_t) *at age t as a function of maximum body mass* (W_∞) *and initial body mass* (W_0)

Growth curves

Gompertz: $W_t = W_0 \exp(G^{(1-e^{-gt})})$

Logistic: $W_t = W_\infty W_0\, e^{gt}/(W_\infty - W_0 + W_0 e^{gt})$

Von Bertalanffy: $W_t = W_\infty(1 - be^{-gt})^3$

Curves from early growth:

Exponential: $W_t = W_0 e^{kt}$

Power: $W_t = W_0 t^k$

Growth rates

Gompertz: $dW/dt = W_0 \exp(G^{(1-e^{-g})} + G^{(1-e^{-gt})})$

Logistic: $dW/dt = gW_t(W_\infty - W_t)/W_\infty$

Von Bertalanffy: $dW/dt = W_\infty (bge^{-gt})^3$

Exponential: $dW/dt = W_0 ke^{kt}$

Power: $dW/dt = W_0 kt^{k-1}$

Note: W_0 and $t = 0$ are not the size and age at birth but rather the smallest size and earliest age at which the equations apply. Each of the three equations for the growth curve requires three constant (W_0, W_∞, and g or k), which need not be identical among different formulations. Two formulas to describe early (for mammals, largely prenatal) growth are also given. Growth rates are the first derivatives of these curves with respect to time (Ricklefs 1968).

(W_0 or, in von Bertalanffy's equation, a function of b), the maximum size toward which growth tends (W_∞ or, in the Gompertz equation, $W_0 \exp G$), and an exponential rate constant of growth (g), which describes the effect of age and size on growth rate. The first two constants are properties of the organism under study and are necessary to specify application of the equation. The last constant reflects the fractional change in size or growth rate over the life span. This last constant may be predicted from other variables, among them size (Ricklefs 1969, 1979).

The equations in Table 8.1 condense much of the information that appears in Figure 8.1. Unfortunately, the analyses required to define such equations are tedious, and few taxa have been appropriately analyzed. One cannot compare the allometry of individual growth within or among larger taxa using the relations in Table 8.1, because too few relations are available.

Fortunately, another, cruder approach does permit these comparisons. Ricklefs (1973) suggested that the average growth rate as an animal grows from 10 to 90% of its adult mass can serve as a suitable index for comparison of complex growth curves. Case (1978) used this parameter extensively in his comparative study of vertebrate growth. For mammals, his analyses show that growth rates (in Watts) rise as adult mass to the power 0.72. The mean exponent for all mammalian subgroups ($\bar{X} = 0.64$, $SE = 0.03$, $N = 18$; Appendix VIIId) is lower because growth rate increases slowly with adult mass among primates and ungulates ($\bar{X} = 0.46$, $SE = 0.05$, $N = 5$).

Reproductive growth and reproductive effort. Reproductive growth rate, or the rate of laying down of fetal tissues, presents a problem analogous to that of somatic growth rates. Ideally, reproductive growth rate would be determined as a function of both adult mass and fetal age or size. This involves fitting fetal size-at-age data to a fetal growth model (Table 8.1), regressing the fitted constants against adult size, introduction of the regression equation into the growth model, and then calculating the first derivative of this relation with respect to time. This intensive analysis has not yet been performed.

A simpler estimate of reproductive growth rate for mammals is sometimes calculated as litter mass divided by gestation time (Bartels 1970; Blueweiss et al. 1978; Payne & Wheeler 1967b, 1968; Appendix VIIId). This index underestimates the total tissue growth associated with mammalian reproduction, for it ignores growth of the uterus. At term, the litter accounts for only 60% of the total uterine mass (Robbins & Robbins 1979), so the placenta and other tissues represent a sizable share of this aspect of parental investment. This calculation of fetal growth rate also misrepresents an increasing value as a constant. The production and demands associated with pregnancy reach a maximum just prior to birth. The calculation, therefore, shares many of the shortcomings of Case's (1978) representation of individual growth. Nevertheless, both parameters have the double appeal of simplicity and, in the context of this book, existing allometric relationships.

Defined in this way, reproductive growth rate scales only as female mass (W_f) to the 0.5 or 0.6 power (Blueweiss et al. 1978; Payne & Wheeler 1967b, 1968). Average rates of fetal growth, therefore, rise more slowly with maternal size than most other physiological and ecological rates. The elevations of these relationships (0.068 to 0.38 Watt for a 1-kg animal) are somewhat less than those for individual growth (0.20 to 0.91 Watt; Case 1978) and would remain so even if corrections for uterine growth were applied. This difference between prenatal and postnatal growth increases among larger mammals.

The eccentric slopes for reproductive growth have implications for the demands that reproduction places on the mother. Reproductive costs are considered critical determinants in both individual survival and genetic continuance; they provide a major focus in contemporary discussions of life history. If reproductive effort is defined as the proportion of available energy and material devoted to reproduction, then reproductive effort declines with body size. In other words, larger mammals devote a smaller proportion of ingestion, assimilation, or assimilation less respiration to reproduction than do small mammals. Since all terms describing the rates of energy gain rise as $W^{3/4}$ and that for reproductive costs only as $W^{0.55}$, then reproductive expenditure relative to supply must decline as $(W^{0.55}/W^{0.75} =) W^{-0.2}$.

The elaboration of fetal tissue is not the only, nor the major, cost associated with mammalian reproduction. However, a similar reduction in reproductive costs is suggested by several other rates (Table 8.2). A consideration of these rates will complete this section on individual growth.

In mammals, the highest additional energy demands associated with reproduction occur just before weaning when the increased demands of large offspring must be met by the mother's capacity to produce milk. Millar (1977) calculated the metabolic rate of litters at this time and concluded that maximum reproductive cost (calculated as the ratio of offspring to maternal respiration) fell as maternal mass to the power -0.2. However, the observed scaling of peak rates of milk production is proportional to maternal metabolism (Hanwell & Peaker 1977; Linzell 1972; Payne & Wheeler 1968; Appendix VIIId), which suggests that reproductive effort associated with peak milk production is unaffected by size.

Sadleir (1980) suggests that, in deer, total milk production over the nursing period also varies as $W_f^{3/4}$. If this is generally so, then the average rate of milk production during nursing will rise only as $W_f^{0.60}$, because the period of lactation is longer in larger species (Table 8.2). Perhaps peak rates of milk production are equally onerous in large and small mammals, but small animals may need to maintain such rates for a greater proportion of the lactation period. Lactation would then place a heavier burden on smaller species.

Table 8.2. *The effect of maternal size* (W_f, *in kg*) *on components of mammalian reproductive effort defined as the ratio of rates of biological expenditure associated with reproduction to maternal metabolic rate* (R_f)

From Kleiber (1961),

$$R_f = 3.28 W_f^{0.76}$$

I. From Blueweiss et al. (1978),

reproductive growth rate $(G_r) = 0.16 W^{0.57}$

reproductive effort $= G_r / R_f$

$$= 0.049 W^{-0.19}$$

II. From Millar (1977),

reproductive effort $=$ metabolic rate of litter at weaning$/R_f$

$$= 0.96 W^{-0.17}$$

III. From Hanwell & Peaker (1977),

peak rate of milk production $(G_{milk\,(max)}) = 6.16 W^{0.75}$

reproductive effort $= G_{milk\,(max)} / R_f$

$$= 1.9 W_f^{-0.01}$$

IV. From Sadleir (1980),

total milk production $(\sum G_{milk\,(ave)}$, in J$) = 17.2 \times 10^6 W_f^{0.75}$

From Blaxter (1971),

Time to weaning $(T_w$, in s$) = 2.92 \times 10^6 W_f^{0.15}$

∴ average rate of milk production $(G_{milk\,(ave)}) = \sum G_{milk\,(ave)} / T_w$

$$= 5.89 W^{0.60}$$

reproductive effort $= G_{milk\,(ave)} / R_f$

$$= 1.8 W^{-0.16}$$

V. From Payne & Wheeler (1968),

milk protein concentration (protein) $\propto W^{-0.28}$

From Hanwell & Peaker (1977),

$G_{milk\,(max)}$ (in kg d^{-1}) $= 0.0835 W^{0.75}$

rate of production of milk protein $\propto W^{-0.28} \times 0.835 W^{0.75}$

$$\propto W^{0.47}$$

reproductive effort $=$ rate of milk Protein production$/R_f$

$$= W^{-0.29}$$

Table 8.2. (*cont.*)

VI. From Brody (1945),

total metabolic cost associated with reproduction (C_r, in J) $= 1.84\ W_l^{1.2}$

where W_1 is litter mass in kg.

From Blueweiss et al. (1978),

$$\text{litter mass} = 0.16W_f^{0.82}$$
$$\text{gestation time } (T_g \text{ in s}) = 5.46 \times 10^6 W_f^{0.258}$$
$$\therefore\ C_r = 1.84 \times 10^7\ (0.16W_f^{0.82})^{1.2}$$
$$= 2.04 \times 10^6 W_f^{0.984}$$

Average increase in metabolic rate through reproduction $= C_r/T_g$
$$= 0.362W_f^{0.726}$$

$$\therefore\ \text{reproductive effort} = (C_r/T_g)/R_f$$
$$= 0.11W^{-0.024}$$

Of six indexes of mammalian reproductive effort, two (cases III and VI) suggest that effort is independent of size and four suggest that effort falls with increasing size. Since each index is independent of the others, the table suggests that reproductive effort is greatest among small mammals.

Note: Similar estimates of reproductive effort would be obtained if maternal ingestion or assimilation rates were substituted for respiration rate as a measure of resource availability. Rates in this table are expressed in Watts unless otherwise noted.

Lactational costs will be greater for small species in any case, because small, fast-growing species produce milk that is higher in protein. Payne and Wheeler (1968) reported that milk protein concentration declines as maternal mass to the power -0.28. Since milk production rises only as $W_f^{3/4}$, milk proteins need only be produced at a rate that rises as $W_f^{0.47}$.

Finally, Brody (1945; Appendix IIIf) described the total increase in metabolism associated with reproduction as a function of litter mass. With appropriate manipulation, it can be shown that the total metabolic cost of reproduction is directly proportional to maternal size and the average increase in metabolic rate as a result of reproduction, therefore, varies as $W_f^{3/4}$ (Table 8.2). This is an exception to the general pattern in Table 8.2, which shows that most indexes of reproductive efforts decline with maternal size. Since most of these measures are mutually independent, there is little contradiction in the observation that some indexes of reproductive effort are independent of size, whereas others decline among larger mammals. The net effect is a reduction in reproductive effort as size increases.

The allometry of life history in other homeotherms

Marsupials differ from eutherians in the details of fetal and prenatal development. Nevertheless, much of the eutherian model also applies to marsupials. Data presented by Russell (1982) shows that litter mass at the age of permanent exit from the pouch and at weaning rises more slowly than maternal mass (W_f) and approximately as $W_f^{3/4}$. The age at pouch exit and weaning rise as $W_f^{1/4}$. In consequence, reproductive effort will rise more slowly than metabolic rate. Marsupials differ from placental mammals of similar size in that individual offspring are larger at pouch exit and weaning than are eutherians at birth and weaning. Moreover, the size of marsupials at these stages is a nearly constant fraction of maternal mass. Perhaps marsupial reproduction permits more extended and more intensive parental care by avoiding the bottlenecks of placental size or pelvic passage, which may set a maximum size for the offspring of eutherians.

The eutherian model also applies to birds with very few modifications (Appendix VIII). Birds, like mammals, have determinate growth but reach a given proportion of adult mass somewhat earlier (Taylor 1968) than mammals of the same adult size. Growth rate increases approximately as $W^{3/4}$ (Ricklefs 1968, 1973) and ontogenetic periods, such as incubation time (Blueweiss et al. 1978; Rahn & Ar 1974), maturation time (Blueweiss et al. 1978; Taylor 1965), and life span (Blueweiss et al. 1978; Lindstedt & Calder 1976), all increase as the fourth root of adult mass. Time to first flight may increase more rapidly (Ricklefs 1968). Various masses (egg, hatchling, and total clutch mass; Appendix VIIIa) increase as $W_f^{3/4}$. Like mammals, birds show no trend in the number of offspring per clutch, but the mean of eggs in a clutch is greater ($\bar{X} = 4.85$; Blueweiss et al. 1978) than the average number of neonates per mammalian litter ($\bar{X} = 2.7$).

Estimates of avian reproductive effort, comparable to mammalian indexes discussed above, cannot be teased from the information in Appendix VIII. However, the close similarity in scaling of both periods and masses associated with avian and mammalian ontogenetics suggests that avian reproductive effort will also decline with size. Pearson (1968)

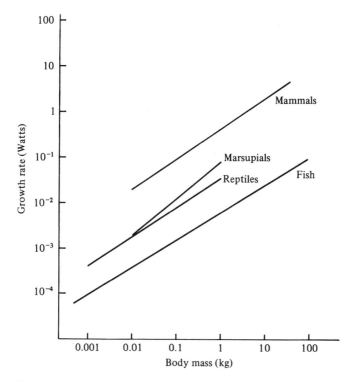

Figure 8.2. The effect of adult size on the somatic growth rate of the individual calculated as the average rate of increase as the animal grows from 10 to 90% of adult body size. Modified from Case (1978).

provided some circumstantial evidence for this when he showed that larger seabirds devote less time in searching for food for their nestlings. Despite the obvious differences in reproductive physiology of the two groups, one model applies fairly well to both.

Poikilotherms and the eutherian model

Many of the basic patterns in the mammalian model apply well to poikilotherms, but important differences exist too. Unlike homeotherms, many poikilotherms do not have a fixed maximum size but continue to grow throughout their lives. Even insects, which have determinate growth, usually grow for a much larger fraction of their lives than do mammals and birds. Rates of growth attained over this extended period are lower than those of homeotherms of comparable size (Figure 8.2). Despite this slower and indeterminate growth, somatic growth in

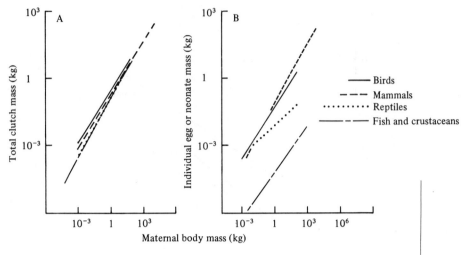

Figure 8.3. The relation between adult body mass and (A) clutch mass or (B) neonate mass in various taxa. Most animals make similar total reproductive investments in each clutch but differ in the allotment per individual neonate. Modified from Blueweiss et al. (1978).

poikilotherms appears sigmoid and can be described by the growth models in Table 8.1. This S shape reflects a marked decline in somatic growth once sexual maturity is achieved. Thereafter, reproductive growth rapidly assumes an increasing share of the available resources. The intervals between events in poikilotherm life history again scale roughly as $W^{1/4}$ (Appendix VIIIb). Blueweiss et al. (1978) suggest that maturation times and maximum longevities are similar in homeotherms and poikilotherms of the same size. Fenchel (1974), citing data compiled elsewhere (Altmann and Dittmer 1968), argued that maturation times in poikilotherms were an order of magnitude greater. Independent regressions for fish (Ware 1980) and mammals (Western 1979) suggest a much smaller difference. More striking similarities and differences are seen in comparing patterns of reproductive investment in the two metabolic groups. Unfortunately, the only available indexes are the masses of the clutch or egg and fecundity, since rates of reproductive growth among poikilotherms have been little studied.

Total mass of the litter does not vary much among vertebrates or, perhaps, invertebrates of similar size (Figure 8.3). This remarkable constancy across very different taxa suggests that total investment in a single reproductive event is a highly conservative property. Relative investment tends to decline with size but absolute investment rises as about $W^{3/4}$. This conservatism is not imposed by constraints of space, since most anatomical volumes rise as $W^{1.0}$ (Appendix IV). The exponent instead suggests some unidentified metabolic constraint. The elevations of clutch mass–maternal size relations are also similar among different taxa. Regardless of taxonomic affiliation, animals invest similar amounts of tissue in each reproductive bout.

Different organisms do differ remarkably in the amount of this investment directed to each offspring. Homeotherms have a small number of young. Each of these few receives both a relatively large prenatal investment and a considerable postnatal investment in care and feeding.

Poikilotherms usually provide less parental care and investment to their eggs and newborn than homeotherms. Individual offspring sizes increase with adult mass, but less dramatically than among homeotherms (Figure 8.3). Across many taxa, the relation between offspring and parent size may parallel that for homeotherms so that, at the coarsest level, neonate size scales as adult mass to the $\frac{3}{4}$ power (Blueweiss et al. 1978). This similarity is achieved despite wide scatter and consistent departures within even quite large taxa. Within such taxa, offspring size is less dramatically influenced by adult mass. This may reflect a non-allometric constraint on maximum egg size, such as the rate of oxygen diffusion into the egg. Reptiles, which have large eggs, may partly escape this restriction because of higher rates of oxygen diffusion from the air; Seymour and Ackerman (1980) have suggested that very large eggs may still be impractical for reptiles that bury their eggs, because subsurface oxygen levels may be low. This may be reflected in the intermediate position of reptiles in Figure 8.3. Small reptiles have offspring only slightly smaller than mammals of similar size, but large reptiles, like sea turtles and crocodiles, have quite small eggs relative to their size.

Since total litter mass is a similar function of body size for both homeotherms and poikilotherms, these trends in individual size imply that fecundity, the number of offspring per clutch, is higher in poikilotherms and much higher in large poikilotherms. Except for the smallest members of any group, poikilotherms rarely have clutches of one, two, or a few eggs. Apparently, most species either invest appreciably in reproduction or do not reproduce. If this were not the case, the lower limit of litter mass would be identical with individual egg mass, and variation in clutch size would increase markedly with body size.

Higher fecundity in larger poikilotherms implies higher mortalities (Table 8.3). This is most pronounced if mortality is expressed per generation or some similarly scaled unit of time but also holds if mortality

Table 8.3. *The effect of adult size on mortality as suggested by the scaling of fecundity and generation time to size*

I. For poikilotherms

$$\text{fecundity} \propto W^{0.5}$$
$$\text{generation time} \propto W^{1/4}$$
$$\text{birth rate} \propto \text{fecundity/generation time}$$
$$\therefore \quad \text{birth rate per generation} \propto W^{0.5}$$
$$\text{birth rate per unit time} \propto W^{1/4}$$

If population numbers are stable,

$$\text{mortality} = \text{natality}$$
$$\therefore \quad \text{death rate per generation} \propto W^{0.5}$$
$$\text{death rate per unit time} \propto W^{1/4}$$

Mortality rates rise with poikilotherm body mass.

II. For homeotherms

$$\text{fecundity} \propto W^{0}$$
$$\text{generation time} \propto W^{1/4}$$
$$\therefore \quad \text{rates of birth and death per generation} \propto W^{0}$$
$$\text{rates of birth and death per unit time} \propto W^{0}/W^{1/4}$$
$$= W^{-1/4}$$

Homeotherm mortality rates, per unit time, decline with size. Since smaller homeotherms may have more than one litter per year and since litter size may tend to decline with size, it is likely that mortality rates per generation slightly decline with size.

Note: Generation time is here approximated by maturation time (Bonner 1965) and will, therefore, underestimate generation time in animals that enjoy a long reproductive career after attaining sexual maturity; such species tend to be large. Fecundity is here taken as clutch size, not total lifetime fecundity. This will underestimate total fecundity, especially for larger species that reproduce more times per life span than do small species. If similar corrections apply to both generation time and fecundity (the numerator and denominator in this table) the calculations above should be approximately correct. All proportionalities in this table are derived from Blueweiss et al. (1978).

is expressed per unit of absolute time. Larger poikilotherms have higher birth rates and higher death rates than do smaller species. Pauly (1979) has found that fish mortality declines with adult size, but his analysis refers only to postlarval stocks. This must be compensated by proportionally greater larval mortality, otherwise the seas would brim with big fish. Larger homeotherms have smaller birth rates (Western 1979) and, therefore, have lower mortality rates per unit of absolute time but probably not per unit of biological or relative time. However, there is conflicting evidence regarding the effect of adult size on mortality (Banse 1979; Blueweiss et al. 1978; Western 1979).

Population production

Since the production of a population determines its capacity for economic exploitation, recovery, and control, population production is one of the key terms one wishes to predict. Population production is usually expressed as the ratio of total annual production (P) to average biomass ($\bar{B}$) and so is often referred to as the $P/\bar{B}$ ratio. It is equivalent to the average specific production rates of all members in the population weighted for their temporal duration or to the sum of all productivity terms associated with the life history again biased according to the number and duration of the population's members at each ontogenetic stage. This bias of production by population size structure and survivorship may lead to departures from the typical allometric exponent of size in equations describing specific rates, $-\frac{1}{4}$.

Some contradictory relations

The most critical allometric analysis of population production (Banse & Mosher 1980) is based on highly selected data from populations of terrestrial and aquatic invertebrates. Regression of these data on size at maturity (W_a) shows that $P/\bar{B}$ declines as $W_a^{-0.37}$ (Figure 8.4), which is significantly faster than the expected slope, $-\frac{1}{4}$, for specific rates. Apparently, populations of small invertebrates are dominated by highly productive juveniles and those of larger animals by less productive adults and subadults.

  Banse and Mosher (1980) also examined mammals, fish, insects, and other invertebrates (Appendix VIIIc). Although these analyses were more superficial and were based on fewer species, they tended to confirm the steeper slope observed for invertebrates; only fish have a slope close

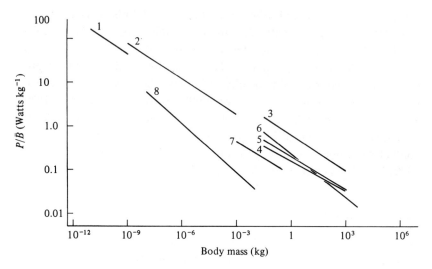

Figure 8.4. Competing relations which describe the ratio of population production to average biomass as a function of adult body size. Equations and sources for this figure are: (1) for unicells $r_{max} = 0.14W^{-0.28}$, (2) for poikilotherms, $r_{max} = 0.28W^{-0.27}$; and (3) for homeotherms, $r_{max} = 0.63W^{-0.28}$ (Fenchel 1974); (4) for tetrapods, $P/\bar{B} = 0.16W^{-0.23}$ and (5) for mammals, $P/\bar{B} = 0.20W^{-0.27}$ (Farlow 1976); also (6) for mammals, $P/\bar{B} = 0.23 W^{-0.33}$; (7) for fish, $P/\bar{B} = 0.074 W^{-0.26}$; and for invertebrates $P/\bar{B} = 0.0067W^{-0.37}$ (Banse & Mosher 1980).

to $-\frac{1}{4}$. These analyses showed marked differences in the elevation of the regression lines among groups (Figure 8.4). Production to biomass ratios in mammalian populations lie more than an order of magnitude above those of fish that are in turn another order of magnitude above those of invertebrates. Very small invertebrates and unicells were not analyzed by regression but appear to have still smaller $P/\bar{B}$ ratios.

These results do not conform to previous analyses. Fenchel (1974) analyzed maximum rate of population growth (r_{max} = maximum $P/\bar{B}$; Appendix VIIIc) of unicells, poikilotherms, and homeotherms. He concluded, like Banse and Mosher (1980), that differences in elevation exist among these groups. However, those differences he found were small, roughly twofold, and the slopes of all three relations were very close to $-\frac{1}{4}$. Fenchel's unicell relation has since been confirmed by other workers (Baldock, Baker, & Sleigh 1980; Finlay 1977). Farlow (1976) examined the effect of adult size on $P/\bar{B}$ for tetrapods. He proposed that, for homeotherms, a significant relation exists with a slope of -0.27 and a somewhat lower elevation than that of Banse and Mosher (1980). The two curves differ markedly only for small homeotherms. A significant relation also exists for all species of terrestrial tetrapods in Farlow's collection and again this curve is scarcely different from that for homeotherms. Equations describing the doubling time ($= \bar{B}/P$) for marine animals (Parsons 1980) suggest that the ratio of production to biomass falls as adult mass to the power $-\frac{1}{4}$ and this ratio is not different from that for mammals reported by Banse and Mosher (1980). Clearly, there are grounds for different opinions as to the effect of size on population production.

At present, these differences cannot be resolved. Indeed, the same data may sometimes be used to support different opinions. Blueweiss et al. (1978) treated all of Fenchel's (1974) data relating r_{max} to body size as a single curve and obtained a satisfactory fit. The different groupings promoted by Banse and Mosher (1980) could be treated similarly. Since small poikilotherms, like rotifers, and unicells lie below an extrapolation of the poikilotherm line and large poikilotherms, like fish, lie above it, one might suspect the existence of a general line with a flatter slope. Reanalysis of their poikilotherm data (Appendix VIIIc) yields a regression with a slope of -0.18. Inclusion of Arctic, tropical, and eccentric invertebrate points, which Banse and Mosher (1980) excluded, yielded a relation with a body mass exponent of -0.23 for invertebrates and regression of all homeotherms and poikilotherm data together gave a slope of -0.17. This lumping of animals substantially reduced differences in elevations between curves based on these data and others in Appendix VIIIc and substantially increased residual variation.

It is not yet possible to select among these interpretations of the data. Statistical selection requires extensive regions of overlap in size among the proposed groupings, but little overlap exists (Figure 8.4). Increased variance in the more general lines may indicate that such lines are less appropriate, but they may show that the phenomenon is actually highly variable. As more data become available, it should be possible to disconfirm one or more of the relations in Appendix VIIIc, because it describes the new data less well. Predictive power is the ultimate criterion by which a theory is judged.

This controversy lies at one extreme of a recurrent dichotomy in biology. Biologists are continuously impressed both by the scientific need for generality and the individuality of each organism under study. Each of us, therefore, strikes some balance in the amount of precision one is willing to sacrifice to gain generality. Since the value of the losses and gains are often subjective, this dilemma has long fostered bitter debate among "lumpers" and "splitters." This same division reappears in the contradictions noted above, although the current division exists between lumpers and even greater lumpers.

Some applications for production relationships

Population productivity is such an important aspect of a population's ecological role that testing and, if necessary, recasting of these theories in the light of new data are well worth the effort required.

Harvesting our natural resources. The potential utility of equations that predict $P/\bar{B}$ can be demonstrated using either the mammalian relationship of Farlow (1976) or that of Banse and Mosher (1980) for these are in general agreement for larger mammals. The North Atlantic harp seal fishery has long been criticized as placing the herds in danger of extinction. Winters (1978) used an extensive population study to estimate that annual pup production is about 335,000 from a population of $1\frac{1}{4}$ million. He estimated a total allowable catch of 220,000. A very similar estimate of pup production (315,000 to 333,000) may be made (Table 8.4) from Winters' estimate of population size, adult body mass, and either $P/\bar{B}$ relationship for mammals. Using an identical approach, one can calculate the probable production rates for Canadian game mammals and compare these to anticipated hunting mortality (Figure 8.5). For most mammals, hunting kill is close to average production rates. Heavily exploited populations, like these, probably have higher than average production rates, and this increase must offset nonhunting mortality. The low position of caribou with respect to the predicted production rates likely reflects government attempts to protect the declining herds and increased natural mortality in their rigorous environment.

The apparent predictive power of the mammalian equations contrasts with the doubtful nature of relations describing the scaling of $P/\bar{B}$ for invertebrates. For vertebrate poikilotherms, the available relations (Banse & Mosher 1980; Farlow 1976; Sheldon, Prakash, & Sutcliffe 1972) are in reasonable agreement. All relations also agree that populations of large poikilothermic invertebrates should be less exploited than populations of small invertebrates; unfortunately, the various equations also show very large quantitative differences. For example, Farlow's tetrapod data, if extended to invertebrates, would suggest that the production-to-biomass ratio of lobsters (mature body mass = 500 g) is about 1.1 year^{-1}. The equation of Banse and Mosher (1980) solves to 0.065 year^{-1} and that calculated from their data for all invertebrates prior to selection to

Table 8.4. *Calculation of harp seal production using Farlow's (1976) allometric relation describing the ratio of production to average biomass* ($P/\bar{B}$, *in year*$^{-1}$) *in mammalian populations as a function of adult body mass* (W_a, *in kg*)

From Winters (1978),

$$\text{population size } (N) = 1.25 \times 10^6$$

If adult body mass $(W_a) = 100\,\text{kg}$

$$\text{biomass of population } (B, \text{ in kg}) = W_a N$$
$$= 1.25 \times 10^8$$

From Farlow (1976),

$$P/\bar{B} = 0.906 W_a^{-0.266}$$
$$= 0.266$$
$$\therefore \quad \text{total production } (P, \text{ in kg year}^{-1}) = B \times P/\bar{B}$$
$$= 33.3 \times 10^6$$

If all production is represented by the growth of pups from 0 to 100 kg, then annual production of seal pups (individuals year^{-1}) $= P/W_a = 333,000$. This compares favorably with Winters' (1978) direct estimate of 335,000 pups and with an estimate (315,000) calculated as above but using the mammalian $P/\bar{B}$ relation of Banse & Mosher (1980) instead of Farlow's equation.

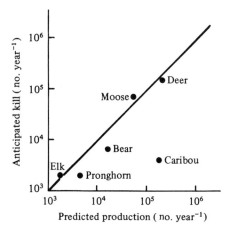

Figure 8.5. A comparison of the anticipated kill (Collins 1979) of large game in Canada and annual production estimated from Farlow's (1976) equation for mammals. With the exception of caribou, the allowed annual take is close to predicted production.

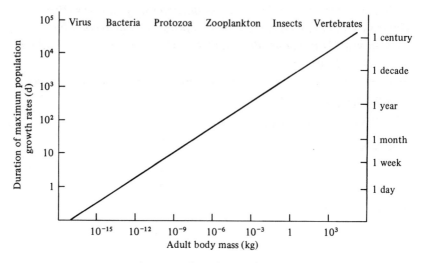

Figure 8.6. The effect of body size on the potential duration of maximum rates of population growth. This figure assumes that population growth is described by an exponential function such that population at time t, $P_t = P_0 \exp(r_{max}t)$, where P_0 is initial population density, t is time, and $r_{max}(d^{-1}) = 0.0041W^{-0.26}$ (Blueweiss et al. 1978). To complete the table, it was assumed that few populations ever expanded to more than 10^5 times initial density; the time required to do so (t_r, d) is thus $t_r = (\ln P_t - \ln P_0)/r_{max} = \ln(P_t/P_0)/r_{max} = 2{,}500\,W^{0.26}$.

0.29 year^{-1}. The equation listed in Parsons (1980) solves to 3.4 year^{-1}. Obviously these differences would dramatically affect the management of the lobster stocks. Some alternatives might extinguish the population, others might lead to serious underexploitation of a valuable, renewable resource. Although the latter choice is preferable because it is ecologically conservative, it may be needlessly expensive and morally untenable on a hungry planet.

The scaling of population dynamics. Estimates of the maximum rate of population growth can be used to calculate the amount of time required for a population to establish itself after a catastrophic decline or after colonizing a new habitat (Figure 8.6). To produce Figure 8.6, I assumed that some catastrophe had reduced all organisms to the extremely low density of 1 g km^{-2}. Thereafter, populations were assumed to grow at the maximum rate given by an equation describing intrinsic growth rate (r_{max}) as a function of size (Blueweiss et al. 1978). This rate was maintained until the population biomass was 100 kg km^{-2}, a density selected on the basis of average density observed for almost 300 animal populations (Chapter 10) rounded up from $32\,kg\,km^{-2}$ to the nearest order of magnitude. This great range in density implies that the durations of maximum growth in Figure 8.6 are upper estimates of values in nature. Although the population limits used in this calculation are rather arbitrary, the calculation will be approximately correct, because exponential growth of the population renders these limits relatively unimportant. If the population could only increase 100-fold instead of the assumed 100,000-fold, the duration of the maximum growth phase would only be halved.

Figure 8.6 has important implications for ecological investigations. It shows that the temporal scale of population change increases with animal size: What is a long time in entomology may be a short time in mammalogy or ornithology. Thus, populations of large animals may appear more stable simply because they are physiologically incapable of great change within the time frame of our observations. The rapid response of small organisms also indicates that monthly, weekly, or even biweekly observations of, for example, phytoplankton populations may be inappropriately long, whereas effective studies of elephant or moose populations may require decades. Similarly, significant fluctuations in the environmental parameters of microorganisms may occur in the space of a few hours, but those of large fish over months or years. Furthermore, the examination of community response at a fixed period after perturbation is inappropriate because small organisms may have overcome any effect of perturbation long before larger species approach equilibrium. If community equilibrium is influenced by its largest members, almost all experimental studies involving community manipulation are impractically short. Biological time scales as $W^{1/4}$. This fundamental relationship should be remembered in planning any programs of ecological research that involve populations of different-sized organisms.

An individual production term for the balanced growth equation

The allometric description of the balanced growth equation that is slowly being developed in this book is based on individual rates. This is the most common, most general, and most easily used of several allometric approaches. One should be able to apply these descriptions to any animal once body size and approximate taxonomic position are known and so to describe much of the autecology of the animal in question. If estimates of the size of the population are also available, the equations can be used to predict population rates, and if size–structure of the community is available, these relations can be applied to describe processes at the community and ecosystem levels.

In principle, population rates are not so versatile. They describe the behavior, not of an individual based on its size, but of the whole population based on average size at maturity. The population rate need not be equally distributed among the members of the population, so it does not lend itself to autecological use; its application to community processes requires the enumeration and classification of all major populations in the community, a difficult and time-consuming chore. Population equations require considerable effort from the investigator, since they are best used if population size is measured; this alone requires a relatively extensive process of sampling and enumeration. A balanced growth equation for populations would still be very useful for questions at the population level. Some steps have been taken toward this development (Banse & Mosher 1980), but, at present, the population approach has received far less attention than the allometry of individual rates. The only exception to this is the rate of production for which population estimates are more studied. The central problem in this section is, therefore, to develop some approximation of individual growth rate based on the better-established relations for $P/\bar{B}$.

The allometry of individual production: An interim solution

There is no standard definition of individual growth rate comparable to ingestion or respiration rate. The sum of the instantaneous rates of reproductive and somatic growth would be unsuitable, since instantaneous rates vary from zero to maximum values with seasonal changes in growth. Average rates that integrate across seasonal and ontogenetic differences in some manner analogous to average energy expenditure are needed. However, the use of long-term averages introduces the new problem of changing body size. As an approximation, Peters (1978b) plotted the sum of somatic and reproductive growth against average size over the period of growth and used the resultant relationship as an estimate of the effect of body size on individual production rate (Figure 8.7).

This plot shows that homeotherms and rapidly growing poikilotherms produce new tissue at a rate close to or slightly above that calculated from maximum population growth rate (Blueweiss et al. 1978). This correspondence might be expected. When losses to the population are small, all individual growth is added to the population, and, if individual produc-

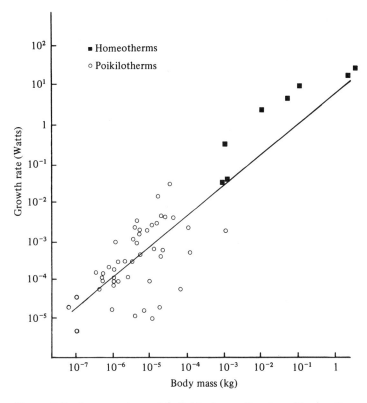

Figure 8.7. A comparison of individual growth rates of homeotherms and poikilotherms (data points) with the maximum rate of tissue production calculated from the allometric relation for r_{max}, the intrinsic rate of population growth. Individual growth rates were calculated as the sum of reproductive and somatic growth over a certain period divided by elapsed time; individual size is the average body mass over this period. The figure suggests that individual growth rate may be approximated by curves describing maximum population growth rates (Blueweiss et al. 1978; Peters 1978b).

tion is high, population growth will be too. In other words, a population can grow only as fast as the summed productivities of its individual components. Since no population can be entirely free of all losses, it is not surprising that the maximum rate of population growth is somewhat smaller than individual production rates.

Figure 8.7 shows that homeotherms exploit a very narrow, but high, range of individual productivities. Apparently, if homeotherms cannot grow near their maximum rate, they will not survive. This is obvious in considering homeotherm reproduction. Since their fecundity is so low, marked reductions in rates of individual reproduction can be achieved only by reducing birth rate to zero. Apparently, young homeotherms must also grow or perish; intermediate and arrested growth rates ("stunting") are not among their capabilities. Poikilothermic production is a more flexible process. Maximum values can be as high as those of homeotherms of similar size, but minimum values can be much lower. Pough (1980) has developed the theme that poikilothermy is an adaptation to fluctuating or chronically low food levels.

The relationship in Figure 8.7 applies only during periods of active growth. If one wishes a less restricted equation, the curve must be reduced to allow for nongrowing periods. Assuming that growth is restricted to the mild part of the year, such a relation might suggest rates only $\frac{1}{3}$ to $\frac{1}{2}$ that calculated from r_{max} (Figure 8.7). If the exponent of mass is unaffected, this average rate might be approximated by Farlow's (1976) equation for $P/\bar{B}$ modified to yield rates in Watts. For the present, Farlow's equation for homeotherms will be used to describe individual production for birds and mammals and that for tetrapods will be used for poikilotherms.

These are clearly only interim solutions, but available evidence suggests that they are not far wrong. Both Parsons (1980) and Sheldon, Prakash, and Sutcliffe (1972) present relations for the productivity of marine poikilotherms that, on conversion, are similar to Farlow's curves. Similar values have been reported for unicells (Baldock, Baker, & Sleigh 1980; Banse 1976; Eppley & Sloan 1966). Nevertheless, the lower relations and anomalous slopes reported by other authors should remind one that poikilothermic rates may be more varied. Systematic departures from "expectations" based on Farlow's curve may be encountered both with size and in certain communities. Banse and Mosher (1980) suggest that larger invertebrates could have lower productivities than expected; Båmstedt and Skjoldal (1980) found that marine zooplankton are less productive than would be predicted from Farlow's equations. Such variation may reflect the great flexibility of poikilothermic patterns of production. Banse and Mosher (1980) suggest that the steeper slope in their equation may reflect changes in population structure that ought not be mirrored by individual rates of production.

Ecological efficiencies

The combination of individual equations for respiration, ingestion, and production permits determination of the effect of body size on several

Table 8.5. *A comparison of various ecological efficiencies calculated as the ratios of allometric equations for rates of respiration (R), growth (G), defecation (D), assimilation (A), and ingestion (I)*

Term	Definition	Poikilotherm	Homeotherm
Respiration efficiency	R/I	$0.48W^{-0.06}$	$0.77W^{0.05}$
Growth efficiency	G/I	$0.21W^{-0.05}$	$0.019W^{0.03}$
	D/I	$\sim 0.31W^{-0.03}$	$\sim 0.22W^{0.01}$
Assimilation efficiency	A/I	$\sim 0.69W^{-0.06}$	$\sim 0.78W^{0.04}$
	R/A	~ 0.70	~ 0.98
Tissue growth efficiency	G/A	~ 0.30	$\sim 0.024W^{-0.02}$

Note: For homeotherms, $R = 8.2W^{0.75}$ (Hemmingsen 1960) $G = 0.20W^{0.73}$, and $I = 11W^{0.70}$ (Farlow 1976). For poikilotherms, $R = 0.38W^{0.76}$ (Robinson, Peters, & Zimmerman 1983); $G = 0.16W^{0.7}$, and $I = 0.78W^{0.82}$ (Farlow 1976). For both groups, D and A were calculated by difference as $A = R + G$ and $D = I - A$; R was taken as twice the standard metabolic rate: All rates are expressed in Watts and body size (W) in kg. Ecological efficiencies are defined following Kozlovsky (1968). It seems likely that all of the residual mass exponents are negligible.

ecological efficiencies and simultaneously checks the validity of the approximation used for production. As indicated in the previous chapter, efficiencies provide easy comparisons of resource use among animal groups. Table 8.5 indicates that there exist marked differences among the major metabolic groups. Within these groups, size has a negligible effect on all energetic efficiencies. Homeotherms of all sizes produce much less for each unit of energy ingested or assimilated than do poikilotherms. This is primarily because homeotherms burn most of their food to maintain their high metabolic rates.

Kleiber (1961) has given these efficiencies more reality with a practical example. If one feeds 10 tons of hay to two half-ton steers and an equal amount to five hundred 2-kg rabbits, both will reduce their food resource to 6 tons of manure while producing 0.2 ton of new tissue. In other words, assimilation efficiency and production efficiencies are independent of size. The remaining 38% of the energy in the feed will be lost as respiration. The major difference in the two species is that 1 ton of rabbits will eat all their food and produce all their growth in only 3 months, whereas 1 ton of cattle will require 14 months. If this analysis were extended to poikilotherms, Table 8.5 suggests that 10 tons of hay could support a population of 1 million 1-g grasshoppers for 9 months, but at the end of that time they would produce 2 tons of new grasshoppers, leaving behind 6 tons of manure and burning off the energy in only 2 tons

Table 8.6. *Comparison of observed regression between population production and respiration (McNiel & Lawton 1970) with similar equations calculated from allometric relations for individual respiration and production*

	Allometric relations		Production–respiration relation	
	Respiration	Production	Calculated	Observed
Homeotherms	$R = 7.78W^{0.79}$	$P = 0.20W^{0.73}$	$P = 0.030R^{0.93}$	$P = 0.017R^{1.01}$
Long-lived poikilotherms	$R = 0.40W^{0.76}$	$P = 0.16W^{0.77}$	$P = 0.40R^{1.01}$	$P = 0.58R^{0.82}$
Short-lived poikilotherms	$R = 0.40W^{0.76}$	$P = 0.34W^{0.74}$	$P = 0.83R^{0.97}$	$P = 0.80R^{0.83}$
Unicells	$R = 0.055W^{0.83}$	$P = 0.34W^{0.74}$	$P = 4.5R^{0.89}$	—

Note: Respiration for unicells is calculated as standard metabolism at 20°C, for poikilotherms as twice standard metabolism at 20°C, and for homeotherms as twice standard metabolism at 39°C (Robinson, Peters & Zimmerman 1983). Production for homeotherms and long-lived poikilotherms is calculated from Farlow (1976) and for others as maximum production rate (Blueweiss et al. 1978). Sample calculations for homeotherms: Given $R = 7.78W^{0.79}$, $W = (R/7.78)^{1/0.79} = 0.074R^{1.27}$. Given $P = 0.20W^{0.73}$, $P = 0.20 (0.074R^{1.27})^{0.73} = P = 0.030R^{0.93}$.

of the food. For both the farmer and the ecosystem, it is less energetically expensive to raise poikilotherms than homeotherms. Unfortunately, a scientific entrepreneur may expect stiff market resistance to the substitution of grasshopper, chuckwalla, carp, or other poikilothermic herbivore for a more traditional diet based on lamb, pork, or beef. However, the figures also suggest that fish meal might be fed more efficiently to trout than to battery chickens.

The relation between respiration and production. Efficiencies may also be presented as plots of one flux against another. If an arithmetic plot is used and a straight line produced, efficiency is represented by the slope. If a double-logarithmic grid is used, efficiency is given by the intercept, and any change in efficiency with flux is reflected in departures of the slope from unity. McNiel and Lawton (1970) regressed population production against population respiration. This analysis identified significantly different elevations (i.e., P/R efficiencies) for homeotherms, short-lived poikilotherms, and long-lived poikilotherms.

Similar equations may be calculated from allometric relations (Table 8.6, Figure 8.8) even though the units of respiration and production in

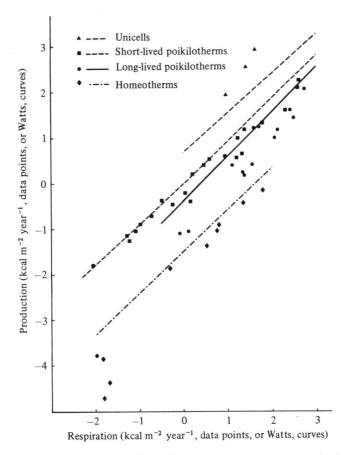

Figure 8.8. A comparison of the relations between respiration and production for homeotherms, short-lived and long-lived poikilotherms, and unicells. Observed values are shown as points (Finlay 1978; McNiel & Lawton 1970). Curves are calculated from relations scaling respiration and production to size (Table 8.6).

allometry (Watts animal^{-1}) differ from those used by McNiel and Lawton (cal m^{-2} year^{-1}). Apparently, the P/R ratio is unaffected by changes in the absolute rates of production or respiration regardless of whether these changes in rate reflect variations in population density or animal size. Humphreys (1979) identified five statistically different groups on the basis of P/R regressions like those of McNiel and Lawton (1970). These are less effectively imitated by calculations like those in Table 8.6. Either Humphrey's relations are not representative or calculations based on allometry do not apply to more closely defined groups.

Both Table 8.6 and Figure 8.8 show that homeothermy is more expensive than poikilothermy and groups with high respiration rates generally have lower production efficiencies. If, as data from Finlay (1978) suggest, the unicell curve is even approximately correct, protozoans, algae, and bacteria must be more efficient producers than either homeotherms or poikilotherms.

Prediction of other rates. The above comparisons suggest that the relation between production and respiration is similar for individuals, populations, and, presumably, communities. Such relationships are useful because respiration is easier to measure than production so that, at any given level of organization, one need only measure respiration to estimate production. If respiration is difficult to measure, one could calculate it from any correlate of respiration, like electron transport activity (King & Packard 1975) or body size. Allometric curves for respiration and production can be used to generate the relationships as in Figure 8.8. This raises the possibility that equations relating respiration to other physiological rates could also be approximated from allometric relationships. If this is so, one could calculate community or population rates of ingestion, defecation, nutrient turnover, and so forth from measurements or estimates of respiration rate.

This chapter addressed three different aspects of growth and production. It began with an examination of the effects of size on eutherian life history in order to provide a standard for comparison with other groups. Other homeotherms largely conform to this model and even poikilotherms show many similarities. The major differences between these two metabolic groups lies in the parental effort directed toward individual offspring and the greater flexibility of poikilotherm growth patterns. The second section examined the productivity of populations, an immensely important topic that is confounded by contradictions among existing allometric relations. Relations for both population production and events in an individual's life history are not homologous to most others in these pages, because the independent variable is the mass of one individual (usually the adult) and the dependent variable is often a characteristic of some other animal or animals (the offspring or the population). Most previous and subsequent equations relate an individual's characteristics to its own mass. The chapter concluded by constructing more typical relations to predict an individual production rate that represents the sum of somatic and reproductive growth. It is these last equations that can be used in the balanced growth equation.

9

Mass flow

To this point, my account of the balanced growth equation has centered on energy flow, but ecological fluxes involve more than the transfer of energy. When an animal eats or breathes, defecates or grows, a quantity of material with a definite chemical composition accompanies each energy flux. One could produce budgets for each compound and element in animal tissues and so repeat the foregoing chapters considering balanced growth equations for phosphorus (Peters and Rigler 1973), nitrogen (Nagy & Shoemaker 1975), nickel (Hall 1978), and so forth. These flows are the autecological expression of nutrient turnover, mineral utilization, and toxicant accumulation, areas of prime environmental concern that are not addressed by ecological energetics.

Whatever advantages such material budgets might offer, they have been far less extensively analyzed than energy budgets, and the treatment here will be correspondingly brief. The prime reason for this asymmetrical development is theoretical. Beginning with Lindeman's (1942) seminal paper and culminating in the International Biological Programme (1964–1974), ecologists hoped that patterns of energy flow would provide the basis for ecological generality. This theoretical attraction gave impetus to the development of techniques and traditions that emphasize energy transfers. Finally, the number of possible compounds that could be studied ecologically is so vast that biologists must hope that the basic proportionalities of energy flow will also apply to mass fluxes. For example, it is not unreasonable to suppose that if 20% of ingested energy is used in growth then about 20% of ingested nitrogen or phosphorus will be similarly used.

The highest hopes for ecological energetics have been deluded. More recent approaches to ecology integrate physical and chemical factors with energetics. Thus, Mattson (1980) shows that the utilization of ingested energy by insects rises with the nitrogen content of the food. In lakes, both physical (Straškraba 1980) and chemical factors (Dillon & Rigler 1974; Smith 1980) influence plankton standing stock and productivity.

147

Nevertheless, the search for ecological generality in energy flow was far from fruitless; the many relations described in earlier chapters show that energetic generalities exist. Moreover, without the direction provided by Lindeman and his successors, our research efforts might have been so dissipated over the gamut of energetic, elemental, and molecular budgets that perhaps no generality would have emerged at all. Unfortunately, energy flow is not the whole of ecology. If it were, energy flows would be sufficient to predict mass flow, and no new information would be gained by considering material flux in addition to energy flow.

The autecology of material flows

Since it is rarely possible to provide allometric descriptions of balanced mass budgets, this chapter instead presents selected sketches to show how material flux rates might be developed and used. First, order-of-magnitude estimates of ingestion rate are established from average tissue concentrations. After a tangential note on defecation, the economy of water use is briefly examined. A treatment of nutrient flow concludes the chapter.

Ingestion and nutrient requirements

Ideally, the ingestion of a substance is determined from the total mass of food ingested and its chemical composition. Often this information is not available, and one must estimate nutrient ingestion from average figures for composition and some estimate of ingestion rate in terms of mass. For example, food usually contains between 17,000 and 34,000 kJ kg^{-1} dry mass. Therefore, the elemental ingestion rate can be approximated from average element concentration in 1 kg dry matter and the energy ingestion rate predicted by allometry (Appendix VIIa). A similar approach could be applied to most natural constituents of living tissue (e.g., vitamins, protein, carbohydrate, etc.) but not to contaminants or to any substance that varies greatly among living species.

Small to moderate differences in the chemical composition of an animal and its food may be nutritionally irrelevant, because ingested nutrients can be used more effectively than ingested energy. All organisms act as nutrient distillers since, through respiration, they burn the food's organic matrix and concentrate the mineral nutrients imbedded there. Those materials that are in excess in the concentrate are excreted and the

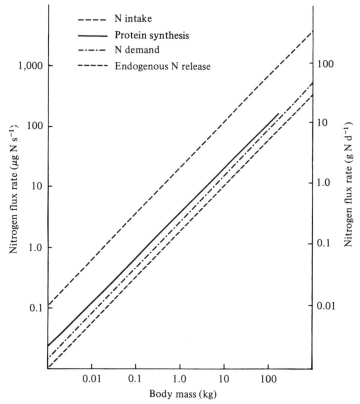

Figure 9.1. A comparison of nitrogen supply through voluntary intake (N_i, Evans & Miller 1968), by mammals with total N demand (N_d, Miller & Payne 1964), endogenous N excretion (N_e, Brody 1945), and protein N synthesis (N_p, Munro 1969). Equations used to construct these curves are $N_i = 1600W^{0.75}$, $N_d = 250W^{0.75}$, $N_e = 146W^{0.75}$, $N_p = 367W^{0.75}$, where flux rates are in milligrams of nitrogen per day, and W in kilograms.

remainder used by the predator. Still more effective use of ingested materials is achieved by the reuse of minerals within an animal's body. Energy may only be used once, but an atom may be recycled indefinitely.

The allometry of nitrogen intake and use by mammals has been sufficiently documented and will serve as an example (Figure 9.1). Total nitrogen requirement can be calculated as the sum (Miller & Payne 1964) of nitrogen loss in feces (22%), sweat (12%), hair and skin (8%), and *endogenous nitrogen execretion* in urine (58%). This last is an analog of basal metabolic rate and represents the minimal rate of loss of an animal's constituent protein. Usually, more nitrogen is ingested than is necessary to meet the basic total nitrogen requirement. Part of this is used for growth, and, in animals, the excess is excreted. Evans and Miller (1968) compared endogenous N excretion with N ingestion by zoo animals and found that supply exceeded demand by eleven times. Total N requirement was one-sixth of supply. Zoo diets may be more mineral rich than those in nature. This is offset, in part, because voluntary food intake and, so, nitrogen intake is probably reduced in zoo animals (Chapter 7). Moreover, the concentration of macroconstituents (C, H, N, O, S, but not P) of organisms is remarkably constant (Bowen 1979), so that wide imbalances are unlikely. The total N requirement shows how much new N must be ingested to replace losses but not how much N is metabolized and reused in the body; the total amount of N used in protein synthesis exceeds total N supplied to the body (Munro 1969) because N is recycled. Nevertheless, N deficiencies can occur, especially among animals that depend largely on plant food (Mattson 1980). Many herbivores have especially efficient mechanisms for N reutilization (Schmidt-Nielsen 1979), and most include some animal food in their diets, especially when N demands increase with growth or reproduction. At least among homeotherms, omnivory is far more common than strict herbivory (Peters 1977) and even the mildest herbivore, such as deer (Krausman & Bissonette 1978), antelope (Leuthold 1977), and manatees (Powell 1978), may take meat if available.

The exponents in Figure 9.1 suggest that the scaling of nutrient flux size parallels that of metabolic rate. This often holds: The ratio of endogenous N excretion to basal metabolic rate is approximately $0.5\,\text{mg}\,[\text{N}]\,\text{kJ}^{-1}$ in mammals (Brody 1945; Mitchell 1962), and a similar value (0.7) has been observed for tortoises (Brody 1945). Brody (1945) reports a somewhat higher value ($1.7\,\text{mg}\,[\text{N}]\,\text{kJ}^{-1}$) for frogs. This is still not remarkably different from those of mammals and tortoises. The requirements for the essential amino acids, methionine and threonine (Munro 1969), and for B vitamins rise as $W^{0.75}$ (Brody 1945); since sulfur excretion rises as $W^{3/4}$ (Brody 1945), one supposes that sulfur requirement does too. This parallel with metabolism is not complete, for the requirement for A vitamins rise as $W^{1.0}$ (Brody 1945).

In summary, most energetically sufficient natural diets meet the minimum requirements for a given nutrient. If we wish to know rate of ingestion of a given element, this may be calculated from food demands (Appendix VIIa) and food composition. The latter should be obtained by direct analyses of the food but can be approximated from average figures like those in Bowen (1979) and Vinogradov (1953) or in their sources.

A comment on defecation

It is scarcely surprising that crude approximations, such as those outlined in the previous section, should be necessary in building mass budgets. Even the allometric descriptions of the balanced energy equation are incomplete; the efficiencies calculated in Chapter 8 were based only on measured values for ingestion, respiration, and growth. Defecation is usually calculated by difference, because fecal production has been little studied.

One might allow scientists a certain disinclination toward the quantitative study of defecation. However, when one considers the vast range of disgusting phenomena that have been examined – for example, some parts of mycology and bacteriology; large areas in anatomy, nematology, and heminthology; and most of teratology, oncology, and necrology – this lacuna is less forgivable.

I have found only two equations (Appendix IXa) describing the allometry of defecation. One, based on the data of Hargrave (1972), predicts that the mass of feces produced by benthic detritivores rises almost as rapidly as body size; Cammen (1980) found a similar trend in substrate ingestion rate and showed that large detritivores eat a less energy-rich food than do small ones. Thus, larger benthic detritivores ingest and defecate large amounts of material but have normal rates of energy consumption (Appendix VIIa). It is, therefore, probable that the defecation rate of the other poikilotherms will be overestimated by this relation. Blueweiss et al. (1978) list an equation for the defecation rate of mammals that is about 40% of ingestion rate (Farlow 1976) and is in reasonable agreement with the difference between ingestion and the sum of growth and respiration (Chapter 8). For a 1-kg animal, defecation is predicted to be 3.8 Watts and calculated, from ingestion rate minus growth (Farlow 1976) and twice basal metabolism (Kleiber 1961), as 3.9 Watts. Defecation rates can often be calculated from ingestion rate and assimilation efficiency, since unassimilated ingesta are defecated. Unfortunately, assimilation rates vary with food, the species under study, and the assimilated material. In general, the nature of the food is the most important factor: Highly digestible foods, like flesh and seeds, may be 80 to 100% assimilated, young vegetation may be 40 to 70% assimilable, mature vegetation and wood perhaps 10 to 40%, and detritus even less. The wide range of reported assimilation efficiencies (Table 9.1) reflects this source of variation.

It is difficult to make empirical generalities about the proportion of the ingested energy or mass that is assimilated or defecated. Logically,

Table 9.1. *Some estimates of assimilation efficiency*

	Assimilation efficiency (%)				
	Detritus	Plants	Animals	Overall	Reference
Nitrogen	3–20	30–90	60–90	3–90	Mattson (1980)
Phosphorus	—	—	—	8–100	Peters (1972)
Energy	1.3–64	13–73	25–100	1.3–100	Schroeder (1981)
Energy	15	30–90	70–90	15–90	Ricklefs (1974)
Radiochemicals	—	—	—	3–100	DiGregorio, Kitchings, & Van Voris (1978)

Note: The assignment to various food catagories is very approximate. More precise information, if required, should be sought in the original sources.

defecation must range between 0 and 100% of ingestion, at least in the long term. The average assimilation efficiency of the ranges listed in Table 9.1 is 51% and an assumed value of 50% is probably a reasonable approximation, especially if applied to material ingestion rates based on allometric equations and average tissue compositions (Bowen 1979). In such cases, the uncertainty from assumed assimilation efficiency is probably small compared to other sources of error and only order-of-magnitude estimates are sought. If this is an acceptable approximation in calculating the rate of defecation of some element, a similar value should apply to order-of-magnitude estimates of excretion rate.

Such estimates of defecation or excretion are of use only when our knowledge of a given process is very weak, as is too commonly the case. Estimates of this type may be useful in comparisons of very different phenomena that cannot all be measured by a single investigator. Such comparisons may indicate which of several possible directions for future research is likely to prove worthwhile (and also those that are likely worthless). For example, nutrient budgets for whole lakes have proven powerful tools in the prediction of eutrophication (Dillon & Rigler 1975). Such budgets are usually determined by measuring nutrients in rainfall, inflowing rivers, and the lake outflow while ignoring other nutrient sources. An aquatic biologist engaged in such a study might wonder if migrating wild fowl or herd animals have any influence on this budget. It is far easier to calculate the possible contribution of these animals from allometry than to measure it directly. If the calculation suggests this is a significant source, then further study may be profitable. If not, there are many other problems of greater importance. Similarly, if the direct

Table 9.2. *Lethal amounts of desiccation in different animals expressed as the percentage of normal body mass lost by death*

Group	%
Mammals	20–36
Birds	44
Reptiles	33–42
Amphibians	40–50
Insects	30–50

Source: Adolph (1943).

absorption of a contaminant from the medium has been measured, a toxicologist might wish to approximate probable rates of uptake from the food to compare both routes. Such approximations can also provide a useful check (but certainly not a touchstone for "true" values) when reading the literature. In all cases, the calculations are coarse; but there is still a place for them in a scientist's repertoire if one does not forget their limitations.

Water economy

For many terrestrial animals, desiccation is a more chronic problem than starvation. Since animals are largely composed of water, and since the relative humidity in most terrestrial environments is less than 100%, water diffuses out of the body. In addition, water serves as the medium for the excretion of potentially toxic nitrogenous wastes, and many animals avoid overheating by the evaporation of body water. Since most animals die if water loss exceeds 20 to 50% of body mass (Table 9.2), these losses must be replaced in food or drink.

The vast majority of animals do not drink for pleasure, and water intake is largely set by need and availability. Water intake by terrestrial animals must offset both evaporation and urination. The former is a function of body size, body permeability, and habitat, whereas the latter is affected by size, habitat, and levels of nitrogenous waste. If fluid intake is excessive, urine volume is increased to maintain water balance. Water consumption (Adolph 1943), urination (Adolph 1943; Brody 1945; Edwards 1975), evaporative water loss (Altman & Dittmer 1968; Calder & King 1974, Claussen 1969; Gordon 1977), and total water turnover (Altman & Dittmer 1968; Eberhardt 1969) all vary approximately as $W^{3/4}$, although some exceptions have been recorded. At normal

Table 9.3. *Calculation of evaporative water loss necessary to offset
metabolic heating in a 1-kg mammal*

From Boddington (1978),

$$\text{basal metabolism } (R_b, \text{ in Watts}) = 3.61W^{0.73}$$
$$\text{average daily metabolic rate } (J\,d^{-1}) = 2 \times R_b \times s\,d^{-1}$$
$$= 6.24 \times 10^5$$
$$\text{heat required to evaporate } 1\,ml \text{ of water at } 40°C = 2,500\,J$$
$$\therefore \quad \text{total water required} = 6.24 \times 10^5/2,500$$
$$= 250\,ml$$

From Adolf (1943),

$$\text{average daily intake of water (ml)} = 105\,ml$$

**A mammal may offset metabolic heating by drinking somewhat more
than twice its average rate.**

temperatures, urine production accounts for about two-thirds of total water output. At very high temperatures, evaporative water loss is
both relatively and absolutely greater.

Evaporative water loss in terrestrial vertebrates. Most terrestrial vertebrates cannot long survive a body temperature of 45°C or more. However, given sufficient water, large animals can indefinitely resist even
higher ambient temperatures by evaporative cooling. At a temperature
of 40°C, the evaporation of 1 g of water requires 2,500 J of energy so that
excess heat may be very effectively dissipated by evaporating body water.
For example, a 1-kg mammal can release all its metabolic heat by
evaporating just a $\frac{1}{4}$ liter of water a day (Table 9.3). This is probably well
within its physiological capacity (Calder & King 1974).

Figure 9.2 compares evaporative water losses in the major groups of
terrestrial vertebrates at 20 to 25°C. The slopes in such relations (Appendix IXc) usually range between 0.6 and 0.8, although the mean slope
($\bar{X} = 0.59$; $SD = 0.22$; $N = 17$) is reduced by several very low slopes
reported for frogs (Claussen 1969) and passerine birds (Altman &
Dittmer 1968). The elevations of these curves hold some surprises. Over
the range from 10 g to 1 kg, birds and mammals have similar low rates of
evaporative water loss. The higher ventilation rates and body temperatures of homeotherms do not lead to higher rates of evaporative water

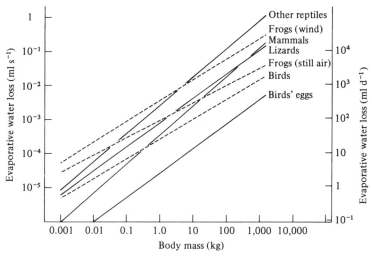

Figure 9.2. A comparison of the effect of body size on rates of evaporative water loss in terrestrial vertebrates (Altman & Dittmer 1968).

loss; apparently, these animals are very conservative in water use at moderate temperatures. It is equally surprising that the "dry, scaly skin" of reptiles does not offer an advantage in water economy. In fact, snakes, turtles, and crocodiles lose water almost as fast as do amphibians of similar size in still air. Wind increases the rate of water loss from frogs almost five times. This is the highest curve in Figure 9.2.

The equations for evaporative water loss in Figure 9.2 are derived under mild conditions and would misrepresent the extent of evaporative water loss in heat-stressed animals. Calder and King (1974) estimated that the rate of evaporation from birds at ambient temperatures above body temperature is about ten times higher than rates at 20 to 25°C. This increase in evaporative water loss with ambient temperature reflects both greater heat gain from the hot environment and reduced losses through conduction, convection, or radiation. Since the problem becomes much more intense when the environment is at, or above, body temperature, this rise in evaporative water loss is very sharp near the upper limit of biological tolerance. Plots of evaporative water loss versus ambient temperature (Weathers 1981) suggest an exponential rise, but no multiple-regression equation is available to summarize the combined effects of temperature and body size on evaporation rates of either poikilotherms or homeotherms.

Table 9.4. *Total water loss from birds'*
eggs over the incubation period expressed
as a fraction of initial mass

	Loss	SD	N
Altricial birds	0.162	0.026	32
Semialtricial birds	0.152	—	—
Semiprecocial birds	0.150	—	—
Precocial birds	0.142	0.022	13
Together	0.154	—	—

Note: All birds' eggs lose about 15% of their initial
mass through evaporation during incubation, but
species with well-developed hatchlings lose less than
those with very dependent offspring.
Source: From Ar and Rahn (1980).

Allometric implications of evaporative cooling. The combined effects of
size and permeability have been analyzed for the avian egg (Ar & Rahn
1980; Hoyt 1980). Evaporative water loss represents a special problem
for birds' eggs, because all water demands are supplied from the egg's
initial internal stores. Although evaporative water loss from eggs is low
(Figure 9.2), such losses are major determinants of incubation time. The
interval from laying to hatching can be predicted more effectively from
both egg size and evaporation rate than from egg size alone (Hoyt 1980).
The proportion of the initial water lost from eggs (Table 9.4) during
incubation is independent of size but varies somewhat with the precocial-
ity of the hatchling (Ar & Rahn 1980). Birds with long incubations
produce eggs that are less permeable to water so that longer incubations
imply reduced rates of evaporative loss. If the amount of water available
in eggs is closely balanced by evaporative loss, it is possible that particularly dry weather may significantly stress the egg. This may change
parental brooding behavior or increase egg mortality, particularly in
smaller birds.

Evaporative water loss plays different roles among large and small
animals. For small animals, evaporative cooling is possible only as a
short-term response to an acute heat load. The surface area of small
animals is large relative to their body volume so that absorbed heat leads
to rapid heating (Chapter 5) unless offset by evaporation of relatively
large volumes of water. Since available body water is correspondingly
limited, such expenditures can only be tolerated briefly. Small species,

therefore, thermoregulate behaviorally by exploiting microclimatic differences: They alternate periods of exposure to hot microhabitats with periods of cooling in shade or burrow. Because the body cools rapidly and isolated cooler areas are readily available to small animals, this strategy is quite effective. In kangaroo rats, laboratory mice, and presumably other small rodents, evaporative cooling appears important only under acute heat stress when these animals may slaver on their chests (Schmidt–Nielsen 1964, 1979).

Evaporative cooling is more attractive for large animals. They can employ fewer microclimatic differences because their size excludes them from burrows, crevices, and the shade of small plants. Even larger, cool sites, like caverns and the shade of trees, are less useful because the large thermal inertia of large animals requires prolonged periods of cooling. However, their small ratio of surface area to body volume ensures a reduced rate of heat gain (a surface phenomenon) and a large reserve of water for evaporative cooling (a volume phenomenon). At mild temperatures, evaporative water loss still rises more slowly with body mass (Appendix IXc) than does sustainable water loss (Table 9.2) so large animals will be more resistant to prolonged drought, regardless of temperature.

Even in temperate climates, animals lose body water through evaporation. All terrestrial animals, therefore, run some risk of desiccation. This is most pronounced in small animals with moist skins (Figure 9.2) and should restrict such animals not simply to cool habitats but also to humid ones or to areas where a ready supply of water is available. For example, Eviator (1973) related the size of adult frogs to water availability in various parts of their range.

In short, evaporative water loss should have marked effects on animals of different size. In hot dry climates, like deserts but also grasslands and many urban environments, small species are obliged to exploit microhabitat differences and should demonstrate marked spatial and temporal patterns in activity. They should not stray far from a source of water, although the water sources they use may be unconventional (Schmidt–Nielsen 1964; Wigglesworth, 1974). This suggests that large species are better suited to hot, dry climates. In hot, humid climates, large size may be disadvantageous, because evaporative cooling is less effective. Behavior of small species in such areas would be little affected, but large species may be forced to moderate their activity especially in the warmest part of the day and year. In cooler environments, water budgets are of less concern. Of course, these hypothetical patterns might be modified, enforced, or effaced by other prevailing ecological factors. Apparently, no overview of the effects of climate and size on animal behavior is available with which these suggestions may be compared.

Nutrients and nutrient turnover

In many ecosystems, nutrient concentration is a prime determinant of plant production and biomass and, therefore, of secondary production and animal standing stock. This has been especially well documented in aquatic systems (Dillon & Rigler 1974; Hanson & Leggett 1982; McCauley & Kalff 1981), but the widespread use of fertilizers suggests that it also applies on land. Except in a few special cases, animals can only play a minor role in setting the ultimate nutrient content of their community for this is largely the result of climate, geology, and time. However, animals may play a significant proximal role in nutrient availability by liberating minerals bound to living and dead tissue.

Animals can promote nutrient release in at least three ways. Most obvious is the direct release of minerals in the feces, urine, and, after death, decomposition products. Second, by grinding and fragmenting large detrital fragments, animals enhance microbial or fungal decomposition and mineral release. Third, because predation is invariably directed toward some portion of the plant or animal community, animals are vehicles for nutrient release from living tissue of selected parts of the ecosystem.

Little can be said about the two latter processes. Case (1979) and Mattson (1980) found that larger animals enjoy a greater range of potential prey and larger herbivores can effectively utilize coarser, larger, or more mature vegetation than can small herbivores. The predator–prey size relations (Chapter 7) indicate that larger predators eat a wider range of prey, and larger detritivores also probably eat and partially digest larger, less digestible (Cammen 1980) food items than small species. These trends suggest that the presence of larger predators reduces the refuge that large size and unpalatability offer to plant and animal prey and speeds the release of nutrient from both living tissue and detritus. If an ecosystem is dominated by small predators, nutrients could eventually be sequestered by large (or otherwise unavailable) prey and so lead to a long-term decline in production. Larger predators may prolong periods of relatively high production by increasing mineral turnover; but this productivity will be associated either with small organisms that can escape predation long enough to reproduce or with still larger or more unpalatable species. Nutrients sequestered in very large species are released only physiologically through excretion and senescence or after death following changes in the physical environment, such as the onset of winter or a forest fire. Thus, if prey can find refuge in still larger size or greater unpalatability, the productivity of the system will eventually decline both because larger organisms are less productive and because some of the nutrient will be bound up in nonproductive organs and mature organisms. Small size and rapid growth are effective long-term strategies if some factor other than food supply holds the predator in check. Systems dominated by these smaller organisms will maintain productivity.

Estimation of nutrient release rates

Direct release rates of nutrients have been studied more quantitatively and a number of allometric descriptions are available (Appendix IXc). As one would expect, the rates of releases of S (Brody 1945), N (Brody 1945; Miller & Payne 1964; Stahl 1962), and P (Hargrave & Green 1968; Johannes 1964; Peters & Rigler 1973) rise approximately as $W^{3/4}$, although the exponents are quite variable ($\bar{X} = 0.688$; range $= 0.33$–0.85; $N = 10$) probably because many analyses are based on a limited range of size and species. Unfortunately, relationships do not exist for many materials or many animal groups. In those cases, only an approximation is possible.

I noted earlier that the ratio of endogenous nitrogen release to basal metabolism in mammals is relatively constant at 0.5 mg kJ^{-1} and similar figures have been reported from tortoises (0.7) and somewhat higher values from frogs (1.7). Thus, one could estimate N excretion as the product of this ratio and the appropriate metabolic rate selected from Appendix III. Similar approaches could also be used whenever estimates of the ratio of excretion per unit of metabolism are available (Table 9.5). Naturally, such calculated values should be approached with caution. For example, the ratio of endogenous nitrogen excretion of fish (Brett & Groves 1979) to standard metabolic rate (Winberg 1960) is only 0.022 mg kJ^{-1}, more than an order of magnitude below the values for other vertebrates. This may reflect a remarkable difference between the physiologies of fish and other vertebrates but could also reflect some unidentified systematic error hidden in the determination of fish metabolism or N excretion. A series of such ratios are listed in Table 9.5. Perhaps

Table 9.5. *Some calculated estimates of nutrient turnover by homeotherms, poikilotherms, and unicells*

Element	Measured ratio	Estimated equations		
		Homeotherms	Poikilotherms	Unicells
Ca	32	$130W^{0.75}$	$4.4W^{0.75}$	$0.6W^{0.75}$
P	72	$300W^{0.75}$	$11W^{0.75}$	$1.3W^{0.75}$
N	500	$2,000W^{0.75}$	$68W^{0.75}$	$8.8W^{0.75}$

Note: This table uses measured values of the ratio of nutrient released per J of metabolism in man (Mitchell 1962) to estimate release $(ng\,s^{-1})$ for each metabolic group as metabolic rate (Hemmingsen 1960) ×ratio.

Table 9.6. *Calculated estimates of nutrient turnover*

	Measurements		Calculated intercept		
Element	W (kg)	Excretion $(ng\ s^{-1})$	Homeotherms	Poikilotherms	Unicells
Mg	70	4,300	180	6.4 ˙	0.77
I	70	0.81	0.034	1.2×10^{-3}	0.14×10^{-3}

Note: These calculations were made by assuming slopes of 0.75 and taking the elevation of the curve from one measured value of excretion rate for man (Mitchell 1962). Thus, the intercept is calculated as human excretion rate$/W^{0.75}$. Poikilotherm and unicell equations were calculated by assuming the elevations of these curves bear the same proportions that Hemmingsen (1960) reported for metabolism (1:0.034:0.0044).

even cruder relations are calculated in Table 9.6. These assume a slope of 0.75 and take elevations from a single measured value.

Both Tables 9.5 and 9.6 assume that excretion rates vary among metabolic groupings with the same proportionalities as metabolic rates. This seems likely; but some evidence both for and against this proportionality in nitrogen excretion has already been noted (Brett & Groves 1979; Brody 1945). This uncertainty should be removed. If a single regression line describes excretion for unicells, poikilotherms, and homeotherms, the specific rates of nutrient regeneration of unicells will be immense and would overwhelm any contribution by larger organisms to nutrient flux (Johannes 1964). However, if unicell mineralization rates are lower, as is unicell metabolism, then other organisms may play an important role in nutrient cycling.

There is evidence from algae that their nutrient requirements and uptake capacity also scale to body size. In aquatic systems, nitrogen and phosphorus are the elements most likely to limit growth. Shuter (1978) collected estimates of the *minimum cell quota* for N and P, the minimum amounts of each nutrient required per cell, and concluded that these rise approximately as $W^{3/4}$ (Appendix IXb). This implies that the internal nutrient concentration of growing cells declines as $W^{-1/4}$. This higher requirement would put small cells at a disadvantage except that their capacity to take up nutrients, the *nutrient affinity*, is also higher than that of larger cells. This has been shown *in situ* by examination of rates of uptake of radioactive P by large and small cells (Burnison 1975; Friebele, Correll, & Faust 1978; Rigler 1956) and in the laboratory by examining the cells' affinity for nitrogen (Eppley, Rogers, & McCarthy 1969; Malone 1980) and phosphorus (Smith 1981) as a function of cell size. These patterns are very much what one might expect from general trends in scaling: Small species meet their greater demands by increased abilities whether the demand be for oxygen, water, food, or minerals.

Nutrient turnover in lakes. The complex effects of nutrient turnover and utilization cannot be described with assurance for any community. Further discussion of the interaction of various components is, therefore, limited to an area with which I have some experience, phosphorus movement in fresh water. Although this account is, of necessity, speculative, it is hopefully not fanciful. In the absence of contrary evidence, I suppose that other ecosystems may function similarly.

In most lake systems, the rate of external phosphorus supply to lakes is far less than that required for observed levels of primary production. These levels are maintained by turnover within the water column. Since the specific rate of P release falls with size (Johannes 1964; Peters & Rigler 1973), the most important animals in this process will be the smallest. This follows because the biomass of different size classes seems to be constant in pelagic systems (Sheldon, Prakash, & Sutcliffe 1972; Sheldon, Sutcliffe, & Paranjape 1977; Chapter 10). Consequently, regeneration rates, calculated as the product of biomass and specific rate, decline with size. Thus, tiny zooplankton, like rotifers, will be more important than the larger crustaceans and crustaceans more important than fish. However, the major contributors to P turnover are phytoplankton and bacterioplankton, which are usually treated together for practical reasons. Like other organisms, unicells excrete a certain proportion of their internal P, apparently as an inevitable cost of existence. The amount of P released by this process is an order of magnitude greater than that released by zooplankton (Lean et al. 1975) and two orders of magnitude greater than that supplied by fish or the watershed (Nakashima & Leggett 1980).

Zooplankton may still play an important role in remineralization, but this is subordinate both to the role of other plankton in nutrient excretion and to the role of the zooplankton in modifying the size structure of the plant community. Excreted phosphorus is rapidly accumulated by planktonic algae and bacteria (Peters 1975a; Peters & Lean 1973; Rigler 1973). Since small algae have faster rates of uptake and can take up nutrients at lower concentrations, they will accumulate a higher proportion of the excreted nutrient. Each time the nutrient is recycled, a greater proportion would pass to small algae until these dominate the system. Zooplankton represent a countervailing force, since they selectively prey on the smaller algae and bacteria (Gliwicz 1969a,b; Morgan et al. 1980), releasing their nutrients. Larger or unpalatable forms that are spared this predation, therefore, are able to conserve their nutrient stores and are presented with an opportunity to absorb the breakdown products of palatable species (McCauley & Briand 1979). Lakes dominated by such forms will recycle nutrients more slowly both because of the phytoplankton's larger size and reduced remineralization by animals. Such systems should decline in productivity until some factor, either external (such as declining light levels or thermal destratification) or internal (such as cell senescence), leads to the release of internal nutrient stores.

The role of zooplankton in structuring the community may be more intense in eutrophic lakes, where high animal concentrations can mineralize the entire store of nutrients in prey species in a single day (Haney 1973; Peters & Rigler 1973). This should hasten the decline of edible algae and promote dominance of eutrophic lakes by larger, inedible species. By definition, oligotrophic lakes have lower levels of zooplankton and phytoplankton. In such lakes, the algal prey are dispersed through a much greater volume of water, and the smaller zooplankton population may be unable to affect nutrient distribution among unicellular competitors. In this case, algae will be dominated by the more nutrient-efficient nannoplankton. Pavoni (1963) and, more quantitatively, Watson and Kalff (1981) have shown that nannoplankton dominate oligotrophic sites.

This section describes a verbal model of the interaction among size, competitive ability, prey vulnerability, and nutrient supply. The extent to which this model applies in other systems is not known. The basic elements are certainly present. In most ecosystems, plants probably dominate nutrient exchange either because their biomass is large, as in forests, or their size is small and biomass similar, as in grassland and pelagic systems. Although I know of no comparison of the allometry of nutrient turnover and uptake in vascular plants, small plants grow and reproduce faster than large ones and should, therefore, dominate, unless more subject to herbivory or other mortalities. Herbivores prefer young, small, nutrient-rich plants (Case 1978; Mattson 1980; Scriber & Feeney 1979), and terrestrial herbivores can certainly change plant community composition (Harper 1969). Thus, trees may be seen as refuges from predation analogous to larger algae. Fire, senescence, and perhaps leaf fall would release their bound nutrients to begin a new succession. Although size, time scales, specific constituents, and external agents certainly differ among systems, each ecosystem may still operate through fundamental processes that are present throughout the biosphere.

10

Animal abundance

Almost all relations in this book are derived from and applied to individual animals. The ecological implications of biological scaling are, therefore, primarily autecological. Such information is valuable and important, but many processes depend on other members of the population or the community. For example, excretion, ingestion, growth, and reproduction are each physiological processes that scale to body size. These autecological rates are given new significance when multiplied by the number of individuals in the population or community. Excretion becomes nutrient regeneration, ingestion might represent prey mortality, and growth and reproduction become production. Each of these population and community processes, and many others, can be calculated from allometric equations if we know population size and community size structure (i.e., the number of organisms of each size, regardless of species, in the community). Such composite values may even be better than individual predictions, because they average over many different members of the population or over many different species and individuals.

Ideally, figures for abundance would be measured accurately and precisely, but such estimates are often unavailable for financial, practical, and biological reasons. One then searches the literature for approximations, values from similar sites and similar organisms. This book is dedicated to the premise that size is a major criterion of similarity; consequently this chapter summarizes the literature relating animal abundance to individual size as a necessary step toward the application of allometry to higher ecological levels.

The key to such applications is the relation between animal number and individual mass. If such a relationship exists, we can then address more intriguing questions: What are the implications of the abundance–body mass relations?, Are there regular shifts in community size structure?, and How can these shifts be predicted? This chapter addresses such questions by first establishing how population density scales to individual size and then examining the implications of such relations for the ecology of populations. Comparison of these relations with allometric descriptions of home range area provides an independent check on the validity of these density relationships. The chapter ends by looking at community size structure, the factors that influence it, and its relevance for community ecology.

The numerical density of individual species

The prediction of animal abundance is so central a problem in ecology that it can be used as a definition of the science (Andrewartha 1961; Andrewartha & Birch, 1954; Krebs 1978): Ecology is the study of those factors that determine distribution and abundance. We are intensely interested both in the factors that control mean population densities and in those that contribute to fluctuations around these means. Only the former aspect, mean densities, is discussed here, but relations for minimum and maximum density are listed in Appendix X in order to represent variation in animal density.

Patterns in animal abundance

Mean density. Population density is so frequently measured that an appropriate allometric should be easy to develop. Unfortunately, few authors have felt the exercise interesting enough to undertake. Mohr (1940) believed that the relations between density and body mass of North American mammals might prove a useful rule for game managers. He found that, per unit area, the number of animals of a given species declines as the inverse of body mass (i.e., as W^{-1}), thus biomass per unit area is unaffected by size. Herbivore biomass ranged from about 100 to 1,000 kg km^{-2} and carnivore biomass only from 1 to 10 kg km^{-2}. Omnivores and seedeaters were intermediate. Mohr (1947) later confirmed this work with a further collation of data. Based on a more thorough statistical analysis and more data, Damuth (1981) proposed that the worldwide density of herbivorous mammalian populations declines as $W^{-0.75}$ and, on average, biomass should range between 25 to 600 kg km^{-2}. Clutton-Brock and Harvey (1977a) found that the density of primate populations declines with individual size and extent of carnivory. Ghilarov (1967) noted that the density of soil animals falls with animal length. With these exceptions, I have been unable to locate relations between size and density. However, a number of authors have reported appropriate data (e.g., Eisenberg 1980; Mohr 1947; Schaller

1972; Sinclair 1972), and this section is based largely on unpublished analyses of that literature (R. H. Peters and J. V. Raelson unpublished; R. H. Peters and K. Wassenberg unpublished).

In general, these analyses show highly significant trends despite substantial residual variance. Body size relations provide one of the most general tools for the prediction of animal abundance. It is a tool that should be more widely available.

The best relations describe mammalian abundance. Figure 10.1 shows obvious differences in density between temperate and tropical mammals and between herbivores and carnivores. On average, temperate mammalian species maintain higher population densities then tropical species. This is especially pronounced among herbivores; Temperate species are ten to twenty times more numerous, per unit of habitat, than are tropical herbivores. This difference is decreased to fivefold, or even less, among carnivores. Herbivorous species are more abundant than carnivores, but the proportion varies because carnivore density declines faster with size than does herbivore density and herbivores are more affected by latitude than carnivores. In contrast to the results in Figure 10.1A, Damuth (1981) found no latitudinal effect among herbivores.

The relations for temperate birds are less reliable because they are heavily influenced by data for wintering avifauna in one area (Emlen 1972), censuses of highly mobile, conspicuous animals, like birds, may reflect density in a restricted area rather badly, and only one set of relationships is available (Appendix Xa; Figure 10.2). These show that bird populations are usually less dense than those of mammals. Densities of carnivorous and insectivorous birds fall with size. Herbivorous bird densities appear to rise with size, but the relation is not significant at the 5% level. The curve for all birds (Figure 10.2) probably represents herbivorous birds nearly as well as the herbivorous line.

Temperate vertebrate poikilotherms (Appendix Xa) appear to follow a relation similar to that for mammals, but too few data (11 species) were available to build this relationship; other poikilotherms appear to be more densely packed. Temperate invertebrate populations (Figure 10.2) were divided into two groups by habitat: aquatic or terrestrial. Although aquatic populations are denser than terrestrial ones, both groups appear to be at least an order of magnitude more abundant than vertebrates, once size effects are removed. Slopes for the invertebrate relationships may be less steep than those for vertebrates. Unfortunately these comparisons may be inappropriate for the region of size overlap between vertebrates and invertebrates is small, and differences among sites from

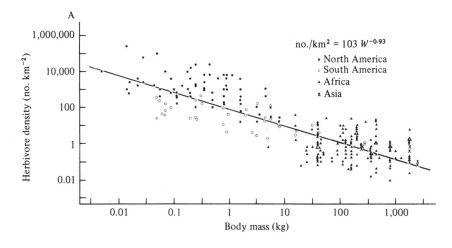

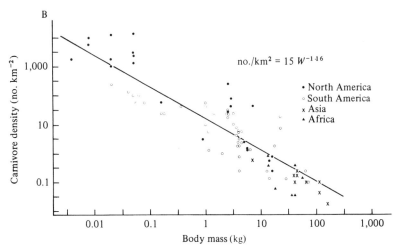

Figure 10.1. The relationship between body size and abundance of different species of mammals. Panel A shows estimates for herbivores, panel B for carnivores. Although the data in each figure can be described with a single line as shown, better descriptions are achieved by separate treatment of tropical and temperate species.

which data were collected may be large. Although all differences mentioned are statistically significant, this may only reflect heterogeneity of the data and not consistent differences among the groupings used.

Indeed, all data can be described by a single curve with a slope of −1 (Figure 10.3), and one cannot, without further tests, argue that this one general line will necessarily prove less effective in prediction than the

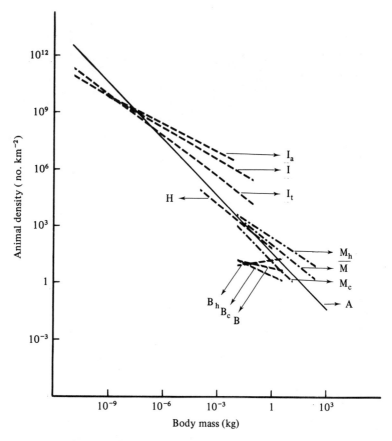

Figure 10.2. Regression lines relating body size to the population densities of different temperate animals. The curves indicated are from R. H. Peters and K. Wassenberg (unpublished; Appendix Xa). A, all animals; B, birds; B_h, herbivorous birds; B_c, carnivorous birds; H, vertebrate poikilotherms; I, invertebrates; I_a, aquatic invertebrates; I_t, terrestrial invertebrates; M, mammals; M_h, herbivorous mammals; M_c, carnivorous mammals.

more specific relations. Certainly, the confidence limits of the general line are broader, but this only indicates that the general line gives a less precise description of data in hand. It is entirely possible that more specific relations provide better fits to available data but make less accurate predictions. One cannot decide if a precise description of a small data set is a better predictor than is an imprecise description of a larger, more heterogeneous set. This choice requires further tests of both relations:

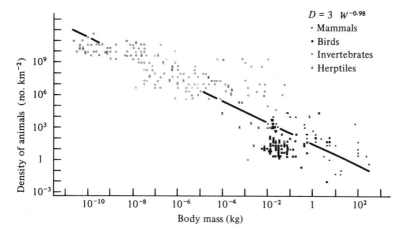

Figure 10.3. The general relationship between animal body size (W, in kg) and population density (D, in no. km^{-2}). Although statistically significant differences exist among the curves in Figure 10.2, such tests are often inappropriate because the ranges in body size overlap little and the data themselves are very heterogeneous. This figure shows that one curve can be fitted to all data.

Some ecological applications of these trends

Figures 10.1–10.3 show that body size allows prediction of one of the most fundamental ecological parameters: animal density. Such relations have such broad significance that only a sampling can be discussed here.

The role of different populations in energy and material flux. In this context, a population's role is defined as the product of density and individual flux rate. In general, population densities decline with size. For herbivores, the slope of such relations is about -0.66, which suggests that a species' role in the flux of its ecosystem increases very slightly with body size. If we accept -0.66 as an approximation of -0.75 (Damuth 1981), then role is not affected by size, and one might argue that each species is equally important in most community rates. This holds for both poikilotherms and homeotherms, because the slower rates of individual activity of poikilotherms are compensated by higher abundance. This generalization may not hold for production. Since poikilotherm production per individual can be as high as that of homeotherms (Chapter 8), poikilotherm production per species or population is greater than that of homeotherms. Larger carnivorous species play smaller roles in a given system because of their lower densities.

The allocation of community resources. If the slope in size–abundance relations for mammalian herbivores is −0.75, it might suggest that densities are scaled to achieve similar flux rates for all populations (Damuth 1981). The most likely external factor to impose this scaling is food supply operating through ingestion rate. A similar rate of food supply to each herbivore population implies that, on average, an equal share of community primary production is allocated to each population of primary consumers. Given average herbivore densities in Figure 10.1 and estimates of net primary production (1 to 50 tons dry mass ha^{-1} y^{-1}; Peters 1980b) the amount of food consumed by each population is only about 0.1 to 1 % of community primary production; so it seems that no single population monopolizes community resources at average population densities. The reader will recognize that this bioenergetic argument for a slope in density–size relationships of about $-\frac{3}{4}$ is very weak, because it must appeal to other, unidentified forces in community structure to explain the elevation of the line. Scaling can explain only the slope.

The abundance of carnivorous mammals similarly will depend on the production of their prey. Because the number of prey species decline with size, both the biomass of large prey and the number of large predators a community can support must also fall (Table 10.1). The increased slope of the equations for carnivorous mammals may reflect this shift in prey species number and the decreased elevation may reflect energy losses between a homeothermic predator and its prey.

If population density, instead, declines as $W^{-1.0}$ (Figure 10.3), this would suggest that communities are structured so that, on average, each population maintains an equal biomass. This might be imposed by numerical and facultative responses of the predators to crop the most abundant population driving it down to average density. In this case, the entire community provides a refuge for rare populations by offering more abundant prey. Again, this explanation is only partly satisfying because it does not address the question of what determines the elevation of size–abundance relations.

Home range area

Animal densities calculated from these equations certainly require further testing. One simple test is to compare these densities with reported home range areas. If the population densities are approximately correct, one would expect that home range would approach the average

Table 10.1. *Calculated relation between the number of prey species* $(N_{s(prey)}$, *in* km^{-2}) *and body mass* $(W_{pred.}$, *in kg) of mammalian predators*

From Appendix Xa,

$$N_{prey} = 214W_{prey}^{-0.61}$$
$$N_{pred.} = 15W_{pred.}^{-1.16}$$
$$\therefore \quad B_{pred.} = 15W_{pred.}^{-0.16}$$

From Appendix VIIc,

$$W_{prey} = 0.11W_{pred.}^{1.16}$$
$$B_{prey} = N_{s(prey)} \times N_{prey} \times W_{prey}$$
$$= N_{s(prey)} \times 214W_{prey}^{0.39}$$
$$= N_{s(prey)} \times 214 \ (0.11W_{pred.}^{1.16})^{0.39}$$
$$= N_{s(prey)} \times 90W_{pred.}^{0.45}$$
$$\therefore \quad N_{s(prey)} = B_{prey}/90W_{pred.}^{0.45}$$

Given prey production rate, predator ingestion rate (Farlow, 1976) and predator: prey size relations (Appendix VIIc), Vézina (unpublished) showed

$$B_{pred.} : B_{prey} = 0.01W_{pred.}^{-0.04}$$
$$\therefore \quad B_{prey} = B_{pred.}/(0.01W_{pred.}^{-0.04})$$
$$\therefore \quad N_{s(prey)} = B_{pred.}/[(0.01W_{pred.}^{-0.04}) \ (90W_{pred.}^{0.45})]$$
$$= 15W_{pred.}^{-0.16}/(0.90W_{pred.}^{0.41})$$
$$= 18W_{pred.}^{-0.57}$$

Substitution of other empirical relations will change the values of slope and intercept somewhat but will always show a pronounced decline in the number of prey species with size. The calculation assumes equal prey vulnerability. If species number declines faster with predator size, larger predators must compensate by taking a greater proportion of the prey. If species number declines more slowly, predation will be a less significant source of mortality in larger animals.

In any case, larger predators depend on fewer prey species.

Note: $B_{pred.}$ and B_{prey} are the biomasses of a given predator and all of its prey species (both in kg km^{-2}). W_{prey} is the mean body size of the prey. $N_{pred.}$ and N_{prey} are the numerical densities of the predator and each of its prey species (km^{-2}).

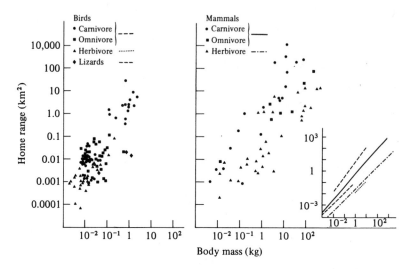

Figure 10.4. The relationship of body size to home range among birds (Schoener 1968), lizards (Turner, Jennrich, & Weintraub 1969), and mammals (Harestad & Bunnell 1979). The inset shows regression lines fitted to these data. In regression, omnivorous birds were omitted and omnivorous mammals lumped with carnivorous species.

area of land available per animal calculated as the inverse of the species population density. This inverse is here termed *individual area.*

Empirical descriptions

Available relations for home range (Appendix Xb) apply almost exclusively to birds or mammals. As one would expect, home range increases with body size. The rate of this increase is more pronounced in carnivores than in herbivores, and carnivores have larger home ranges than herbivores of similar size. A unique study of reptile home ranges suggests that lizards have smaller home ranges than homeotherms (Figure 10.4). These trends parallel those that scale individual area to size. For most animals, home range exceeds individual area so its home range or territory must be shared with conspecifics. Despite this difference, there are many similarities in the scaling of individual areas and home ranges to body size.

The last column in Table 10.2 makes these same comparisons quantitatively. This shows that, at 1 kg, 1 to 20 individuals occupy a given home range. The slopes of these equations are positive; therefore, among smaller animals, fewer individuals occupy the home range. Larger ani-

Table 10.2. *A comparison of home range and "individual area," the inverse of population density, in Figure 10.2*

Group	Diet	Home range	Individual area	Number/home range
Mammals	Plants	$Y = 0.032W^{1.00}$	$Y = 0.0046W^{0.61}$	$Y = 7W^{0.39}$
	Animals	$Y = 1.39W^{1.37}$	$Y = 0.77W^{0.94}$	$Y = 18\,W^{0.43}$
Birds	Plants	$Y = 0.026W^{0.701}$	$Y = 0.04W^0$	$Y = 0.6W^{0.7}$
	Animals	$Y = 8.3W^{1.37}$	$Y = 0.6W^{0.52}$	$Y = 15W^{0.85}$
Lizards	Animals	$Y = 0.12W^{0.95}$	$Y = 0.016W^{0.77}$	$Y = 7.5W^{0.18}$

Note: The nonsignificant exponent of the herbivorous bird relation for individual area was arbitrarily set to zero. Equations describing the number of individuals per home range (= home range/individual area) show that home ranges are usually occupied by several individuals and that this multiple occupancy increases with body mass (W, in kg, home range and individual area in km^2).

Sources: Equations for home range are from the work of Turner, Jennrich, and Weintraub (1969), Schoener (1968), or Harestad and Bunnell (1979) (Appendix Xb).

mals of all types share their home range with more conspecifics. Larger animals, therefore, have more opportunity and need for social behavior. This may contribute to increased frequency and size of social groups in larger species (Clutton-Brock, Albon, & Harvey 1980; Jarman 1974). Generally, carnivores share their larger home ranges with more conspecifics than do herbivores. Since avian population densities appear low, birds rarely have to share home ranges, and exclusive domain may be more common than among mammals or lizards. At small sizes, exclusive domain may be found in some other groups as well.

The overall effect of these comparisons is to support relations describing abundance and individual area. Reported differences in elevations should be expected, since mating and family units are only possible if more than one individual occupies a given home range. Differences in slopes are certainly not negligible but seem plausible given the increased sociality of larger species.

Community size structure

The aggregation of populations into mixed-species assemblages of similar body size is a form of community analysis and should prove useful in treating processes, such as bioaccumulation, site productivity, energy flow, and nutrient processing, which involve all or most members of the ecosystem.

Few analyses of the structure of a full community have been attempted. Instead, typical analyses include only a portion of the animal assemblage defined as much by our abilities as by our goal. One, therefore, speaks of communities of birds, soil organisms, or insects on the leaves of a certain plant. Such divisions are arbitrary. They incur the double jeopardy of isolation of the part from its major environmental determinants and misrepresentation of the unstudied whole because of the anomalous behavior of the much studied part. Because full community analyses are so difficult and time-consuming, partial analyses must be substituted. There is no intrinsic flaw in the partial approach, providing we bear its limitations in mind. The part that we can analyze does not necessarily form a natural entity and need not behave like the entire community.

Pelagial marine communities

The open water of the ocean represents an important methodological exception, because fuller community analyses are possible. For example, large volumes of the ocean can be quickly strained through nets, sieves, and filters of different mesh size. The amount of living material retained on each screen could then be used to build a size frequency distribution of the community. For smaller classes, it may be difficult to distinguish biotic and abiotic material. This problem may be solved by analyzing for some chemical that is biologically bound in a set relation to biomass. ATP is the current choice (Burnison 1975). Even chemical analysis may become obsolete. A variety of particle analyzing systems, such as the Coulter counter, can already provide detailed size analyses of particles in the range from 1 μm to 1 mm (Sheldon & Parsons 1967). In the future, we can expect to see the upper limit extended by developments in sonar and remote sensing (Silvert & Platt 1978). Perhaps we will someday be able to measure the size spectrum of a pelagic site simply driving across it in a boat. Even with our present capabilities, several attempts have been made to analyze the size structure of the full pelagic community.

The size frequency distribution of the pelagic community has been described in different ways. Body size has been measured as mass, length, chemical content, *equivalent spherical diameter*, and *radius*. These last two refer to the diameter or radius of a sphere with the same mass as the organism. In other words, if the animal were a sphere, its diameter would be the equivalent spherical diameter. The quantity of living organisms at a given size or in a given size class can be expressed in numbers, biomass, or chemical composition. Often, the only measure of abundance is number of species. Since different workers have used a variety of these descriptors, quantitative comparisons of community size structure presuppose an ability to convert among notations. This section introduces a series of interconversions and equations that describe the size structure of the most studied community. These quantitative descriptions then serve as standards (if only by default) in comparisons with other communities and in discussions of the role of different size classes in community processes.

Descriptions of pelagial size structure. Electronic particle analyzers like the Coulter counter count particles by logarithmic size class. Sheldon and Parsons (1967) recommend use of a base 2 logarithmic scale in pelagic systems. The scale groups organisms into size categories from 1 to 2, 2 to 4, 4 to 8, 8 to 16, grams, millimeters, or tons, and so forth. This divides the community into a large, but not overwhelming, number of size classes and has an intuitive appeal. Surprisingly, there is no trend in the biomass of each such size class with size (Sheldon, Prakash, & Sutcliffe 1972; Sheldon, Sutcliffe, & Paranjape 1977). If the pelagic community is divided into logarithmic size classes, the amount of matter in each class is approximately constant over the size range from bacteria to whales. This is an extraordinary and audacious claim.

Since the pelagic system is our standard, it would be well to examine some of the implications of this purported regularity. First, the regularity is not dependent on the use of binary logarithms; it applies to any sequence of equal logarithmic classes. A constant biomass over logarithmic size classes is achieved most simply if biomass of organisms of a given body mass, W, declines as a function of W^{-1}. This implies in turn that the number of individuals declines as W^{-2} (Table 10.3). The elevation of these relationships must depend upon the total biomass in the study area. This may be determined by measuring or predicting the concentration of some large part of the spectrum and extrapolating to the remainder. Sheldon, Sutcliffe, and Paranjape (1977) estimate that, in the Gulf of Maine, the concentration of particles in each $\log_2$ size class is about $5,000 \text{ kg km}^{-2}$. This yields (Table 10.3) a relation between biomass (B_i) and individual body mass (W_i) in each size class (i);

$$B_i = 7,200W_i^{-1} \tag{10.1}$$

and between number of individuals (N_i) and body mass,

$$N_i = 7,200W_i^{-2} \tag{10.2}$$

Table 10.3. *Relationships among individual body mass (W, in kg) or radius (r, in m) and biomass (B, in kg km^{-2}), total number of individuals (N, in km^{-2}), number of individuals per species (N$_{s(i)}$, in km^{-2}), and number of species (N$_s$, in km^{-2}) in pelagic marine systems*

I. If $B \propto W^{-1}$, then

$$\int_{W_1}^{W_2} B\, dW = a\, \ln W \Big|_{W_1}^{W_2}$$

$$= a\, \ln(W_2/W_1)$$

if $W_2/W_1 = k$, then

$$a\, \ln(W_2/W_1) = a\, \ln k$$

If successive size classes increase by a constant proportion (i.e., logarithmically) and if biomass falls as W^{-1}, then the biomass in each class will be a constant.

II. From Sheldon, Sutcliffe, & Paranjape (1977), in the Gulf of Maine,

$$a \int_{W}^{2W} B\, dw = 5,000$$

$$= a\, \ln(2W/W)$$

$$= a\, \ln(2)$$

$$\therefore \quad a = 5,000/\ln 2$$

$$= 7,200$$

$$B = 7,200 W^{-1}$$

III. $B = NW$

$$\therefore \quad N = B/W$$

$$= 7,200 W^{-2}$$

$$\therefore \quad \int_{W_1}^{W_2} N\, dw = \int_{W_1}^{W_2} 7,200 W^{-2}\, dW$$

$$= 7,200 W^{-1} \Big|_{W_1}^{W_2}$$

$$= 7,200\ (W_1^{-1} - W_2^{-1})$$

Number of organisms in logarithmically increasing size classes declines as the inverse of body mass.

Table 10.3. (*cont.*)

IV. From Figure 10–3

$$N_{s(i)} = 32W^{-0.98}$$
$$= 32W^{-1}$$
$$\therefore \quad N_s = N/N_{s(i)}$$
$$= 225W^{-1}$$
$$\int_{W_1}^{W_2} N_s dw = 225 \, \ln(W_2/W_1)$$

Species number in logarithmically increasing size classes is constant.

V. Since $W = 4/3\pi r^3 = 4.2r^3$

If W is expressed in kg and r in m,

$$W = (4.2 \times 10^3) r^3$$
Since $\quad N = 7{,}200W^{-2}$
$$N = 7{,}200 \, ([(4.2 \times 10^3) \, r^3])^{-2}$$
$$= 4.08 \times 10^{-2} \, r^{-6}$$

Particle number declines rapidly with particle radius.

===

In the marine literature, particle number is often expressed as a function of particle radius, r. Working from the premise of a constant biomass per logarithmic size class, this suggests that particle number should fall as r^{-6} (Table 10.3). This has been observed for foraminiferans and diatoms, but the sum of all living and nonliving marine particles falls more slowly, as r^{-4} to r^{-5} (Lal 1977). This implies that the N_i would fall as $W^{-1.5}$ and B_i as $W^{-0.5}$. It is possible that such departures would not be observable within the wide scatter in the work of Sheldon, Prakash, and Sutcliffe (1972) and Sheldon, Sutcliffe, and Paranjape (1977).

If we assume that the general relation between species' size and numerical density per species (Figure 10.3) holds in the sea (few of the data used in that relation are for marine animals), then we can calculate the effect of species size on species number. This shows (Table 10.3) that number of species (N_s, km^{-2}) varies as

$$N_s = 230W_i^{-1} \tag{10.3}$$

If we accept instead the lower exponent (-1.5) based on particle distributions, the number of species should fall as $W_i^{-0.5}$.

Equations 10.1–10.3 provide the basic relationships describing community size structure. These can be compared with available information from other systems to see if the hypothesis of constant biomass over logarithmic size classes applies more widely.

Other communities

Descriptions of the size structure of other communities are rarer, less complete, and less quantitative than those of the marine pelagial. Nevertheless, scattered observations permit some probing of the hypothesis that, for all communities, biomass is constant within logarithmically increasing size classes.

Biomass and abundance. Equation 10.2 shows that the total animal number drops off rapidly with increasing size. This observation has been made repeatedly in many systems (Elton 1973; Ghilarov 1967; Hutchinson & MacArthur 1959; Janzen 1973; Janzen & Schoener 1968; Mittelbach 1981; Pough 1980; Thiel 1975) but is probably too crude to be termed a confirmation. If individual number falls faster than W^{-2}, biomass will decline in large logarithmic size classes; if it falls more slowly, biomass will rise.

Schwinghamer (1981) tested the applicability of the pelagic model to marine benthic communities. He was able to identify three biomass peaks, which correspond to bacteria, meiofauna, and macrofauna. His work would suggest that although benthic communities do not have constant amounts of material in logarithmically increasing size classes they do show predictable patterns. Schwinghamer (1981) suggests that these coherent patterns may reflect opportunities provided by the physical structure of the environment. Bacteria utilize the surface of small particles, meiofauna, the pore space, and macrofauna, the sediment–water interface. It is plausible that the unique characteristics of a given environment may perturb community structure over some parts of the size spectrum while maintaining a similar distribution of biomass (i.e., no trend) over all logarithmic size classes.

Janzen and Schoener (1968) provide data on the biomass in different length categories for insect communities. If data from their four sites are combined into logarithmic length classes, the biomass in each of the size classes is remarkably constant: The ratio of maximum to minimum biomass observed in six size classes was only 2.5. This may be taken to extend the trends observed in the marine pelagial.

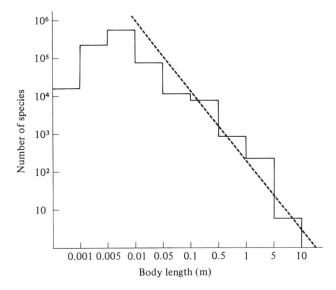

Figure 10.5. The size distribution of animal species (May 1978). The dotted line drawn through the histograms forming the upper tail of the distribution approximates the decline of species number with length. Species number falls approximately as L^{-2} or $W^{-0.67}$.

Species density. A number of authors have prepared histograms relating species number to body size (Fleming 1973; Hutchinson & MacArthur 1959; Lindsey 1966; May 1978; Pianka 1970; Stanley 1973; Van Valen 1973). Typically, these curves approximately normalize the data by using logarithmic size intervals. The histogram (Figure 10.5) developed by May (1976) to treat all species is typical of both general and more taxonomically restricted treatments. Usually, the logarithmic treatment does not succeed in a perfect normalization and the right tail is longer than the left. As May (1978) points out, the exact shape of the left tail is in some doubt because small species are not so well classified as large ones.

Often an approximate *log normal* distribution is enough to describe the effect of size on species number for larger species. If the length frequency is log normally distributed, so is the mass frequency, but variance is increased (Schoener & Janzen 1968). If we look only at the upper tail of the frequency distribution and the large animals, it seems that species number falls approximately as $W^{-0.67}$. This slope may be somewhat less than would be predicted (-1) from the pelagic standard (Equation 10.3). This tendency to underestimate would be enhanced, because small species have smaller geographic ranges than large ones (Brown 1981; Van Valen 1973). Thus, the histogram in Figure 10.5 overestimates the number of small species at a given site. I have no estimate of how significant such an effect might be.

Most of these observations are not in sharp contradiction with the marine pelagic model, but enough differences occur that we must doubt the general applicability of the observation that logarithmic size classes have constant biomasses. Although this generalization must be viewed with reservation, no alternative can be offered at this time. The approximation may stand until further studies indicate a preferable relationship. Until sufficient tests are made, the speculations outlined below must be treated skeptically.

Some possible implications of community size structure

This is clearly a highly speculative account, and much more support is required before we can build on the proposed relations. Since this support is not yet available, detailed discussion of their implications is probably premature. The discussion here is correspondingly shortened.

Species density. One can use the species number relationship to provide a further check on the validity of the estimates. One should be able to solve Equation 10.3 to yield the number of species in a given size interval. For example, Elton (1966) estimates that there are approximately 5,000 animal species in Whytham Woods, England. Assuming these range in size from 1 mg to 1 kg, we can also estimate the number of species using Equation 10.3. In this case, we would calculate a satisfying figure of ($230 \ln(1/10^{-2}) =$) 5,400. However, the same calculation could be used to suggest that, on average, 2,600 species between 10 g and 1,000 kg should be found (per km^2) in North America. Whatever the merits of Equation 10.3, it clearly overestimates the number of species of large animals. Possibly, species number is not related to size by a power formula but by some other relation that leads to a faster decline in species number with size. For example, the distribution favored by Schoener and Janzen (1968) requires that species number decline as $W^{-k \ln W}$. This would decrease the number of species in larger size classes.

Community rates of energy or material flow. The effect of size structure on the rate of ecological processing may not be immediately apparent. Table 10.4 shows that importance of individual animals and size classes

Table 10.4. *The effect of community size structure on flux rates (F) of energy, nutrients, and so forth*

From Table 10–3

$$N = aW^{-2}$$

Since most fluxes vary as $W^{3/4}$, then

$$F = a' \ W^{3/4}$$

$$NF = a' \ a \ W^{-1.25}$$

$$\therefore \int_{W_1}^{W_2} NF dw = a' a \ (W^{-0.25}/-0.25) \ \Big|_{W_1}^{W_2}$$

$$= 4a'a \ (W_1^{-0.25} - W_2^{-0.25})$$

The ecological significance of logarithmic size classes declines with body size.

Note: For this calculation, it is assumed that logarithmic size classes have equal biomasses.

declines with size. Community processes are dominated by smaller species. Therefore, removal of larger species and individuals may have little effect on the rest of the community, and destruction of the smallest size classes could be disastrous. Nakashima and Leggett (1980) have given empirical support to this argument by examining the role of different size classes on nutrient regeneration in lakes. In lakes, fishes apparently excrete only one-tenth as much phosphorus as zooplankton, and zooplankton only one-tenth as much as algae and bacteria, although the standing crop of these three classes is remarkably similar. The minor role of larger organisms may be indicated more generally by the observation that widespread extinction of larger forms has not markedly changed the forests of Europe or North America.

Even if the hypothesis of constant biomass in log size classes is an acceptable approximation, this should not obscure departures from that norm in certain size categories or certain communities. Any such regularity would be particularly important to the application of scaling equations to the prediction of the community phenomena. For example, if a given factor leads to shifts in the community size spectrum to smaller animals, then mass specific rates, such as $P/\bar{B}$ or $I/\bar{B}$, will increase. Assuming that the total biomass of the animal community is determined by the balancing of a given level of primary production by animal demands, then a shift to smaller animal size implies that animal biomass must be reduced. If it is not reduced, then the system's demands will be higher than in the unperturbed system and food must be cropped more heavily. This would lead either to a reduction in producer biomass and subsequent malnutrition of the consumers or possibly to increased production as a result of downward shifts in the size structure of both animal and plant producers. Communities dominated by smaller animals must process energy, nutrients, and contaminants more rapidly; they will show more dramatic spatial and temporal changes in population size; and they will be less integrated and interactive because of the restricted existence of their members in space and time. Community size structure is one of the fundamental characteristics of an ecosystem. It is a pity that it should be so little studied.

Factors influencing community size structure

The most general theory of community size structure is that of Thiel (1975). He examined the literature concerning the size structure of deep-sea benthos and found that deeper communities were dominated by smaller organisms. He then hypothesized that reduced levels of *resource supply* lead to communities with smaller individual size.

There is some, usually qualitative, literature in support of the effect. M. J. Coe (personal communication) found small herbivores to be more dominant in drier East African sites; in East Africa productivity is closely related to rainfall (Coe, Cumming, & Phillipson 1976; Sinclair 1975). Pavoni (1963) suggested that tiny nannoplankton dominate oligotrophic lakes and larger phytoplankton are more common in eutrophic lakes. This has been confirmed and quantified by Watson and Kalff (1981). Davies (1980) has found that freshwater midges, one of the major components of the lake benthos, increase in mean adult size in more productive waters. Finally, environments that are absolutely small, like islands, have less total resource and generally have smaller-sized animal communities than corresponding mainland communities (Brown 1971). The small area of an island may not provide enough material to support a viable population of large animals. The loss of the larger animals may permit some smaller species to increase their size, but rarely enough to compensate completely.

There is also said to be a general trend to increased size and to higher rates of production with succession (Odum 1969). This is obviously the case in terrestrial plant communities and has also been demonstrated in marine phytoplankton (Kilham & Kilham 1980; Margalef 1963). Exceptions are not uncommon (Smayda 1980), and various physical and physiological factors limit or thwart the process. For example, even in very stable aquatic habitats, phytoplankton do not grow as large as grapefruits. The effect of succession on animal communities has been little studied, but we can assume that mean animal size also tends to increase. There is an element of circular reasoning in this, because increase in size is one of the prime indications of succession (Peters 1976).

There are, however, some possible exceptions to these patterns. Boreal species, which are usually associated with less productive sites, tend to be larger than their southern counterparts (*Bergman's Rule*). This pattern is also reflected in the species size distribution of the community. Northern communities of American mammals (Fleming 1973), frogs, salamanders, and marine and freshwater fish (Lindsey 1966) all contain proportionately more large species than do tropical communities. We can only suspect that similar patterns also apply to the size structure of the community, since this has not been studied. In the oceans, phytoplankton tend to be smaller in more productive polar sites and coasts and larger in the open ocean at midlatitudes, which may have very low productivity (Kilham & Kilham 1980; Semina 1972). Nutrient-rich upwellings in the ocean may have smaller phytoplankton communities (Semina 1972), perhaps because they are often so transient that large species have insufficient time to become established.

Unfortunately most of these trends are qualitative and debatable. Too often, they could be derived from a selective reading of the literature, and, because they are qualitative, they are difficult to test. Terms like *resource availability* or *succession* are so vague as to be inapplicable and may only serve by inspiring real theories. Qualitative terms cannot be combined with allometric relations to build further predictive tools. Perhaps the strongest possible conclusion to this discussion is that quantitative study of community size structure needs and merits further study.

11

Other allometric relations

The body of this book is broadly restricted to an allometric analysis of the balanced growth equation and its extensions. This equation provides a convenient framework upon which many diverse relations may be organized. However, the validity of the individual relations is not dependent on the concept of energetic balance, and not all allometric relations can be arranged in that context. For example, larger animals are less sensitive to high-frequency sounds (Heffner & Heffner 1980). I am prepared to believe that this has ecological implications and should be mentioned but I see no place for it in the balanced growth equation. The interest an equation holds is in no way diminished (and is arguably increased), because it bears little relation to the central theme of the balanced growth equation.

This chapter briefly examines the role of body size in three areas: animal behavior, ecological economics, and evolution. In general, these areas rarely use body size as an independent variable in interspecific comparisons, and a consideration of these topics carries me further from my area of expertise. Consequently, the discussions below are briefer and more speculative than many earlier arguments. I hope points raised here will be sufficiently interesting that more capable workers will pursue them, if only to offer falsifications.

Animal behavior

All patterns of animal behavior exist within the range of possibilities defined by physiology and ecology. Since these scale with size, the range of behavior must also do so. A knowledge of allometry might allow animals behaviorists a greater definition of the probable behaviors a study animal might exhibit. For example, large herbivorous mammals sleep much less than small ones (Zepelin & Rechtschaffen 1974). Although this says nothing about an animal's waking behavior, it clearly has implications for the study of time budgets and may be of some use when framing a research program. In the ignorance of such knowledge, researchers might only restate a general rule in specific terms.

Behavioral studies also provide especially attractive and convincing tests of allometric theories because of the distance between the two fields. If body size relations can provide accurate predictions for so plastic a phenomenon as animal behavior, they should be even more powerful when applied to more physiological variables, such as birth rate and nutrient turnover. Partly for this reason, I have frequently alluded to the behavioral implications of scaling in writing this book.

Some examples of behavioral scaling

This chapter will add several more examples of behavioral scaling: intellectual ability, social dominance, and sociality.

One would expect that the larger brain size of bigger animals would permit them to learn better and to develop more complex and varied behavior. The longer life span and prereproductive period of large animals may render such a capacity both more useful and more necessary; for a big animal may have enough time to learn from its mistakes. This capacity will vary among taxa: Small primates are probably more intelligent than large ungulates. However, within taxa, larger species have greater abilities to learn and memorize than small ones (Passingham 1975a,b; Rensch 1956; Riddell & Corl 1977).

A number of authors have reported that larger species are normally socially dominant. Morse (1974) reviewed the literature and found this to be so in 31 of 35 cases of interspecific conflict among birds, mammals, fish, insects, and crustaceans. Foster and Tate (1966) described interspecific social hierarchies associated with sap feeding at sapsucker holes. Larger animals are dominant among vertebrates and frequently so among invertebrates (Remmert 1980). Among the exceptions are ants, which maintain a higher position among invertebrate visitors than their individual size would warrant, and sepsid flies, which achieve dominance over larger flies by a display that may give the appearance of greater size. Remmert (1980) reports that normally (and not surprisingly) elephants have precedence at water holes, followed by rhinoceroses and hippopotamuses, and finally zebras and antelopes, in that order. These orders, however, are subject to rearrangement. For example, sable antelope have been seen to drive away zebra and even elephants. This effect of size on social dominance may also be reflected in predation. Predators are normally larger than their prey, and the prey rarely offer even token resistance. However, if the potential prey is considerably, perhaps three times, larger than the predator it may actively and successfully defend itself (Schaller 1972).

Several trends should contribute to increased intraspecific sociality in larger mammals. Larger mammals have longer periods of maternal dependence so that family units must persist longer. The home ranges of larger mammals are generally so large that they must be shared with other conspecifics, and, among predators, the kill is often too large for one individual to kill or to eat easily. Each of these trends should promote sociality, but none imposes group behavior on large animals. Nevertheless, although some large species do not form herds or packs, group behavior generally increases among larger mammals. This has been shown for antelope (Jarman 1974), deer (Clutton-Brock, Albon, & Harvey 1980), and primates (Clutton-Brock & Harvey 1977b). A similar trend must also exist among predators, but I am not aware that this has appeared in the literature. Sociality does not appear to be affected, either positively or negatively, by size in nonmammals.

These trends can scarcely be called an ethogram, but they may provide a basic, admittedly hazy, outline within which an ethogram can be formed. Table 11.1 summarizes such an outline for African antelopes. Two reviews by Clutton-Brock and Harvey (1977a,b) give a similar overview of the behavioral ecology of primates. Many trends are similar. Smaller primates are more cryptic and nocturnal; they feed more selectively on richer, but more monotonous, diets. Larger primates feed in larger groups and have larger home ranges and greater daily ranges; they are more folivorous and spend more time eating. Larger species tend to show greater sexual dimorphisms and more skewed sex ratios within the social group. Wider study of behavioral scaling should permit the identification of similar generalities and the extension of the approach to other animal groups.

Difficulties in scaling behavior

Allometric trends in animal behavior are not common and are often only semiquantitative. Clutton-Brock and Harvey (1977a) suggest five reasons for this: (1) methodological differences may prevent close comparisons among studies; (2) realistic species averages are difficult to obtain because intraspecific variation is high; (3) field research concentrates on a few taxa to the exclusion of others; (4) only gross behavioral traits can be treated; and (5) the lumping of different groups may obscure trends that exist within taxa. These problems may be more acute in animal behavior, but it is difficult to believe that they are more acutely felt. All biologists tend to be "splitters" in their fields of expertise, and all are suspicious of "lumpers" for the reasons noted here. Animal behavior has only very recently emerged from a nonquantitative phase of ethological description and natural history in which the search for generality had been held in abeyance. The work of Clutton-Brock and Harvey (1977a,b) and others cited above suggests that this phase is now over and we may expect rapid progress in this area.

Ecological economics

In succession, individual quality is said to increase (Odum 1969) and since *r*-selected species (Pianka 1970) have weedy characteristics, *K*-selected species may be thought of as more highly desirable. Both successional stage and position on the *r* and *K* continuum are heavily influenced by body size. Indeed, increasing size is a major element in the definition of succession and, through the correlates of size, of increasing *K*-selection. One can, therefore, argue that larger size is correlated with higher quality and desirability.

The meaning of quality and desirability in the ecological literature is vague. Frequently, this is related to "competitive ability," but the substitution of one vague phrase for another offers no solution. An economist would have no difficulty in isolating a quantitative index of quality and desirability: price.

This definition is anathema for most conservationists and many ecologists. Certainly, current market value cannot be the sole guide in environmental decisions. However, *quality* is a subjective term depending entirely on the point of view. If we accept that price is a reflection of human values, we can use currency as one measure of quality. Monetary worth rises with body size.

Size and price

Good evidence supports that conclusion. In pet stores, larger birds, mammals, fish, and reptiles fetch higher prices; in lumber yards, thicker and wider boards from larger trees cost more than the equivalent weight of smaller lumber; the cost of hunting licences rises with the size of game; governments are willing to invest more to protect larger species and naturalists will pay more to see them; conservationists find that funding is

Table 11.1. *The summary of relationships between body size, ecology, spatial organization, and social behaviour of African bovidae*

	Class A	Class B	Class C	Class D	Class E
Number of species	26	11	20	12	8
Mean size (kg)	13.5	40.5	62.0	165	310
$\pm SD$	± 0.36	± 0.27	± 0.38	± 0.16	± 0.33
Range (kg)	3–64	15–120	22–220	80–250	102–820
Main habitat	Forest or dense bush (closed)	Bush, tall grass (fairly closed)	Medium to open bush or grassland (very variable)	Savanna (mainly open) and grassland	Savanna; mostly open
Feeding style	"Delicate" browsing; highly selective	Browsing or grazing; fairly to highly selective	Browsing or grazing; fairly selective, marked seasonal changes	Mainly grazing; selection for plant parts	Grazing or browsing; relatively unselective
Social groupings	Solitary or pairs; rarely over 3 in group	Commonly 3–6; adult males solitary, young males in small groups	Very variable: 5–50 to several 100; adult males often solitary; young males in herds[a]; seasonal changes	Very variable: 5–50 to several 1,000; adult males often solitary; young males in herds[a]; seasonal aggregations	+permanent[b] herds of 50–500 to 2,000; adult males often solitary or in small groups
Group stability	No groups or strong pair bond	Generally high	Variable; groups often ±open	Variable; groups often ±open	Relatively high[b]

Home range	Small, stable	Small, relatively stable	Medium to large; some seasonal movements	Large to very large; sometimes large seasonal movements	Large to very large; mobility quite high
Territory	++	++[c]	++[c]	++	—
	Individual or pair; "resource" type; approximates home range	Male only; may encompass home range; resource type	Male only; part of home range only; mating territory	As Class C	Absolute dominance hierarchy; several adult males may associate
Antipredator behavior	Concealment freezing	Concealment, freezing; but also flight	Flight (defense of young only)	Flight; occasionally group defense	Flight; individual and communal defense
Representative	Duikers, dikdik, klipspringer	Reedbuck, bushbuck, sitatunga, lesser kudu	Impala, waterbuck, kob, gazelles, greater kudu	Wildebeest, hartebeest	Buffalo; eland, oryx (?)

Note: Mean body sizes were calculated as the geometric mean of the average species mass and standard deviations refer to variation around the decimal logarithm of mean size.

[a] Male herds also include nonterritorial adult males.

[b] Substantiated only for buffalo; eland may be quite different.

[c] Territoriality not confirmed in tragelaphines.

Source: After Jarman (1974) and Leuthold (1977).

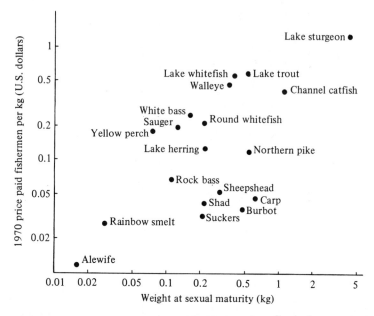

Figure 11.1. The effect of individual fish size on 1970 price for various species of Great Lakes fish. In general, big fish command higher prices, but the scatter in this plot indicates the involvement of other factors. Modified from Regier (1973).

found more easily to defend the whale than the snail darter. Even antivivisectionists seem to prefer pictures of suffering large species to suffering small ones: Lobotomized apes draw more sympathy and bigger support than pithed frogs. A number of factors must underly price but one of the more significant is surely availability or supply (J. A. Downing unpublished). Larger animals are worth more because they are rarer and less productive than small ones.

Fisheries and fish size

These consistent economic trends suggest that further analysis may prove profitable. As an example, consider a mixed-species fishery. In general, the price a fisherman receives per unit mass of fish catch rises with individual fish size (Figure 11.1). If total catch is not affected by size, then the gross for each fishing run must also rise with fish size. However, the fisherman's behavior is determined, not by gross, but by net earnings, by profit. Successful fisherman are those who pursue the most profitable catch within the limits of their abilities. The fisherman's profit is deter-

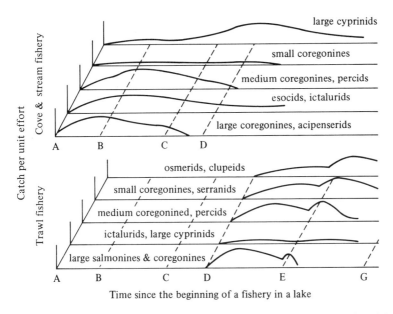

Figure 11.2. Decline of inland fisheries: Reduced levels of exploitation associated with cove and stream fisheries extends the fishery longer in time than trawling. In each case, selective fishing sequentially replaces a large-fish fishery with one for small fish. A, B, C, D, and so forth, indicate temporal sequence. Modified from Regier and Loftus (1972).

mined as the difference between the price his fish fetch and his cost in their collection. First, assume that the fishing boat is filled to capacity on each run. In undisturbed aquatic systems, the biomass of larger fishes is equal to the biomass of small ones (Sheldon et al. 1972, 1977). Thus, the effort, time, and costs involved in catching a boat load is probably unaffected by fish size. Since there is little difference in the cost of collection between large and small fish, a wise fisherman will collect the largest fishes he can. Only when the stocks of large fish are so reduced that the time and costs in their collection has risen to offset their higher price does it become profitable to fish for smaller species. Unfortunately, declining availability of larger fishes increases their market value so that larger fishes can be profitably fished even at quite low population densities. The net effect of these processes is that uncontrolled fisheries result in the sequential destruction of stocks of the largest available fish (Figure 11.2).

One role of fish stock management is to avoid this scenario by scaling catch to production. This can be achieved by limiting catch or by an appropriate system of taxation and subsidy. The latter should seek to equalize profits over the range of fish sizes either by raising the costs of catching large fish through taxation and licence fees or by reducing the costs of small-fish fisheries by subsidation. If the system has equal biomass among size classes, then catch should fall, like production, as $W^{-1/4}$. Large fish must be allowed to escape. This could be achieved by a system in which taxation rises as $W^{1/4}$. Naturally, if larger fish stocks are already reduced, they must be given greater protection and the body size exponent of taxation must rise faster.

Without government intervention, small-fish fisheries are feasible only if costs are reduced. Large-scale fisheries may achieve this by fishing very large amounts of the small species from large, mechanized vessels. Sometimes, small-fish fisheries are profitable if some characteristic of the fishes allows large numbers to be caught with little effort. Some examples are the Peruvian anchoveta fisheries and the Canadian fisheries for smelt; the latter form dense spawning runs that can be economically fished. Finally, low-cost fleets may be used profitably to pursue smaller prey – this tactic is available to many third-world fisheries because of low labor costs.

This brief examination of the scaling of economics to size is superficial. I have ignored biological differences in species abundance and distribution, social problems in the application of any system of controls, and financial and market problems in setting the level of any control. The goal of this section is to point to the economic significance of body size, not to create a new economic order. The example shows that the economics of exploitation of our biological resources partly reflects the ecology and physiology of these organisms. Because size is so important a biological trait, it is also important in economics.

Evolution

It would be a peculiar book in contemporary ecology that made no mention of evolution. Nevertheless, body size relations exist almost independently of evolutionary discussion. In most areas, the two fields are mutually irrelevant.

The major bridge across this gulf is Cope's law: Large species tend to appear later in a group's phylogeny. This is an historical pattern with considerable empirical support in the fossil record. As an empirical phenomenon, this trend merits notice in the current account. Originally, support for Cope's law was drawn from mammalian lines: Horses, elephants, titanotheres, and many other fossil sequences suggest that those mammalian lines that developed large species began as small species (Rensch 1960). Newell (1949) demonstrated phyletic increase in size for a variety of lower organisms, including echinoderms, corals, brachiopods, mollusks, and foraminiferans. Naturally, exceptions have occurred; both birds and amphibians appear to have begun at body sizes that are large relative to their modern relatives and the largest species in many animal lines – birds, terrestrial mammals, insects, scorpions, reptiles – are extinct (Stanley 1973).

Explanations of Cope's law

Evolutionary theory does not lack for explanations of this trend. Most assign a selective advantage to large size. It has been suggested that an evolutionary trend to larger size results because larger animals have greater control over the effects of their environments (Schoener & Janzen 1968) and, therefore, are less likely to suffer predation, desiccation, heating, cooling, and starvation. Rensch (1960) argues that larger species are more mobile and have greater visual power, higher fecundity (in poikilotherms), larger individual offspring, increased capacity to learn, and increased morphological specialization. Admittedly, some of these advantages are debatable. Larger animals have fewer predators, but these predators have fewer prey; the net result may well be similar predation levels in most species. In addition, larger species may suffer higher levels of parasitization. The relative specialization of large and small species is debatable. Large species may have more spectacular sexual dimorphisms and smaller densities, which Cope saw as evidence of specialization (Rensch 1960), and they may also use a reduced range of habitats (Morse 1974). However, Hutchinson (1959) argues equally convincingly that small species are more specialized. Certainly, small species often have more restricted geographic ranges (Brown 1981; Van Valen 1973) and more restricted diets (Case 1979; Clutton-Brock & Harvey 1977a; Leuthold 1977; Wasserman & Mitter 1978). Most positive advantages in large body size are offset by compensating disadvantages, and principles of similitude suggest that large and small animals normally perform equally well. Arguments for the adaptive advantage of large size are thus distinctly flat. As Gould (1966) suggested, the continued existence of both large and small species argues that both are viable strategies.

Stanley (1973) rephrased Cope's law to great advantage. Rather than ask why many phylogenetic lines evolve toward large species, he asked

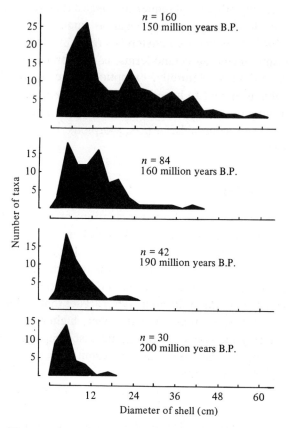

Figure 11.3. The expansion of the size frequency distributions of ammonites through evolutionary time. Although the maximum size increases, the mode and median are little changed. Modified from Stanley (1973).

why many lines begin at a relatively small size. Stanley reexamined data showing phyletic increase in size and pointed out that, for broader taxa, the maximum size does increase with evolution but the median and minimum sizes are little affected (Figure 11.3). This argues against a general selective advantage for larger size. On the basis of Stanley's figures, one might ask why so few species become large rather than why phyletic size increase should be a common evolutionary feature. The likely reason most lines arise at relatively small body sizes is that most potential ancestors were relatively small.

If we look to any available size frequency distribution for modern (Figure 10.5) or extinct (Figure 11.3) taxa, it is obvious that small species are overwhelmingly abundant. Stanley (1973), following Rensch (1960), suggest that the left-hand tail of these distributions is normally truncated, because some irreducibly small feature sets a lower size limit for each basic animal body plan. For example, the minimum size of mammals and birds may be set by physical laws governing heat loss or the minimum size at which an adult can pass a relatively large egg or offspring; the insect head capsule must always be large enough to hold a neuron: This limits the lower size of insect hatchlings, insect eggs, and, so, of insect adults. Similar limitations can be proposed to restrict the left-hand tail of the frequency distribution in most groups. The limitations of the right-hand tail are less sharp and so that tail tends to extend over time. Stanley (1973), therefore, argues that large size is one form of specialization that, like most specializations, tends to be an evolutionary cul-de-sac. Large species are not only relatively rare but they are so specialized that they are poor potential ancestors.

A number of traits may ensure that small species are more likely to produce new evolutionary lines than even their greater abundance would suggest. Small species are more easily isolated geographically, because they have smaller geographic ranges (Brown 1981; Van Valen 1973), smaller home ranges (Chapter 10), and lower mobilities (Chapter 6). If species biomass is unaffected by species size (Chapter 10), small species will produce more offspring because of their higher specific production (Chapter 8). Their offspring will also be more heterogeneous genetically, because more parental combinations will be involved in each generation: Among small species, there are more parents and among poikilotherms, each smaller mother produces fewer eggs. Small species may also be able to present more genetic combinations to natural selection by expanding rapidly during brief favorable periods (Chapter 10), thereby accumulating greater genetic variability. Finally smaller species may be more intensively selected: They experience greater absolute changes in population size and have less capacity to withstand adversity: Because their populations produce more offspring per generation, they necessarily suffer greater mortality.

Stanley (1979) stresses that large species should evolve more slowly because of their low rates of speciation and high rates of extinction. He argues that large species are, therefore, trapped by their longer generation times, high vagility, low population numbers, and essentially fine-grained use of their habitat, which lessens their likelihood of finding some refuge from environmental change. These are the inverse of those characteristics that ensure rapid speciation in small species. I cannot claim that these are the only or the most important mechanisms underlying evolution and speciation. I do hope that this form of argument will render the empirical evidence more palatable. Evolution has favored smaller species, and large size tends to be an evolutionary dead end. Because evolutionary arguments are normally couched in causal terms, the foregoing points should lend support to the arguments based on the size frequency distributions of modern and fossil faunas.

This is not conclusive evidence that small species evolve faster than large ones. The same result would be obtained if all species were produced at similar rates, but large species were extinguished more rapidly. Stanley (1973) lends support to this view by suggesting that small species have more available niches. Unfortunately, a niche can only be identified if occupied by a species, and the argument that more niches imply or are implied by more species is tautological (Colinvaux 1973). Simpson (1953) argued that no correlation exists between evolutionary rate and generation time. Since generation time scales with size, this would suggest that size does not influence the rate of speciation. However, Simpson does not indicate the extent or existence of negative evidence, and his argument must be taken only as an expression of opinion. In any case, the argument for the scaling of speciation depends on more characteristics than generation time.

Finally, one can refer to experimental evidence that small species have more rapid rates of evolution. Small species are preferred for research in genetics and evolution because their rates of genetic change are more rapid. It is surely acceptable Darwinism to suggest that also this aspect of artificial selection parallels natural selection.

12

Allometric simulation models

Introduction

The allometric approach is often unsatisfying, because it draws far more from physiology than ecology. Thus, bare scaling relationships (Table 12.1) suggest that all animals of the same size will behave in the same way, despite differences in their biotic and physical environments. Clearly, this is not the case; but empirical relations rarely exist that would permit descriptions of the habitat's modifications of allometric equations. Just as we cannot tell what effect a particular environment will have on a given organism, we also have no way of predicting what effects the organism will have on its ecosystem.

Simulation models address this problem, because the models permit the investigator a larger creative role than do empirical relations alone. If a necessary piece of information is unknown, the modeler can supply a "reasonable assumption" (a good guess) instead. Because such models treat whole, if hypothetical, ecosystems, they allow a further extension of body size relations to include the effect of size on community level processes, like succession and material flow. Most of this chapter discusses the implications of putting a single allometric organism or population into the biotic environment provided by a community of such organisms. This is a simple demonstration of the role of imagination in extending and connecting empirical relationships. No attempt is made to embed this community in a physical environment, because such embellishments would probably confuse the issue rather than illustrate the principle that computer models may serve when empirical knowledge fails.

The computational power of computers also makes simulation an attractive tool. Previous discussion of the ecological implications of body size has been limited to comparisons of pairs or triplets of allometric relations, because more extensive treatments are difficult to present and, therefore, hard to follow. In most cases, these several equations were reduced to a single formula. This is a useful process for demonstration, but it is unnecessary for calculation and tedious for complicated

Table 12.1. *Body size relations for ingestion (I_i), respiration (R_i),*
production (G_i), and defecation (D_i), which define the balanced growth
equation for the poikilothermic allometric animals used in the simulation
models of this chapter

$I_i = 0.0059\,W_i^{-0.25}$	$G_i = 0.0018\,W_i^{-0.25}$
$R_i = 0.0018\,W_i^{-0.25}$	$D_i = 0.0023\,W_i^{-0.25}$

Note: The equations expressed in g fresh mass $g^{-1}\,d^{-1}$ are modified from Farlow (1976) and Hemmingsen (1960) by standardizing the exponent of mass to $-\frac{1}{4}$ and assuming that 1 g fresh mass = 7,000 J. Defecation is determined by difference; W_i is in kg. Population rates would be determined on multiplication by biomass.

problems. Simulation models may appear more difficult, because they usually involve larger numbers of equations. However, since such models leave calculations to a computer, they do allow more ambitious treatments than have been possible in the preceding chapters.

Simulation modeling is attractive, because it makes both biological and mathematical complexities more tractable. This chapter demonstrates the process by building an allometric community and examining its behavior in succession and nutrient enrichment. Those examples suggest that simulation can be a promising approach.

A complete description of an ecosystem is impossible since the number of components and potential interactions are infinite. We must abstract those features of the complex system that we feel most significant to our own predictive goals. The goal of the model outlined below is to predict the movement of material through an ecosystem as a function of time and ecosystem size structure. The features of the ecosystem that I deem significant in this process are (1) the components of the balanced growth equation, (2) the initial size structure of the system, and (3) the linkages for material transfer among the ecosystem's components (Figure 12.1). This treatment shows the logical possibility of an allometric model to predict redistribution of mass through time and simplifies a real ecosystem to tractable terms.

The basic model

This chapter is not a review of applications of scaling in simulation. The basic model in Figure 12.1 was selected because it is simple, it can be applied to several ecosystem level problems: Succession, variations in resource level, and contaminant flow, and I am most familiar with it. The

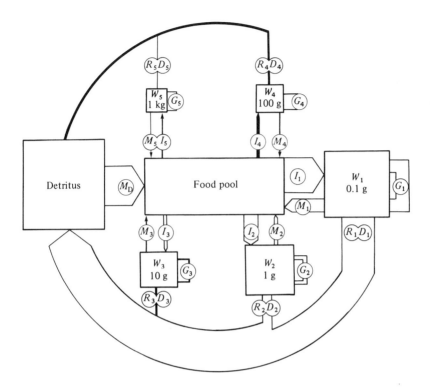

Figure 12.1. The basic model of material transfer. This is a simulation model of an abstract ecosystem. It consists of five size classes of poikilothermic organism distinguished by their average individual masses (W_i) and biomasses (B_i). Ingestion (I_i), respiration (R_i), production (G_i), and defecation (D_i) are defined in Table 12.1. Mortality losses (M_i) are defined by an arbitrary function (Equation 12.1). Losses from the system through respiration and defecation are made good by equal additions to a free resource or detrital pool (B_0), and all mortality is summed in a food pool before being distributed to the various populations. With the exception of this food pool, which is drawn disproportionately large, the size of each compartment in this diagram is proportional to the equilibrium biomass achieved when $F = 2$. The widths of connecting arrows indicate the rates of flow. Modified from Griesbach, Peters, and Youakim (1982).

model serves as one example of how scaling equations can be combined in ecosystem models.

Description of the model

The basic model is exceedingly simple. There are five classes of poikilothermic animals each of different body size (W_i) – 0.1, 1, 10, 100, and

1,000 g – and biomass (B_i, which is initially only 1 g). The number of size classes, initial biomasses, and individual sizes are arbitrary; different numbers have no effect on the qualitative trends that the model predicts. Ingestion, respiration, defecation, and production rates are semi-empirically defined by reference to the equations in Table 12.1. The system is assumed to be closed; its size is arbitrarily determined by the modeler when assigning initial biomass values to each size class and a pool of free resources or detritus. This compartment (B_0) may be conceived as all the resource that is not biotically bound.

For computation, all components and flows must be expressed in the same units, here in grams of mass and grams per day. However, any consistent units might be used, Joules and Watts, micrograms and micrograms per hour of nitrogen or phosphorus, even picograms and picograms per day of DDT (dichlorodiphenyl trichloroethane) or methyl mercury. In treating mass or energy flows, the model seems particularly unrealistic because it lacks primary producers and would eventually "run down" and disappear through respiration and defecation losses. To avoid this, daily additions to the detrital pool compensate for the summed respiration and defecation losses. Total ecosystem size is, therefore, unchanged. The detrital pool consequently represents not only detritus but all components of the system that are not included in the five size classes: detritus, plants, and animals of other sizes.

The model's most distinctive characteristic lies in its trophic relations. Ingestion rate is determined by individual size and class biomass so that the animal community demands an amount of food equal to its food requirements (ΣI_i). This food is drawn from all classes, including detritus, according to an arbitrary function that apportions this predation mortality (ΣM_i) over all classes as a function of their abundance (B_i):

$$M_i = (B_i^F / \Sigma B_i^F)\, \Sigma I_i \qquad (12.1)$$

where F is an arbitrary constant. Total mortality is, therefore, equal to total ingestion. This formulation implies that all organisms eat the same food. Thus, no differences in trophic level exist among the size classes, or, in other words, trophic position is unrelated to body size. The common food pool in Figure 12.1 from which all classes draw their food indicates this absence of trophic level. The pool is very much a bookkeeper's fiction, and though necessary, it is unlikely to have an equivalent in the real world.

Ecologists may feel uneasy at abandoning trophic structure. The concept has, however, come under increasing attack (Cousins 1980;

Isaacs 1973; Payne 1980; Peters 1977; Rigler 1975), and the possibility of alternatives has become more acceptable. The model does not require that we abandon trophic levels. It suggests only that dietary differences within and between size classes are not so strong as to be included. Dietary preferences are just one of many characteristics of animals that have been ignored. Such simplifications are one of the costs of scientific treatment (Peters 1980a). The model was originally designed to treat community development from a low initial biomass ($B_i = 1$ g; $i > 0$) in all size classes to a nearly stable system in which biomass is set by the resources that were initially available ($\Sigma B_i = 100$ g; initial $B_0 = 95$ g). Given the basic structure of Figure 12.1, the empirical relations in Table 12.1, and the arbitrary linkage among classes described by Equation 12.1, net daily gain (or loss) from each size class was calculated and added to (or subtracted from) the biomass at the beginning of the computer day. This process was repeated for a period of 2 computer years by which time the community size structure was nearly stable.

Examination of the model's properties

The first step in using a simulation model is generally an exploration of its inherent characteristics. The model's dependence on its structure and the values of its parameters are investigated by systematically changing both. Ideally, the model will be little affected by variations in elements of little immediate interest and will indicate which of the more relevant properties have the greatest effect on the output. In the present example, one would hope that the model's output is unaffected by changes in Equation 12.1, because that equation is the model's most fantastic element. On the other hand, one might wish to know if system behavior depends on the number of size classes or on the presence of homeotherms in the system that would change the constants in the allometric relations. These investigations are termed *sensitivity analysis*.

The stable biomass distribution of this model is highly sensitive to the value of F in Equation 12.1. When $F = 1$, mortality losses are directly proportional to abundance, and small species rapidly dominate the system (Figure 12.2). This is scarcely surprising for production is a declining function of size (Table 12.1). If mortality is unaffected by size, then the low production of large species is not compensated by a lower death rate, and larger animals decline in abundance. Such a situation must be rare in nature or large animals would not persist.

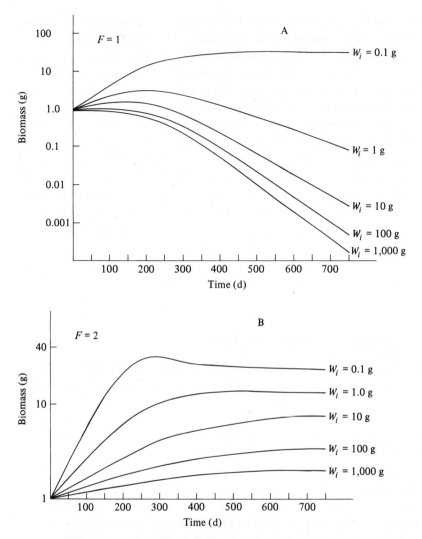

Figure 12.2. Change in biomass of different classes as a function of time and F (Equation 12.1). As F increases (A, B, and C) predation falls more heavily on the most abundant size class. This allows populations of small organisms to increase more rapidly, producing more abrupt growth early in succession when the detrital pool is large and a more even distribution of biomass at equilibrium. Modified from Peters (1978b).

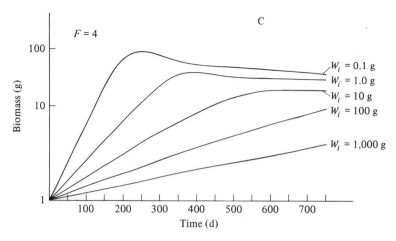

Figure 12.2. (*cont.*)

Discontinuous size-independent mortality is sometimes imposed on the community by catastrophic changes, such as the drying of a pond, the burning of a field, and the flooding of forests. Continuous size-independent mortality is less common in nature but can be imposed in laboratory systems, such as chemostats that continuously flush away a certain proportion of the planktonic ecosystem they contain. Similar effects may be imposed on the plankton of lakes by entrainment of large volumes of water for cooling, hydroelectricity, and other industrial uses. In oceans, downwellings may have similar effects. In such cases, the model anticipates dominance by small, rapidly growing species.

Persistence of all size classes is achieved only when $F > 1$ (Figure 12.2). In these cases, small populations receive some respite from predation, whereas large populations provide disproportionately more to the food pool. Initially, most food is provided by the large detrital pool, and animal populations grow almost as fast as their physiologies permit. Small species, therefore, dominate because of their higher growth rate. Once the detrital supply is reduced, these larger populations of small species must bear the brunt of predation, whereas populations of larger species continue to grow. Eventually, an effectively steady state is reached in which biomass still declines with individual size although large species have substantial representation.

Larger values of F lead to larger representations of big animals and higher total biomasses at equilibrium. This is indicated in Figure 12.3, which shows the average body mass of all animals over 750 d of system development. However, in all cases, small species dominate the system as

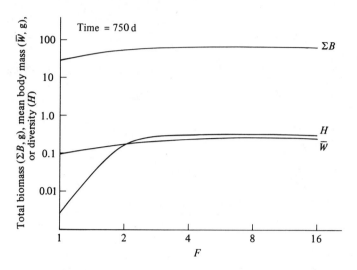

Figure 12.3. The effect of F (Equation 12.1) on mean body size in the basic model (Figure 12.1), after 750 d. Increasing values of F lead to higher final biomass, higher average size, and an increase in size class diversity. Diversity is calculated using the Shannon–Wiener index, $H = -\Sigma P_i \log P_i$ where $P_i = N_i/\Sigma N_i$ and $N_i = B_i/W_i$.

indicated by the low average size. Other formulations of the mortality function (Peters 1978b) may weigh more heavily on the smaller size classes; these formulations lead to larger populations of big animals and to larger average sizes.

Simulation models are highly seductive, at least for their creators. Once the model is constructed, we can change and adjust variables easily, thereby performing a type of thought experiment. In the present example, one could add more compartments: This increases final biomass somewhat and distributes resources more thinly, reducing the size of all compartments. One could add a class of homeotherms; in this case, behavior is little changed in terms of developmental trend and biomass relative to that of the poikilotherm system. When the system proceeds for much longer periods, larger classes continue to increase their biomass (but at such slow rates that they are unlikely to be significant).

Various alternatives to this model exist in the literature. The mortality function, the most artificial part of the model, can be expressed quite differently (Peters 1978b) without marked effects. Some authors (Kerr 1974; Thomann 1978, 1981) have linked trophic structure to size so that larger animals always feed on the next smallest class. Others (Platt &

Denman 1977, 1978; Silvert & Platt 1978, 1980) have presented models in which trophic interaction has been eliminated. Silvert and Platt (1978) offer a more elegant model that permits continuous size distribution rather than the clumsy discrete classes in Figure 12.1. Jørgensen (1979) provides far more comprehensive models in terms of the biological processes that are represented. I do not advocate any one of these as better, but I hope that a combination of all with appropriate empirical data will permit the development of an improved generation of allometric models.

The logical status of a computer model provides no basis for Pygmalion-like enthusiasm. Computer models are strictly deductive, and a good logician should see their conclusions as transparently necessary products of the model's design. For example, the present model shows that unless small animals suffer losses that are proportional to their high population growth rates, they will dominate an ecosystem. In retrospect, building and running a model often seems superfluous support for some truism. However, even the best of logicians may err. Those biologists who are not among the best logicians can certainly profit by the computer's logical assistance. Ideally, we can find a path between the rejection of all models as elaborations of the obvious and the acceptance of all models as intrinsically appealing.

Succession

One of the major questions facing any simulation model is the description of an adequate test. If these models are to be considered as theories and, therefore, as science, they must yield falsifiable predictions. Simulation models are notoriously difficult to falsify. In part, this is because some minor modification can almost always be found that allows model to mimic observation. Strictly speaking, this adjustment falsifies the original model and entails a new theory; but frequently the model tested is actually an undefined suite of models, sharing some similar form, but differing in detail. Such vague descriptions of the theory to be tested are compounded by equally vague descriptions of the expected results. This imprecision makes most models difficult, if not impossible, to test.

The present model is no exception. The suite of models to be tested is reasonably well defined: all models described by Figure 12.1 in which F (Equation 12.1) exceeds unity. The qualitative trends in succession that the model could emulate are given in Table 12.2; but since the trends were available before the model was built, it is probable that both

Table 12.2. *The direction of some qualitative trends in succession as indicated by Odum (1969)*

	Observed	Model
I.		
Biomass	+	+
Intrabiotic nutrient	+	+
Organic material	+	+
II.		
Body size	+	+
Life span	+	+
Stability	+	+
$P/\bar{B}$	−	−
Yield	−	−
Energy flow/biomass	−	−
Nutrient exchange rates	−	−
J-shaped population growth	−	−
III.		
Size class diversity and evenness	+	+
Spatial heterogeneity	+	+
Food web connectivity	+	+
Diversity	+	+
Information	+	+
Entropy	−	−
IV.		
Symbiosis	+	?
Biochemical diversity	+	?
Species diversity	+	?
Life cycle complexity	+	?
Detrital food chains	+	?
Quality of production	+	?
Ratio of production : respiration	Approaches 1	?
Niche specialization	+	?
Closure of mineral cycles	+	?

Note: These characteristics are arranged in four blocks. The first three are attributes expected as the basis of increased biomass, increased average body size, and increased size class diversity, respectively, and, therefore, correspond to the attributes of succession in the simulation model of Figure 12.1. The remaining block includes trends that cannot be observed in the model, because the observations and the model lack comparable terms of reference

conscious and unconscious bias toward the correct result has influenced the model at all stages. Certainly, the suite of models was selected after some experience revealed their implications. Any test I perform is, therefore, not independent of the theory's elaboration.

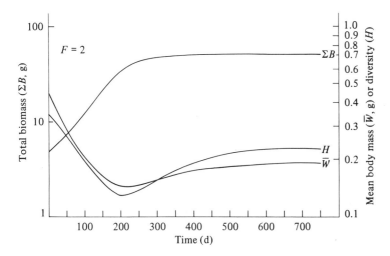

Figure 12.4. The effect of time on mean body size (W), total biomass (B), and size class diversity (H) given the model in Figure 12.1 and $F = 2$. Size class diversity is defined in Figure 12.3. After an initial period reflecting "founder effects," all three parameters increase during ecosystem development.

The present comparison cannot constitute convincing tests of the model. However, no theory can be tested adequately by its originator (Kuhn 1970), for some subjectivity is unavoidable. We are all too fond of our own ideas. At best, we can only help others to criticize them for us. We can do this best by describing our constructs as clearly and unambiguously as possible.

Although the model is very simple, it is sufficient to predict almost two-thirds of the observed characteristics of succession (Table 12.2). The model predicts that biomass, individual size, and size class diversity will increase (Figure 12.4). Since the composition of most living tissue is relatively constant (Bowen 1979), increased biomass implies an increase in both organic material and intrabiotic nutrients. If nothing else, this book has shown that an increase in body size will decrease most specific biological rates, among which are the specific rates of production ($P/\bar{B}$), energy flow, turnover, and yield. Larger species have longer life spans, and their populations respond less dramatically to changes in the environment. Thus, mature systems are more stable in their resistance to perturbation and less J-shaped in response to improved conditions. The increase in evenness of biomass distribution implies, by definition, an increase in size class diversity. In the model, this increased evenness implies an increasingly complex pattern of trophic interactions, for the system moves from one entirely dependent upon the detrital pool to one in which all compartments serve as prey. In this sense, food webs become more richly interconnected in mature systems. Spatial heterogeneity is closely connected to, and possibly interchangeable with, size class diversity, since a greater diversity of sizes implies greater spatial heterogeneity. Similarly both increased information and decreased entropy are implied by increased diversity. Several observations (Table 12.2) are not predicted by the model, because the model provides no common point of reference, even when interpreted very broadly. For the same reason, these observations cannot falsify the model either.

Despite this apparent agreement, these are scarcely strong tests. Frequently, the same observation has been repeated several times in different words so that a single trend in the model can be confirmed in several ways. For example, increased size class diversity implies increased diversity, increased spatial heterogeneity, increased information, and decreased entropy. Often, the observed phenomena and the model's behaviors are only very approximately congruent; for example, the J-shaped curves of population growth from the model refer to the behavior of size classes, whereas observations refer to individual species. Finally, many successional trends are derived from plant communities that are excluded from these models; the correspondencies of model and observation may, therefore, be fortuituous. Nevertheless, predicted and observed trends are indeed similar. If these are not crucial tests of the model's predictive ability, such comparisons do suggest that body size may play an important and tractable role in ecosystem behavior. That conclusion is the aim of this chapter.

Community development in perturbed systems

Not all communities develop in the ordered progression described above, because successional replacement is frequently stopped or redirected. The effect of some types of perturbations on succession can be examined using the model.

Nutrient enrichment. The basic model describes a closed system in which the availability of resources for the community is fixed by total ecosystem size and, for individual size classes, by the allometric equations in Table

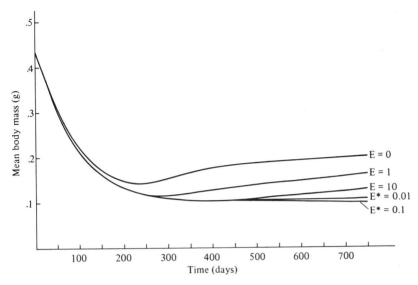

Figure 12.5. Community succession in a system to which a given quantity or proportion of resource is added daily. When the amount of enrichment is fixed (E = constant), the early dominance of small species is increased and extended in time; but eventually the system tends to the same structure as do closed systems, like those in Figures 12.2 and 12.4 and here shown as $E = 0$. If additions to the system are, instead, proportional to total system biomass ($E^* = E/B$ = constant), small organisms remain dominant. Otherwise the model is described in Figure 12.1 with $F = 2$.

12.1. These limitations can be relaxed, permitting either perturbation of the entire system by varying total resource supply or perturbation of particular size classes by modification of the allometric equations. The behavior of the perturbed model can then be seen as a new class of predictions.

The closed system in Figure 12.1 may be opened by daily additions of a given quantity of resource to the free pool. This continued enrichment initially benefits the smallest species that increase and extend their dominance of early successional stages (Figure 12.5). Eventually, however, this enrichment passes to larger organisms. Although the mature system has a greater biomass, the distribution of biomass over the various compartments is similar to that in an unperturbed system. If the daily addition increases in proportion to ecosystem size instead, the dominance of small species is permanent. Apparently, increasing the amount of free resource keeps the system in an early successional state by protecting small size classes from predation and so extending their period of rapid growth.

It is difficult to find observations to compare with this prediction. The modeled enrichment process is not identical with eutrophy of lakes, since, in eutrophication, net mineral additions are small relative to rates of internal nutrient flux and seasonal succession should not show perturbations in size structure. Approximations to the system in Figure 12.5 are achieved only if relatively large amounts of nutrient are continuously added to the system. Such additions may occur in sewage lagoons, large estuaries in lakes and oceans, or the immediate vicinity of marine upwellings. The last could present an interesting example of succession in that plankton communities close to the upwelling should be in an earlier successional state (Table 12.2) than those that have been carried some distance in the plume. Kilham and Kilham (1980) suggest that coastal algal communities do tend to smaller individual sizes than pelagic communities; the size structure of planktonic systems at upwellings is an area of active controversy (Hecky & Kilham 1974; Parsons & Takahashi 1973; Semina 1972). Perhaps the model may be of use there by indicating that cell size need not be constant throughout the zone of upwelling.

The predicted effect in planktonic communities is also observed in the laboratory (Turpin & Harrison 1980), where occasional pulses of nutrient lead to a reduction of individual size in mixed-species cultures. It is difficult to imagine a nontrivial equivalent in terrestrial systems.

Periodic stress. A different effect is observed when different size classes are periodically stressed by reducing their ingestion and growth rates to zero while maintaining normal respiration costs (Figure 12.6). The scheduling of this stress was achieved by assigning each size class a probability of not eating or growing for a day using the formula in Figure 12.6. Under all schedules, large organisms are initially more resistant to stress and dominate early succession in the system. This reflects their lower specific respiration, which reduces losses on lean days. However, over the long term, smaller species establish themselves and dominate the system when they are stressed less than large ones. Under equally frequent stress or when small species starve more often than large ones, both large and small species can survive; the size distribution is often more even and the average size much larger than that in previous models. Total biomass is reduced by stress, especially by frequent stress. I know of no observations that would serve to test these predictions, but they have an intuitive appeal.

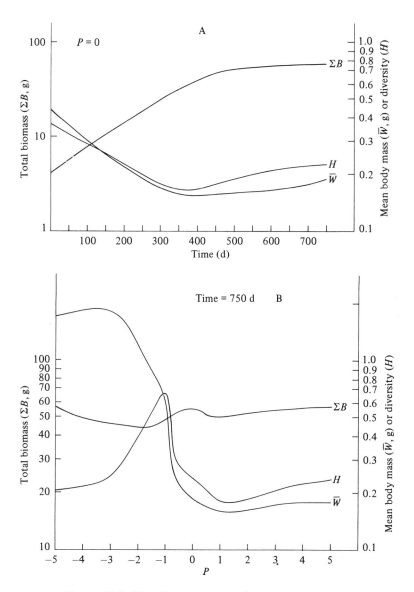

Figure 12.6. The effect of a periodic stress that occasionally reduces ingestion and growth of a particular size class to zero but does not affect metabolic costs. The probability of a hungry day is set by the factor, P, in probability $(I_i = 0;\ G_i = 0) = i^P/\Sigma i^P$, and the scheduling of stress is determined by a random number generator. When $P = 0$, (A) all classes receive the stress with equal frequency (20% of the time) and the general trend is similar to those observed with no stress (cf. Figure 12.4). When $P > 0$, large species are stressed more frequently and, in the long term (B), small species are more abundant; when $P < 0$, small species experience more lean days, and larger animals increase in representation. Diversity is greatest when small species are stressed more frequently, but the frequency of any stress is rather small $(-1 < P < 0)$.

The model certainly suggests that both periods of bounty and deprivation will have marked effects on succession. It is tempting to pursue these analyses to variations in the type of limiting resource or to the scheduling of starvation periods or to other topics. However, exercises with "reasonable assumptions" are too easily carried to extremes. The two extensions described will suffice to illustrate the line of thought.

13

Explanations

Ecologists generally place more importance on science as explanation than science as prediction. In this book, that emphasis was reversed by stressing the predictive and descriptive roles of theory. Nevertheless, one cannot but wonder why the power formula, in general, and the mass exponents of $\frac{3}{4}$, $\frac{1}{4}$, and $-\frac{1}{4}$, in particular, are so effective in describing biological phenomena. This chapter addresses such questions by examining some of the explanations proposed for the $\frac{3}{4}$ law. I do not consider the problem solved and so offer only a review without conclusion.

Two basic components of allometric explanations

Similitude

One of the striking features of allometry is that very different processes show parallel responses to variations in body size. This parallelism is usually referred to as the *principle of similitude* (Thompson 1961) or *similarity*(Kleiber 1961). This may be stated in a restricted form: For example, the size relations of biological rates can be described by a power formula in which the exponent of mass is approximately $\frac{3}{4}$. A wider formulation could be used instead: Over the longer term, the ratio of physiological processes is constant. The second formulation is the more testable for it applies wherever a range of rates are encountered. The first is, however, a more accurate description of the empiricisms on which the principle is founded. Either statement is empirically based and should be viewed as a general theory. Using the theory of similitude, we can explain most body size relations. One answer to the question "Why does metabolic rate rise as $W^{3/4}$?" is "Researchers have consistently found that biological rates rise as $W^{3/4}$; this observation is a further instance of that generality." Under the most sharply drawn and restricted of definitions, this answer is a satisfactory scientific explanation. According to Hempel (1965), an observation (e.g., metabolism rises as $W^{3/4}$) is explained when the observation is shown to be a special case of a more

213

general theory (e.g., all rates rise as $W^{3/4}$). Such explanations share the basic characteristics of predictions but differ in their temporal relation to observation. Explanations are offered after the fact and predictions, before.

If we are not satisfied by this explanation, we might try to explain similitude by reference to a still more general theory. This appears to raise the danger of infinite regress, but, in practice, scientists must stop asking for further explanations rather quickly either because we lack general theories that can predict at the specific level of our observations or because the availability of a highly successful general theory has removed interest in explanation at some intermediate level. Thus, the choice of general theory to which explanatory reference is made is usually clear.

Master reactions

Many authors accept similitude only because they feel it must reflect some underlying causal relationship. In their explanations, they treat one process, usually metabolism, because they believe that process imposes its form on all others. Thus, ingestion and excretion rates follow the $\frac{3}{4}$ law because one process must fuel metabolism and the other must remove metabolic end products. Metabolism seems to control the scaling of other processes. It may, therefore, be termed a *master reaction*. The assumption of similitude is an integral part of explanations invoking a master reaction. Such explanations differ from similitude alone because they center on a single process that is assumed to be fundamental and thereby the problem is reduced to an explanation of the master reaction.

The use of respiration as the master reaction implies an explanation of the form: "All rates vary as $W^{3/4}$ because metabolism does so; metabolism varies as $W^{3/4}$ because" followed by a particular explanation. The choice of metabolism as the controller is not essential. Convincing mechanisms could be built by assigning the central role to longevity (Boddington, 1978), food availability and ingestion, assimilation, and so forth.

Personally, I consider the search for a fundamental or master reaction empty and unprofitable. First, it invokes a concept of dubious merit, causation. Causal connection has been in disrepute as undemonstrable and unnecessary for over 200 years (Hume 1739, 1740, 1748; Kuhn 1977; Russell 1963). Second, the concept of a master reaction leads us to select one essential process as more essential than others. Just as one cannot decide if the head or the heart is more necessary, one cannot isolate any one allometric relation as more important than another. Organisms are interrelated systems in which all processes act in concert. Each allometric relation describes a process that is both cause and effect, master and servant, fundamental and derived. An adequate explanation for the system's scaling should predict at the level of the organism not at the level of some subcomponent.

A series of possible explanations of varying quality has been advanced. Earlier explanations have been reviewed with characteristic style and erudition by Thompson (1961) and will not be examined here. I will begin with Kleiber's (1961) review of several possible explanations: size-related shifts in surface area-to-body volume ratios, compositional change with size, and an attempted argument from genetics. Next, a structural argument based on engineering principles (McMahon 1973) is examined and finally a set of arguments from dimensional analysis is introduced. Each explanation accepts the principle of similitude but adds some further component that is intended to rationalize the observed exponents of body mass. In each case, one must ask if this rationalization can be considered an explanation either in Hempel's sense (in which case, I believe, the answer is "no") or in some other.

Some allometric explanations

The surface law

The surface law is among the oldest explanations for the rise of physiological rate with size. The rationale for such arguments begins by noting that the areas of geometrically similar objects rise not directly with volume but only as $V^{2/3}$. If the rate of any process associated with V is dependent on surface area, then that rate must vary as $V^{2/3}$ and, presuming constant density, as $W^{2/3}$.

The surface law has a number of disadvantages when used to explain the $W^{3/4}$ law. Rates do not rise as $W^{2/3}$ but as $W^{3/4}$, and animals are not geometrically similar. This has been apparent at least since Galileo. Large animals are stockier than small ones (McMahon 1973, 1980) so their body area should rise even more slowly than the surface law would suggest. Finally, most arguments for the surface law refer to the area of an animal's hide. In many species, this bears only a distant relation to the surfaces that are used in material or energy exchange, like the areas of villi of the gut and the alveoli of the lung. The latter (Appendix IV) appears to vary as $W^{1.0}$ in mammals (Tenney & Tenney 1970). The practical difficulties in accurately estimating surface area make any test involving surface area more difficult than theories using body mass. Nevertheless, the simplicity of the surface law as an explanation proved so attractive that over a century of science was distorted by trying to fit observation to this inappropriate model (Kleiber 1961).

Compositional explanations. Kleiber (1961) also considers explanations that are based on changes in body composition with size. Early explanations had made reference to size-related shifts in protoplasmic composition concluding that the concentration of "active protoplasm" declines with size. This ineffective explanation can now be made more convincing because allometric changes in chemical composition have been reported (Appendixes IVb, f). Smaller species have higher concentrations of RNA (Båmstedt & Skjoldal 1980; Sutcliffe 1970), respiratory coenzyme (Munro 1969, King & Packard 1975), and enzymes (Munro 1969). In each case, concentration rises approximately as $W^{3/4}$. Correlation does not indicate causation, and it is difficult to imagine that an organism could scale metabolic rate without appropriate biochemical (or physiological) machinery. This seems another example of similitude, and one wonders if it is considered more fundamental simply because it is chemical. Kleiber makes the same point by saying that the chemical concentration of the cells is probably under central, hormonal control. In other words, cellular composition is not independent of the body but a body product.

The second compositional explanation suggests that scaling results from anatomical change. Different tissues and organs have different metabolic rates (Durnin & Passmore 1967), and anatomical proportions change with size (Appendix IV). This combination could lead to the $W^{3/4}$ law, but these small changes in anatomy are insufficient to account for the observed changes in metabolism. Moreover, this would not explain the adherence of organisms with fundamentally different body plans to the same scaling pattern.

Tissue metabolism. Finally, Kleiber (1961) discusses the suggestion that the metabolic rate of the body is fixed because the tissues and organs have genetically fixed respiration rates. The experimental evidence for this is equivocal. Some reports suggest that excised tissues respire at rates different from in vivo rates and others suggest the opposite. However, if the argument is only that metabolic rate is influenced by an animal's genes, then the suggestion that the body's oxygen requirement is set by the cells is indubitable but unhelpful. If this means instead that the metabolic rate of the cells is independent of factors affecting the body as a whole, then the statement is wrong. A great deal of evidence shows that in vivo oxygen consumption, ultimately an intracellular process, is influenced by both physiology and ecology.

McMahon's structural explanation

McMahon (1973, 1980) offers one of the most ingenious explanations in recent years. His analysis combines engineering principles based on elasticity, buckling, and bending with several biological empiricisms, including the principle of similitude, to deduce that basal metabolic rate should rise as $W^{3/4}$.

McMahon (1973, 1980) begins with recognition that animals are not geometrically similar but become progressively stockier as size increases. Following a number of very early arguments (Thompson 1961), McMahon suggests that a disproportionate increase in girth and cross-sectional area is necessary to support a greater body mass. His quantitative argument (Table 13.1) suggests that the square of limb or trunk diameter should rise as mass does, and that diameter should rise as length$^{3/2}$.

From this conclusion, McMahon turns to a series of biological generalities that allow him to deduce that basal metabolic rate rises as $W^{3/4}$. If diameter rises as length$^{3/2}$, then it can be shown (Table 13.1) that cross-sectional area will rise as $W^{3/4}$. The maximum power a muscle can produce depends on cross-sectional area, since the maximum power development per unit area of muscle is relatively constant (McMahon 1980; Schmidt-Nielsen 1979). If we assume that maximum metabolic rate of an organism is proportional to the maximum power production of its muscles and if we accept that basal metabolic rate may be proportional to maximum metabolic rate (Chapter 3), then basal respiration rates rise as muscle cross-sectional area and as $W^{3/4}$ (Table 13.1).

Despite the rather lengthy list of assumptions, McMahon's argument is neat and explains a number of allometric relations among mass, area, diameter, and circumference; however, it is not unflawed. First, the argument predicts that mass rises as L^4. Although McMahon (1980) offers some support for this from bovids and primates, the bulk of available evidence (Appendix IIa), from a very broad spectrum of animals, shows that W rises as L^3. The average slope of the mass–length relations in Appendix IIa is only 2.83 ($N = 21$, $SD = 0.81$; median = 2.98). The buckling argument would more strictly apply only to

Table 13.1. *A summary of McMahon's (1973, 1980) explanation of the $W^{3/4}$ law in terms of elastic criteria, which hold that all organisms are constructed so that loads supported by the trunk and limbs are proportional to the maximum load supportable without buckling*

For a cone, column, or hollow cylinder, the critical length (L_{crit}) at which the structure buckles under its own weight is given as

$$L_{crit} = kd^{2/3}$$

where k is a constant that reflects the shape, density, and elasticity of the column, and d is the average diameter.

Elastic similarity implies that length (L) is proportional to critical length, thus

$$L \propto L_{crit}$$
$$\propto d^{2/3}$$
$$\therefore \quad d \propto L^{3/2}$$

Also,

$$\text{mass } (W) = d^2 \, L$$
$$\propto (L^{3/2})^2 L$$
$$\propto L^4$$
$$\propto d^{8/3}$$
$$\therefore \quad L \propto W^{1/4} \quad \text{and} \quad d \propto W^{3/8}$$

If animals can be approximated as a cone, column, or hollow cylinder, body size relations describing circumference (C), area (A), and maximum metabolic rate (R_{max}) can be deduced

$$C = \pi d$$
$$\therefore \quad C \propto W^{3/8} = W^{0.375}$$

and

$$A = \pi d^2 / 4$$
$$\therefore \quad A \propto (W^{3/8})^2 = W^{3/4}$$

Since maximal muscle power is a function of cross-sectional area (Schmidt–Nielsen 1979),

maximum muscle power $\propto A \propto W^{3/4}$

Assuming $R_{max} \propto$ maximum muscle power, then $R_{max} \propto W^{3/4}$.
Assuming basal metabolic rate, $R_b \propto R_{max}$, then $R_b \propto W^{3/4}$.

Under the above assumptions, basal metabolism varies as body mass to the three-fourths power.

skeletal elements, but even there, skeletal mass rises too slowly with length (as $(L^3)^{1.1} = L^{3.3}$, not L^4) to support McMahon's theory. Similarly, McMahon's argument is applied indiscriminately to skeletal and muscular components. Even if the supporting skeleton did increase its diameter as $W^{3/8}$, this need not apply to the muscles that generate metabolic power. Finally, the explanation need not apply to organisms with very different organic and structural relations, such as protozoans, trees, and eggs, although the observation to be explained does (Hemmingsen 1960). Ingenious as McMahon's explanation is, it does not explain enough aspects of allometry to be definitive.

Dimensional analyses and arguments from similarity

Both the surface law, which assumes geometric similarity, and McMahon's explanation, which assumes elastic similarity, can be seen as examples of dimensional or similarity analysis. This is a common form of mathematical argument, which has found growing acceptance in biology. Günther (1975) presents the most extensive application of dimensional analysis to allometric problems.

Quantities and dimensions. Every quantity used in science has an associated unit that has been assigned more or less arbitrarily. For example, length may be measured in furlongs, feet, miles, or meters. The confusion of multiple units for the same physical quantity was one of the prime reasons for the construction of the international system of units. This system contains seven base quantities that are, by definition, fundamental and independent. Each of these seven quantities (Table 13.2) represents a physical dimension, and, together, the seven can completely characterize any quantity. Thus, all physical variables can be expressed by some combination of these seven quantities, and a rational system of scientific units should take advantage of this property. Each of these base quantities is measured with a standard unit. Other physical quantities are measured by combinations of these basic quantities and are, therefore, derived quantities. Seventeen derived quantities have special names and units. Some of those that have been used in this book are also listed in Table 13.2. For example, force or pressure is a derived quantity measured with a derived unit, the Newton. Force (F) represents the acceleration of mass (i.e., $F = Ma$), acceleration (a) represents change of velocity (v) over time ($a = vT^{-1}$), and velocity is change of distance (L) over time

Table 13.2. *Fundamental and derived physical quantities (Q), their units (U), and dimensions (D)*

Fundamental			Derived			Derived		
Q	U	D	Q	U	D	Q	U	D
Length	meter	L	Energy	Joule	MLT	Volume	cubic meter	L^3
Mass	kilogram	M	Force	Newton	MLT^{-2}	Acceleration	meters per squared second	LT^{-2}
Time	second	T	Pressure	Pascal	$ML^{-1}T^{-2}$	Density	kilograms per cubic meter	ML^{-3}
Electric current	ampere	A	Power	Watt	ML^2T^{-3}	Longevity	second	T
Thermodynamic temperature	degree Kelvin	K	Electric charge	Coulomb	AT	Production rate	Watt	ML^2T^{-3}
Luminous intensity	candela	I	Electric potential	Volt	$ML^2T^{-3}A^{-1}$	Biological conductance	Watts per degree Kelvin	$ML^2T^{-3}K^{-1}$
Amount of substance	mole	S	Frequency	Hertz	T^{-1}	Concentration	moles per cubic meter	SL^{-3}

Note: Only the list of fundamental quantities is exhaustive. Derived quantities in the last three columns do not have special units in the international system.

$(v = LT^{-1})$. Thus, the quantity force is derived from three base quantities (mass, length, and time), and the dimensions of force are $(F = Ma =)$ MLT^{-2}. By convention, mass, length, and time are measured in kilograms, meters, and seconds, respectively, therefore, 1 Newton (1 N), the measure of force, equals $1 \, \text{kg m s}^{-2}$.

Biological quantities are also expressed in, and derived from, the fundamental quantities. This is usually expressed by stating that any parameter X can be expressed as a function of mass (M), length (L), time (T), temperature (K), amount of substance (S), luminous intensity (I), and electric current (E) such that

$$X \propto M^a L^b T^c K^d S^e I^f E^g \tag{13.1}$$

where a to g are real numbers (Derome 1977). In allometry, electric current and luminous intensity are rarely important, and dimensional arguments in allometry typically ignore temperature and amount of substance. For an exception in which temperature is treated, see Günther and Léon de la Barra (1966b). Consequently, physiological variables, X_p, are usually represented in three dimensions, mass, length, and time, so

$$X_p \propto M^a L^b T^c \tag{13.2}$$

(Derome 1977; Economos 1979a; Günther 1975; Günther & Guerra 1955).

Similarity. The number of dimensions may be further reduced if the systems to be described maintain a fixed relation between any pair of dimensions. For example, if one deals only with systems that have the same density $(D = M/V)$ and that are geometrically similar $(V \propto L^3)$ then mass varies as volume (V) and the cube of length. One of the two dimensions may be removed by substitution in Equation 13.2, so since $L \propto M^{1/3}$, then $L^b \propto M^{b/3}$ and

$$X_p \propto M^{a+b/3} \, T^c \tag{13.3}$$

The time term can also be removed if we could establish a constant that relates time and mass. Günther (1975) suggest that, in some physiological systems, time might vary with length because the speed of propogation (V_p) of electric impulses is a constant. Thus, if V_p is a constant and $V_p = L/T$, then $T \propto L$. Since $L \propto M^{1/3}$, then $T^c \propto M^{c/3}$ and

$$X_p \propto M^{a+b/3+c/3} \tag{13.4}$$

The constants used to reduce three dimensions to one may be seen as constraints that imply that these conversions apply only in those in which length and time scale as mass$^{1/3}$, or they may be seen as universal constants suggesting that one of the basic similarities of organisms is that length and time scale as $W^{1/3}$. Because this analysis depends on some underlying similarities of all animals or of all organisms, it may be called *similarity analysis* (Derome 1977).

Equation 13.4 may prove useful in determining the relation to mass of a given variable. For example, respiration rate, as a measure of power, has the dimensions $M^1 L^2 T^{-3}$ (Table 13.2). The values 1, 2, and -3 may be substituted into Equation 13.4 for a, b, and c, respectively, to suggest that the mass exponent of allometric equations for respiration should be $(1 + \frac{2}{3} - \frac{3}{3} =) \frac{2}{3}$. This, then, is a restatement of the surface law (Economos 1979a). This approach is often called *dimensional analysis*, because it is based on consideration of the physical dimensions of the dependent variable.

Criteria of similarity. The erroneous identification of $\frac{2}{3}$ as the exponent of body mass should not be disturbing. It simply indicates that the proposed similarities in geometry, density, and time were incorrect. Similarity analysis depends on the choice of universal constants – an error there and the analysis will be wrong. For example, McMahon (1980) suggests that we use elastic similarity criteria where (Table 13.3) $L \propto W^4$. By implication (Economos 1979a), $T \propto L$, so Equation 13.4 becomes

$$X_p \propto M^{a + b/4 + c/4} \qquad (13.5)$$

and the allometric exponent for respiration becomes $\frac{3}{4}$, as desired. Other criteria for similarity could also be used. McMahon (1980) shows that if animals were constructed so that static stress was similar, then length would vary as the fifth root of mass. Assuming that $T \propto L$, Equation 13.5 shows that respiration would vary as $W^{0.80}$. Derome (1977) points out that selection of the proper similarity depends upon intuition, judgment, and experience. If one errs at first, one can try again until the "proper" results are obtained.

One disturbing element in similarity analysis is that different approaches can yield the empirically established result. Günther (1975) argues that animals are "mixed regimes" in which a compromise is reached among competing similarity criteria. In particular, he believes that at the biochemical level biological time is largely governed by electrodynamic principles (i.e., $T \propto L$), but at the macroscopic level, time

Table 13.3. *A comparison of the exponents of mass in equations describing respiration rate and other physiological rates among terrestrial and aquatic organisms*

	Respiration			Other processes			All		
	$\bar{X}$	SE	N	$\bar{X}$	SE	N	$\bar{X}$	SE	N
Aquatic organisms	0.754	0.019	54	0.761	0.0235	33	0.756	0.0146	87
Terrestrial homeotherms	0.752	0.0092	126	0.674	0.0167	59	0.727	0.0086	185
Terrestrial poikilotherms	0.737	0.023	22	0.723	0.084	3	0.736	0.021	25
Terrestrial animals	0.749	0.0085	148	0.677	0.016	62	0.728	0.0079	210
Organisms	0.750	0.0079	202	0.706	0.014	95	0.736	0.0071	297

Note: All equations are listed in the appendixes, but no conductances or homeothermic rates outside the thermal neutral zone have been used because these are expected to depart from either $\frac{3}{4}$ or $\frac{2}{3}$ laws. These comparisons lend no support to the hypothesis that different scaling rules apply in terrestrial and aquatic environments.

is related to the gravitational constant g. Since $g = LT^{-2}$, then $T \propto L^{1/2}$ and $T \propto M^{1/6}$. The gravitational constant, g, is independent of animal size only if this is so. Since an organism cannot follow two different similarity criteria simultaneously, it must reach a compromise between them. Günther (1975) argues that the exponent of time in Equations 13.4 and 13.5 should lie between $c/3$ and $c/6$. He introduces an empirical constant to the equation so that dimensional analysis of respiration yields the expected value, following Brody (1945), of 0.73. If we accept this approach but take the expected value as 0.75, then Equation 13.4 becomes

$$X_p \propto M^{a + b/3 + c'/3} \tag{13.6}$$

where $c' = c - 0.25 \, (c/|c|)$. This approach yields values for the exponent of mass that are very close to those observed (Figure 13.1).

Platt and Silvert (1981) use different similarity criteria–the caloric constant and weight density of animal tissue–to show that metabolism ought to vary as $W^{0.75}$. This follows because dimensional analysis shows that one constant, the caloric content of tissue (J kg^{-1}), has dimensions $L^2 T^{-2}$ (Table 13.2) so $L \propto T$; and the other constant, weight density (N m^{-3}), has dimensions $ML^{-2} T^{-2}$, so $M \propto L^4$. This, then, follows McMahon's argument (and is subject to the same flaw: Mass does not

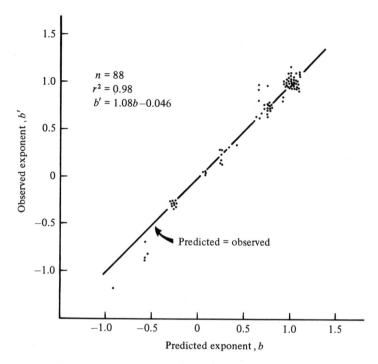

Figure 13.1 A comparison of the measured exponents of body mass for 88 physiological functions with those predicted using the dimensional analysis of Equation 13.6. The measured exponents are those listed by Günther (1975). Each deals with some aspect of homeothermic respiratory and circulatory physiology. Variables that Günther treated with other similarity criteria like the blood-air barrier of the lung (cellular similarity) or diffusion capacity and blood viscosity (transport similarity) have not been used. The solid line indicates perfect correspondence between prediction and observation. The regression line through the points is not significantly different.

vary as L^4 in animals). However, their analysis shows that another approach to the problem can yield the same result.

Platt and Silvert (1981) go one step further. They argue that although weight density is significant for terrestrial animals this may not be so for aquatic animals that live in a weightless environment. For such organisms, the relevant constant is mass density with dimensions ML^{-3}, because they must push water aside to move. Substituting these values in Equation 13.4 yields the surface law. Platt and Silvert (1981), therefore, propose that metabolic rates of aquatic species should vary as $W^{2/3}$ and for terrestrial species as $W^{3/4}$. Unfortunately, there is no evidence to support this distinction. Regressions assembled in the appendixes of this book suggest that an exponent of $\frac{3}{4}$ applies to both terrestrial and aquatic species (Table 13.3).

Dimensional analysis codifies the principle of similitude and is a useful tool in determining the probable exponent in a variety of allometric equations. However, this method of analysis presents so broad an array of possible allometric relations that prior knowledge of the empirically determined exponent is required. One would accept such an ad hoc explanation only if it had surprising implications. Both the explanations of McMahon (1973, 1980) and Platt and Silvert (1981) held such implications in suggesting that mass varied as length to the fourth power. This renders the explanation more interesting because it is more testable, but, unfortunately, this also results in falsification.

Some other explanations

The explanations examined above do not exhaust those available. Blum (1977) sees the scaling of metabolic rate as possible evidence for the functioning of the surface law in four dimensions rather than three – the area of a hypervolume (V) rises as $V^{1-1/n}$, where n is the number of geometrical dimensions on which the hypervolume is measured; unfortunately, he does not identify the additional dimension. Apple and Korostyshevskiy (1980) see the power laws as evidence for central control, which again is not identified. Economos (1979a,b, 1981) proposed that increasing gravitational load may differentially stress larger terrestrial animals. However, this argument suggests that the metabolic rate of aquatic organisms, which ought not suffer such a stress in their weightless environment, should rise as $W^{2/3}$. Table 13.3 shows that this is not so. Gray (1981) proposes that the $\frac{3}{4}$ law results from the relation of local variations in body temperature to metabolism. I am unable to follow his abstruse arguments, but, as they appear to apply only to homeotherms, their scope is insufficient to explain the wide applicability of the $\frac{3}{4}$ exponent.

Allometry and scaling have attracted empiricists rather than rationalists, and the literature of explanation has become peripheral to the mainstream of research in body size. The prime reason for this is that available explanations add little to our knowledge. An empiricist is most interested in an explanation when it implies something new about nature. In other words, explanations necessarily show that the observations could have been predicted, but attractive explanations, like the good theories they are, explain many different phenomena and so can make many different predictions. The quality of allometric explanations does not therefore, rest on their ability to provide a rationale for the $W^{3/4}$ law – all explanations should do that – but on their implications in other areas. So far the implications in other areas are either nonexistent or incorrect. It is hardly surprising that such explanations have not proved popular.

Academics can scarcely cease the search for fundamental scientific regularities, and biologists have long been trained to ask the unanswerable question, Why? However, a century and a half's explanation of the scaling of metabolism has not brought us closer to a definitive answer and may have seriously impeded research by stressing the need to explain a nonexistent regularity, the surface law, rather then the need to test it.

Attempts to explain the $W^{3/4}$ law, which replaced the surface law, have not proven fruitful, whereas empirical rules for biological scaling have developed rapidly and extensively. This dichotomy might suggest that, although we cannot close our minds to the possibility of a general explanation, we might more profitably concentrate our efforts on the improvement of specific size relationships. Indeed, because the power curve is only one of many potential models to describe scaling, the $\frac{3}{4}$ exponent may only be a statistical regularity. If this were so, and there seems no way of identifying such regularities, our time would be wasted trying to find some further 'mechanism' to explain our observations.

14

Prospectus

One of the central tenets in Thomas Kuhn's (1970) famous book on scientific growth and revolution is that normal, vigorous fields of science present many unsolved, but soluble, problems and questions. Pursuit of this premise leads to the paradox that if a scientific monograph is to be a success, it must also be a failure. When all holes and doubts are filled, when the approach has been prodded and viewed from all angles, when it has been pulled apart and put together in every possible way, when the definitive review has been written, then the field is closed to future investigation and withers as a scientific pursuit. The information represented by the science may still be used by technologists in various applications, but scientists show little interest. They have moved on to the next challenge.

This book does not endanger allometry as a scientific field. Instead the book should be a sufficiently successful failure that it encourages others to study and advance our knowledge of biological scaling. There is certainly room for improvement. Even as a review, this book is incomplete because it ignores the rich Soviet literature. Other gaps in our knowledge are indicated throughout the text. These should appear as interesting opportunities for further research.

Although testing proposed relations is an essential part of science, surprisingly few tests are encountered in allometry. This is a major flaw. Allometricians are more apt to provide a statistical description of a new data set than to use this set to test existing relations. The result is a plethora of slightly different equations, all of which claim to describe the same phenomenon and none of which can be objectively judged as superior. We need a scientific housecleaning to identify our best tools by comparing the predictive power of our current theories with new information.

Further work is obviously needed to extend current relations to unstudied phenomena, taxa, and sizes. In scaling, as in so many areas in biology, we know far more about homeotherms than about poikilotherms or unicells. Since most organisms are not homeotherms, a great deal of

227

work is required before our knowledge would be proportional to animal abundance. Of the many variables related to body size, only basal and standard rates of respiration have been extensively studied. We require confirmation of many relations that have been studied little or only by a restricted group of researchers. We would like overviews of the allometries of poikilotherm and unicell physiology (cf. Günther 1975; Lasiewski & Calder 1971), which might indicate the extent of similitude in these organisms. Some processes (e.g., fluxes of materials) have rarely been correlated to size even for mammals and birds. Finally, existing relations are often built from, and, therefore, apply over, a restricted range of animal size. Most relations would be improved if still larger and smaller animals were included.

These suggestions for testing and extending current allometric relations involve the collection of more data; other improvements are possible that may only require reanalysis of data used in existing relationships. Critical reanalysis of the data underlying the current regressions might also identify spurious or biased data. Many allometric regressions have used all available data and must include some information that has subsequently been found inaccurate. Removal of such data or their replacement with better estimates should improve regression.

Last, it is a rare paper in allometry that adequately describes the results of regression analysis. Usually some statistic (e.g., the range of the original data, the mean value of the independant variable, the standard error of the coefficients, the number of individuals, and the number of species) is unreported so that the adequacy of the equation remains in doubt; an important future contribution to scaling could include reanalysis of the current relations and a fuller description of these regressions.

For many, these problems and suggested programs will appear either uninteresting and irrelevant or pedestrian and unnecessary. These value judgments are dangerous and unfortunate. On the one hand, experimentalists sometimes act as though still one more careful measurement of, say, feeding rate of one more organism under yet another set of laboratory conditions is more useful and interesting than the construction of a general, if approximate, biological law. I do not disparage precise experimentalism. Without it, the study of size would be impossible. However, the experimentalist should ask how more detailed information will advance the science. One answer is that this detail will eventually permit better generalizations. On the other hand, many generalists apparently believe that one generality is as good as another. As a result, they show little interest in improving our current general theories.

Neither attitude is appropriate; one denies the possibility of more general relations and the other, the possibility of better ones. Obviously, a balance between detail and generality is desired.

Other ecologists will find the suggested problems uninteresting because they allow little scope for thought and originality. Unlike many parts of ecology, further growth of allometry does not require the creativity of genius, but rather a modicum of originality greatly leavened with hard work, perseverance, and criticism so that new relations can be isolated from the existing literature or established experimentally. Contemporary "theoretical ecologists" regard Charles Darwin as a prototype. Oddly enough, he seems more revered because he, like Wallace, Spencer, and many before, recognized the logical force of natural selection. What distinguished Darwin was not one brilliant idea but a lifetime of dedication to its pursuit and elaboration. The role of hard work is too often ignored in ecology.

Nevertheless, allometry is not an exclusive purview of the scientific drudge. A number of less routine (and consequently vaguer) problems exist within the orbit of body size; these permit more scope for technical innovation, scientific originality, and intellectual daring. Almost all data used in building the reported body size regressions were determined in the laboratory but are treated here as relevant to wild animals. An important step in biological scaling would test this assumption by measuring appropriate rates in free-living animals. Even if this did not confirm the laboratory results, it might lead to a complementary set of ecologically relevant relations. Further, each chapter of this book tries to indicate the ecological relevance of body size through a series of ecological deductions and hypotheses, some of these are only weakly supported by available information and further tests would often require considerable ingenuity and knowledge. The wide scatter in most size relations should provide an additional outlet for scientific creativity by using multiple regressions to identify other powerful independent variables of general biological interest. Finally, I have tried to pursue as many of the ecological implications of body size as possible, but this is obviously limited by my own thinking and experience. Fresh minds will find implications that I have completely overlooked.

The present book is, therefore, weak enough to provide scope for a range of scientific aptitudes and abilities. However, the availability of a poor text does not guarantee future growth for a scientific field, and I have no wish to end on a note of personal mortification. The gaps and shortcomings mentioned above will eventually be filled and corrected.

However, even the present body of allometric relations provides answers to many biological questions. It is a powerful approach that should find wider application.

At the very least, this book represents the largest available collection of body size relationships and should, therefore, provide a clearinghouse for much of the Western literature. The relationships listed in the appendixes are described more completely than in other collations and should permit better, more detailed predictions. The translation of all equations into standard units should facilitate comparison and the identification of inconsistencies in the literature. This identification is the first step to resolution. These are concrete points in favor of the current text about which there can be little discussion.

I would like to think this text offers some less quantifiable advances too. First, I feel it has increased coherence within a growing field of allometric physiology and autecology primarily by collating so many relations but also by directing these toward a related set of ecological questions. The book is more than just an updated extension of the existing series of books on allometry and body size (Bonner 1965, Brody 1945; Huxley 1972; Kleiber 1961; Pedley 1977; Thompson 1961). It is also a point of confluence or hybridization between that school of empirical biology and a growing trend in contemporary ecology toward general theories erected through the statistical analysis of literature data. This approach allows us to use what we have already determined by summarizing vast amounts of experimentation in predictive equations. Body size relations form a major part, but still only a part, of a growing subdiscipline: Predictive Ecology (Peters 1980b). Ecology has survived by trading on the ideas of its pioneers; now it must prove itself useful, not just diverting, by formulating predictions.

Finally, I would stress my belief that the allometric relations listed in the appendixes represent by far the greatest body of quantitative general theories in biology. I sincerely hope that this closing show of arrogance will stimulate others to prove me wrong. This can only be shown by amassing an even greater body of ecological theory.

Appendixes

Appendix contents

I. Units and useful conversions

Appendix Ia. Approximate conversions (the equivalents listed here are not exact; they are determined empirically)

$$1 \text{ kg dry mass} = 3\text{--}10 \text{ kg wet mass}$$
$$1 \text{ kg dry mass} = 22 \times 10^6 \text{ J}$$
$$1 \text{ kg wet mass} = 7 \times 10^6 \text{ J}$$
$$1 \text{ kg fat} = 40 \times 10^6 \text{ J}$$
$$\text{Tissue density} = 1 \text{ kg liter}^{-1}$$
$$1 \text{ kg wet mass} = 1 \times 10^{15} \, \mu\text{m}^3$$
$$1 \text{ kg dry mass} = 0.4 \text{ kg carbon}$$
$$1 \text{ ml O}_2 = 20.1 \text{ J}$$

Appendix Ib. Exact conversions (the equivalents listed here are determined by definition)

Acceleration
$$1 \text{ m s}^{-2} = 0.102 \text{ G}$$
$$9.8 \text{ m s}^{-2} = 1 \text{ G}$$

Force (mass × acceleration)
$$1 \text{ Newton} = 1 \text{ kg m s}^{-2}$$
$$= 0.102 \text{ kg force}$$
$$= 1 \times 10^5 \text{ dynes}$$

Work and energy (force × distance)
$$1 \text{ Joule} = 1 \text{ kg m}^2 \text{ s}^{-2}$$
$$= 1 \text{ N m}$$
$$= 0.239 \text{ cal}$$
$$= 1 \times 10^7 \text{ ergs}$$

Power (energy per unit time)
$$1 \text{ Watt} = 1 \text{ kg m}^2 \text{ s}^{-3}$$
$$= 1 \text{ J s}^{-1}$$
$$1 \text{ ml O}_2 \text{ s}^{-1} = 0.0446 \text{ mMol O}_2 \text{ s}^{-1}$$
$$= 1.43 \text{ mg O}_2 \text{ s}^{-1}$$

Pressure (force per unit area)
$$1 \text{ Pascal} = 1 \text{ kg m}^{-1} \text{ s}^{-2}$$
$$= 1 \text{ N m}^{-2}$$
$$= 9.87 \times 10^{-6} \text{ atm}$$
$$= 1.0 \times 10^{-5} \text{ bar}$$
$$= 7.501 \text{ torr}$$
$$= 7.501 \text{ mm Hg}$$
$$= 0.102 \text{ mm H}_2\text{0}$$

Volume
$$1 \text{ m}^3 = 1 \times 10^3 \text{ liters}$$
$$= 1 \times 10^6 \text{ cm}^3$$
$$= 1 \times 10^{18} \, \mu\text{m}^3$$

Mass
$$1 \text{ kg} = 1 \times 10^{-3} \text{ ton}$$

Note: A bewildering variety of units have been used in allometric relations. To facilitate comparisons, all relations in Appendixes II–X have been standardized by expressing W in kilograms of fresh mass and by expressing Y in comparable units when possible. This standardization involves a number of conversions.

II. Mass relations

Appendix IIa. *Relations between fresh body mass and linear dimensions of wet mass*

Taxon	Independent variable	Range of W (kg)	N	No. of spp.	Length	Reference	Intercept at $L=1$ m a (kg)	Slope $b \pm \bar{S}_b$	r^2	S_{xy}
Ungulates & carnivores	Body mass		113		Snout–vent	Radinsky (1978)	25	3.01±0.08	0.92	
Ungulates & carnivores	Body mass	0.2–3000			Total	Jerison (1973)	24	3.03		
Ungulates	Body mass		59		Snout–vent	Radinsky (1978)	21	3.45±0.11	0.94	
Carnivores	Body mass		54		Snout–vent	Radinsky (1978)	23	2.73±0.15	0.87	
Land mammals	Body mass		240	130	Total	Economos (1981)	14	3.23		
Bats	Body mass			5	Total	Jerison (1973)	5	2.0		
Birds	Body mass	0.008–10.6	118	118	Body length	From Covo (1965)	7.39	2.74	0.92	
Birds (no hummingbirds)	Body mass				Wingspan	Tucker (1973a)	0.75	3.0		
Hummingbirds	Body mass				Wing length	Greenewalt (1975)	0.87	1.8	0.89	
Legged lizards	Body mass	0.0005–6.7	636		Snout–vent	Pough (1980)	28	2.98		
Legless herptiles	Body mass		500		Snout–vent	Pough (1980)	0.72	3.02		
Legless herptiles	Body mass		500		Total	Pough (1980)	0.38	3.02		
Turtles	Body mass	0.006–448	508		Carapace length	Pough (1980)	94	2.69		
Frogs	Body mass		368		Snout–vent	Pough (1980)	181	3.24		
Salamanders	Body mass		645		Snout–vent	Pough (1980)	14	2.94		
Salamanders	Body mass		645		Total	Pough (1980)	12.4	3.64		

Fish	Body mass		60		Total	Vézina (unpubl.)	10.6	2.57		
Fish	Body mass				Total	Webb (1975)	10	3		
World record fish	Body mass	0.6–800	64		Total	Jerison (1973)	13.8	2.53		
Insects	Body mass	20×10^{-9}–48×10^{-6}			Total	Rogers et al. (1976)	8.8	2.62±0.055	0.94	0.644
Insects	Body mass	4.9×10^{-9}–364×10^{-6}	48		Total	Vézina (unpubl.)	1.4	2.35	0.82	
Aquatic insects	Body mass	30×10^{-9}–895×10^{-6}	1,362	43	Total	Smock (1980)	8.8	2.62	0.80	
Planktonic crustaceans	Body mass	0.03×10^{-15}–1×10^{-9}	283		Total	Peters & Downing (unpubl.)	0.08	2.1±0.055	0.84	0.024
Algae	Body mass	5×10^{-15}–1×10^{-10}	48		Diameter	Peters & Downing (unpubl.)	0.0058	1.9±0.22	0.60	0.10

Notes: Conversion from dry to wet mass assume that fresh mass = 4 × dry.

Appendix IIb. *Relations between surface area and body mass*

Taxon	Independent variable	Range of W (kg)	N	Reference	Standardized relation		
					Intercept at $W = 1\,kg$, $a\,(\text{m}^2)$	Slope $b \pm S_b$	r^2
Homeotherms	Surface area			Kayser & Heusner (1964)	0.0824	0.73	
Mammals	Surface area		>100	Stahl (1967)	0.11	0.65	
Mammals	Surface area			Dawson & Hurlbert (1970)	0.10	0.67 ± 0.018	0.98
Mammals	Surface area		37	Martin (1981)	0.109	0.67	
Birds	Surface area	0.03–4.24		Lasiewski & Calder (1971)	0.09	0.74	
Birds	Surface area		114	Calder & King (1974)	0.10	0.67	

III. Metabolic rates

Appendix IIIa. *Allometric relations for metabolic rate* (**R**), *basal metabolic rate* (**R**$_b$), *standard metabolic rate* (**R**$_s$), *resting metabolic rate* (**R**$_r$), *and existence metabolic rate* (**R**$_e$) *of homeotherms.*

Taxon	Independent variable	Conditions	Range of W (kg)	N	No. of spp.	Reference	Intercept at W=1 kg, a (watts)	Slope b±S$_b$	r²	S$_{xy}$
Homeotherms	R$_b$		0.01–1,000			Brody (1945)	3.42	0.734		
Homeotherms	R$_b$		0.01–3,000			Benedict (1938)	4.22	0.65		
Homeotherms	R$_b$	T_b=39°C	0.005–1,000	13		Hemmingsen (1960)	4.10	0.751±0.015		
Homeotherms	R$_s$	T_b=39°C				Robinson et al. (1983)	3.89	0.79		
Homeotherms[a]	R$_s$	exp(0.087±0.019)T_b	0.005–3,000	89		Robinson et al. (1983)	0.131	0.79±0.014	0.77	
Mammals	R$_b$					Boddington (1978)	3.61	0.73±0.02	0.89	
Mammals	R$_b$		0.0035–600	26		Kleiber (1961)	3.28	0.756±0.004		
Mammals	R			349		Stahl (1967)	3.89	0.76±0.013	0.96	
Mammals	R$_b$	W<120 g	0.005–0.12	51		Tracy (1977)	2.23	0.61	0.48	
Small wild mammals	R$_s$		0.003–3.4	55	50	Lechner (1978)	2.84	0.725±0.022	0.96	
Small mammals	R$_s$		0.003–3.4	63	56	Lechner (1978)	3.06	0.727±0.020	0.96	
Hibernating mammals	R	Awake	0.01–5	7	11	Kayser (1950)	2.95	0.62		
Hibernating mammals	R	Awake		7	7	Kayser & Heusner (1964)	3.08	0.62±0.04		
Hibernating mammals	R	Asleep ($T_b \propto W$)		46	16	Kayser & Heusner (1964)	0.105	1.02±0.043		
Hibernating mammals	R	Asleep T_b=18°C			15	Kayser (1950)	0.086	0.728		

Appendix IIIa. (*cont.*)

Taxon	Independent variable	Conditions	Range of W (kg)	N	No. of spp.	Reference	Intercept at $W=1$ kg, a(watts)	Slope $b \pm S_b$	r^2	S_{xy}
Hibernating mammals	R	Asleep $T_b=10°C$		46		Kayser & Heusner (1964)	0.101	0.69		
Hibernating mammals	R	Asleep $T_b=10°C$		46		Kayser (1964)	0.101	0.69		
Hibernating mammals	R	Asleep ($T_b \propto W$)			12	Kayser (1950)	0.095	0.94		
Higher primates	R					Konoplev et al. (1978)	5.66	0.73		
Perissodactyla	R					Konoplev et al. (1978)	5.21	0.73		
Artiodactyla	R					Konoplev et al. (1978)	5.21	0.73		
Carnivora	R					Konoplev et al. (1978)	4.52	0.73		
Rodentia	R					Konoplev et al. (1978)	2.84	0.69		
Lagomorpha	R					Konoplev et al. (1978)	3.33	0.72		
Insectivora	R					Konoplev et al. (1978)	3.22	0.73		
Chiroptera	R					Konoplev et al. (1978)	2.83	0.73		
Chiroptera	R_s					McNab (1969)	3.38	0.75		
Edentates	R					Konoplev et al. (1978)	2.90	0.77		
Marsupials	R					Konoplev et al. (1978)	2.32	0.74		
Marsupials	R_s		0.009–54	8	8	Dawson & Hurlbert (1970)	2.28	0.737	0.99	
Dasyurids	R		0.007–5	12		McMillen & Nelson (1969)	2.26	0.74±0.004		
Monotremes	R					Konoplev et al. (1978)	1.57	0.79		
Birds	R					Zotin & Konoplev (1978)	3.35	0.64		
Birds	R		0.003–10	130	100	Zar (1969)	3.76	0.739±0.0087		1.48

Group	Parameter	Conditions	Body-mass range	n	Source			
Birds	R		0.03–10	35	King & Farner (1961)	3.88	0.66±0.76	
Birds	R		0.003–100	120	Lasiewski & Dawson (1967)	4.19	0.668	0.086
Birds	R		0.03–0.12	30	King & Farner (1961)	3.59	0.744±0.074	
Passerines	R_r	Day		14	Aschoff & Pohl (1970)	6.81	0.704±0.029	
Passerines	R_r	Night		14	Aschoff & Pohl (1970)	5.56	0.726±0.022	0.113
Passerines	R_s			48	Lasiewski & Dawson (1967)	6.26	0.724	
Passerines	R	$T_a=30°C$	0.006–1	49	Zar (1969)	5.47	0.632±0.034	0.38
Passerines	R_e			15	Calder (1974)	5.55	0.75	
Passerines	R_s				Calder (1974)	6.74	0.73	
Passerines	R_b	Summer, night	0.01–1	41	Kendeigh et al. (1977)	5.01	0.688	1.09
Passerines	R_b	Winter, night	0.01–1	35	Kendeigh et al. (1977)	5.07	0.658	1.12
Passerines	R_e	$T_a=30°C$	0.01–1	71	Kendeigh et al. (1977)	7.12	0.66	1.13
Passerines	R_e	Photoperiod=10L:14D / Nesting / Photoperiod=15L:9D	0.01–1	70	Kendeigh et al. (1977)	8.21	0.688	1.11
Nonpasserines	R_r	Day		17	Aschoff & Pohl (1970)	4.41	0.729±0.033	
Nonpasserines	R_r	Night		17	Aschoff & Pohl (1970)	3.57	0.734±0.034	
Nonpasserines	R_s			72	Lasiewski & Dawson (1967)	3.64	0.723	0.068
Nonpasserines	R		0.003–10	81 / 64	Zar (1969)	3.69	0.743±0.013	0.15
Nonpasserines	R_b	Summer, night	0.003–4	30	Kendeigh et al. (1977)	4.20	0.728	1.17
Nonpasserines	R_b	Winter, night	0.003–4	12	Kendeigh et al. (1977)	4.01	0.701	1.13
Nonpasserines	R_b	Winter, 30°C	0.003–4	72	Kendeigh et al. (1977)	4.06	0.735	1.23
Nonpasserines	R_e	Photoperiod=10L:14D	0.003–4	40	Kendeigh et al. (1977)	5.32	0.626	1.19
Nonpasserines	R_e	Nesting, 30°C / Photoperiod=15L:9D	0.003–4	70	Kendeigh et al. (1977)	6.00	0.688	1.11
Nonpasserines	R_e				Calder (1974)	4.8	0.75	
Nonpasserines	R_s	30°C		9	Calder (1974)	4.47	0.73	

Appendix IIIa. (cont.)

Taxon	Independent variable	Conditions	Range of W (kg)	N	No. of spp.	Reference	Standardized relation			
							Intercept at W=1 kg, a(watts)	Slope b±S_b	r^2	S_{xy}
Nonpasserines	R		0.027–1.24	72	24	Prinzinger & Hännsler (1980)	5.39	0.716		0.010
Apodiformes	R		0.003–0.008	9	9	Zar (1969)	4.35	0.724±0.10		0.538
Strigiformes	R		0.04–2	12	10	Zar (1969)	3.03	0.696±0.096		0.117
Columbiformes	R		0.04–0.4	11	5	Zar (1969)	3.95	0.772±0.073		0.727
Galliformes	R		0.04–5	14	8	Zar (1969)	3.75	0.627±0.088		1.23
Falconiformes	R		0.1–10	5	5	Zar (1969)	2.3	0.865±0.119		1.05
Anseriformes	R		0.4–10	9	6	Zar (1969)	4.36	0.685±0.055		0.94
Ciconiiformes	R		0.6–6	7	7	Zar (1969)	4.63	0.658±0.070		0.926
Corvidae	R			8		Zar (1969)	5.19	0.476±0.141		0.066
Ploceidae	R			17		Zar (1969)	5.52	0.698±0.138		0.054
Fringillidae	R			20		Zar (1969)	6.01	0.706±0.070		
Bird eggs	R		0.01–0.17	16		Rahn et al. (1974)	1.08	0.772±0.094	0.94	
Bird eggs	R			34		Hoyt (1980)	0.936	0.714±0.004	0.965	
Duck eggs	R			13		Hoyt et al. (1979)	1.06	0.72	0.92	

Notes: W, wet body mass; T_b, body temperature; T_a, ambient temperature.
[a] Multiple regression to the model $\ln Y = \ln a + b \ln W + cT_b$ or $Y = aW^b \exp(cT_b)$.

Appendix IIIb. *Allometric relations for metabolic rates (R) and standard metabolic rates (R_s) of poikilotherms and unicells*

Taxon	Independent variable	Conditions	Range of W (kg)	N	No. of spp.	Reference	Intercept at $W=1\,kg$ a (Watts)	Slope $b \pm S_b$	r^2	S_{xy}
							Standardized relation			
Poikilotherms	R_s	$T_b=20°C$	1×10^{-6}–100		33	Hemmingsen (1960)	0.14	0.751±0.015		
Poikilotherms	R_s	$T_b=20°C$	1×10^{-3}–30			Robinson et al. (1983)	0.20	0.76		
Poikilotherms	R_s	$\exp(0.051\pm0.0033)\ T_b$	1×10^{-3}–30		729	Robinson et al. (1983)	0.071	0.76±0.0084	0.59	
Reptiles	R	$T_b=18°C$	0.01–100		11	Kayser & Heusner (1964)	0.086	0.728±0.034		
Reptiles	R_s	$T_b=20°C$	0.001–100		73	Bennett & Dawson (1976)	0.14	0.80±0.0075		
Reptiles	R_s	$T_b=30°C$	0.001–100		44	Bennett & Dawson (1976)	0.32	0.77±0.0075	0.69	
Reptiles	R					Zotin & Konoplev (1978)	0.41	0.74±0.015	0.83	
Reptiles	R					Zotin & Konoplev (1978)	0.94	0.77		
Lizards	R_s	$T_b=20°C$	0.001–1		13	Bennett & Dawson (1976)	0.13	0.80±0.018	0.64	
Lizards	R_s	$T_b=30°C$	0.001–7		24	Bennett & Dawson (1976)	0.41	0.83±0.010	0.73	
Lizards	R_s	$T_b=30°C$	0.001–1			Bartholomew & Tucker (1964)	0.378	0.821±0.018		
Lizards	R_s	$T_b=37°C$	0.001–1		24	Bennett & Dawson (1976)	0.68	0.82±0.018	0.67	

Appendix IIIb. (*cont.*)

Taxon	Independent variable	Conditions	Range of W (kg)	N	No. of spp.	Reference	Intercept at $W=1$ kg, a (Watts)	Slope $b \pm S_b$	r^2	S_{xy}
Xanthusids	R_s	$T_b=25$°C			10	Mautz (1979)	0.24	0.79±0.033		
Xanthusids	R_s	$T_b=35$°C			10	Mautz (1979)	0.83	0.88±0.03		
Varanids	R_s	$T_b=30$°C	0.016–4.4		10	Bartholomew & Tucker (1964)	0.333	0.62±0.024		
Snakes	R_s	$T_b=20$°C	0.007–20	35		Bennett & Dawson (1976)	0.137	0.77±0.03	0.64	
Snakes	R_s	$T_b=30$°C	0.007–20	13		Bennett & Dawson (1976)	0.202	0.71±0.035	0.83	
Snakes	R_s		0.350–22.5	83	15	Dmi'el (1972)	0.455	0.60±0.036	0.72	
Turtles	R_s	$T_b=20$°C		10		Bennett & Dawson (1976)	0.137	0.86±0.025		
Turtles	R_s		0.003–0.86	24		Kayser & Heusner (1964)	0.212	0.86±0.032		
Amphibians	R_s					Zotin & Konoplev (1978)	0.296	0.77		
Temperate salamanders	R_s	$T_b=5$°C		43	6	Feder (1976)	0.00724	0.622±0.014	0.929	
Tropical salamanders	R_s	$T_b=5$°C		26	4	Feder (1976)	0.025	0.837±0.026	0.920	
Temperate salamanders	R_s	$T_b=15$°C		83	12	Feder (1976)	0.059	0.823±0.012	0.935	
Tropical salamanders	R_s	$T_b=15$°C		81	13	Feder (1976)	0.046	0.860±0.018	0.884	
Lungless salamanders	R_s	$T_b=15$°C		37		Whitford & Hutchinson (1967)	0.0968	0.72		
Lunged salamanders	R_s	$T_b=15$°C		66		Whitford & Hutchinson (1967)	0.227	0.856		

Temperate salamanders	R_s	$T_b=25°C$	0.0005–0.030	56	9	Feder (1976)	0.135	0.802±0.0095	0.970
Tropical salamanders	R_s	$T_b=25°C$	0.0005–0.030	26	5	Feder (1976)	0.123	0.813±0.030	0.891
Frogs	R	$T_b=5°C$				Hutchinson et al. (1968)	0.0142	0.71	
Frogs	R	$T_b=15°C$				Hutchinson et al. (1968)	0.139	0.71	
Frogs	R	$T_b=25°C$				Hutchinson et al. (1968)	0.394	0.71	
Fishes	R_s	$T_a=20°C$	2×10^{-5}–10	369		Winberg (1960)	0.386	0.78±0.096	
Fishes	R_s	$T_a=20°C$	2×10^{-5}–10	364		Winberg (1960)	0.428	0.81±0.0105	
Teleosts	R	$T_a=25°C$	0.01–0.05	119		Kayser & Heusner (1964)	0.141	0.70±0.01	
Teleosts	R					Zotin & Konoplev (1978)	0.447	0.81	
Teleosts	R					Zotin & Konoplev (1978)	0.273	0.77	
Freshwater fishes	R_s		2×10^{-5}–10	266		Winberg (1960)	0.446	0.81±0.014	
Marine fishes	R_s			123		Winberg (1960)	0.420	0.79±0.014	
Salmonids	R_s	$T_a=20°C$	1×10^{-4}–1	31		Winberg (1960)	0.53	0.76±0.032	0.95
Salmon	R_s	$T_a=15°C$	0.003–1.5		1	Brett (1965)	0.163	0.775	
Cyprinids	R_s	$T_a=20°C$	1×10^{-3}–1	43		Winberg (1960)	0.47	0.80±0.044	
Cyprinodonts	R_s	$T_a=20°C$	1×10^{-5}–2	23		Winberg (1960)	0.14	0.71±0.042	
Sturgeons	R_s	$T_a=20°C$	2×10^{-5}–10	33		Winberg (1960)	0.47	0.80	
Cyclostomes	R					Zotin & Konoplev (1978)	0.494	0.81	0.99
Amphioxi	R					Zotin & Konoplev (1978)	0.854	0.91	
Insects	R					Zotin & Konoplev (1978)	0.600	0.76	
Insects	R					Kayser & Heusner (1964)	0.283	0.62±0.08	

Appendix IIIb. (*cont.*)

Taxon	Independent variable	Conditions	Range of W (kg)	N	No. of spp.	Reference	Intercept at $W=1\,\text{kg}$, a (Watts)	Slope $b \pm S_b$	r^2	S_{xy}
Cryptozoa	R	$T_a = 20°C$	1×10^{-5}–0.015		15	Reichle (1968)	0.196	0.84 ± 0.071		
Spiders	R			47	13	Greenestone & Bennett (1980)	0.0748	0.71	0.91	
Moths	R_r			86	23	Bartholomew & Casey (1978)	0.473	0.775	0.79	
Beetles	R_r	$T_a = 24°C$	1×10^{-4}–0.0063	37	25	Bartholomew & Casey (1977)	0.487	0.86	0.83	
Beetles	R		2.5×10^{-4}–2.5×10^{-3}		17	Kayser & Heusner (1964)	0.57	0.77 ± 0.16		
Insect eggs	R	$T_a = 25°C$				Smith & Kleiber (1950)	0.0286	0.66 ± 0.11		
Crustacea	R					Zotin & Konoplev (1978)	0.275	0.81		
Crustacea	R					Zotin & Konoplev (1978)	0.139	0.76		
Crustacea	R	$T_a = 15°C$	2×10^{-5}–0.6	54	22	Weymouth et al. (1944)	0.180	0.826 ± 0.08	0.94	
Crustacea	R	$T_a = 20°C$	11×10^{-9}–2.5×10^{-3}	247		Ivlea (1980)	0.265	0.783 ± 0.008	0.95	
Crustacea	R	$T_a = 25°C$	16×10^{-9}–2.4×10^{-3}	249		Ivlea (1980)	0.217	0.725 ± 0.043	0.53	
Crustacea	R	$T_a = 29°C$	15×10^{-9}–1.2×10^{-3}	212		Ivlea (1980)	0.123	0.664 ± 0.013	0.93	
Decapods	R					Konoplev et al. (1978)	0.289	0.85		
Decapods	R	$T_a = 20°C$	0.74×10^{-6}–0.28×10^{-3}	53		Ivlea (1980)	0.390	0.806 ± 0.038	0.89	

Taxon		Temperature	Weight range	n	Reference			
Decapods	R	$T_a=29°C$	$95×10^{-6}–950×10^{-6}$	35	Ivlea (1980)	0.163	0.689±0.039	0.68
Decapod larvae	R	$T_a=29°C$	$0.87×10^{-6}–230×10^{-6}$	29	Ivlea (1980)	0.536	0.578±0.020	0.97
Amphipods	R				Konoplev et al. (1978)	0.185	0.79	
Amphipods	R	$T_a=20°C$	$0.5×10^{-6}–45×10^{-6}$	15	Ivlea (1980)	0.102	0.72±0.069	0.88
Amphipods	R	$T_a=25°C$	$0.2×10^{-6}–137×10^{-6}$	30	Ivlea (1980)	0.544	0.83±0.019	0.99
Amphipods	R	$T_a=30°C$	$0.14×10^{-6}–105×10^{-6}$	32	Ivlea (1980)	0.203	0.72±0.053	0.85
Isopods	R				Konoplev et al. (1978)	0.285	0.85	
Copepods	R				Konoplev et al. (1978)	0.229	0.77	
Copepods	R	$T_a=20°C$	$11×10^{-9}–9.9×10^{-6}$	56	Ivlea (1980)	0.0578	0.667±0.036	0.86
Copepods	R	$T_a=25°C$	$16×10^{-9}–6.2×10^{-6}$	98	Ivlea (1980)	0.198	0.721±0.030	0.86
Copepods	R	$T_a=29°C$	$15×10^{-9}–7×10^{-6}$	85	Ivlea (1980)	0.361	0.736±0.030	0.86
Ostracoda	R	$T_a=29°C$	$24×10^{-9}–370×10^{-9}$	15	Ivlea (1980)	0.0026	0.435±0.109	0.52
Cladocera	R				Konoplev et al. (1978)	0.216	0.81	
Phyllopods	R				Konoplev et al. (1978)	0.057	0.70	
Euphausiacea	R	$T_a=29°C$	$0.92×10^{-6}–8.5×10^{-6}$	16	Ivlea (1980)	0.612	0.767±0.083	0.84
Mollusca	R				Zotin & Konoplev (1978)	0.159	0.75	
Meristomata	R				Zotin & Konoplev (1978)	0.126	0.81	
Oligochaeta	R				Zotin & Konoplev (1978)	0.091	0.75	
Nematoda	R	$T_a=20°C$			Klekowski, Wasilewsko, & Paplinska (1972)	0.0236	0.72	
Anthozoa	R				Zotin & Konoplev (1978)	0.0738	0.86	

Appendix IIIb. (cont.)

Taxon	Independent variable	Conditions	Range of W (kg)	N	No. of spp.	Reference	Standardized relation			
							Intercept at $W=1$ kg, a (Watts)	Slope $b \pm S_b$	r^2	S_{xy}
Scyphozoa	R					Zotin & Konoplev (1978)	0.00018	0.15		
Porifera	R					Zotin & Konoplev (1978)	0.00625	0.55		
Marine eggs	R	$T_a = 25°C$				Smith & Kleiber (1950)	0.0317	0.82 ± 0.065		
Unicells	R	$T_a = 20°C$	$1 \times 10^{-15} - 1 \times 10^{-9}$	80		Hemmingsen (1960)	0.0179	0.751 ± 0.015		
Unicells	R	$T_a = 20°C$				Robinson et al. (1983)	0.0552	0.83	0.86	
Unicells[a]	R	$\exp(0.049 \pm 0.016)T_b$	$1 \times 10^{-16} - 1 \times 10^{-10}$	67		Robinson et al. (1983)	0.0207	0.83 ± 0.056	0.86	
Algae	R	$T_a = 21°C$	$30 \times 10^{-15} - 10 \times 10^{-12}$	13	8	Banse (1976)	1.24	0.90 ± 0.058	0.96	
Protozoa	R	$T_a = 22°C$	$1 \times 10^{-12} - 1 \times 10^{-8}$	34	29	Klekowski (1981)	0.0170	0.68	0.94	

Note: N, wet body mass; T_b, body temperature; T_a, ambient temperature.
[a]Multiple regression to the model $1n\ Y = 1n\ a + b\ 1n\ W + cT_b$ or $Y = aW^b \exp(cT_b)$.

Appendix IIIc. Allometric relations for metabolic power production homeotherms outside the thermal neutral zone: Standard metabolism (R_s), existence metabolism (R_e), the maximum rate of heat production by nonshivering thermogenesis (R_{NST}), and summit metabolism (R_{sum})

Taxon	Independent variable	Conditions	Range of W (kg)	N	No. of spp.	Reference	Standardized relation			
							Intercept at $W=1$ kg a (Watts)	Slope $b \pm S_b$	r^2	S_{xy}
Mammals	R_s	$T_a=0°C$ $W<120$ g	0.01–0.11	8		Tracy (1977)	3.24	0.36	0.92	
Shrews	R_s	$T_a=24°C$	0.001–0.02	8		Tracy (1977)	3.87	0.525	0.96	
Mammals	R_{NST}			23	9	Heldmaier (1971)	7.30	0.546	0.93	
Birds	R_e	Nesting, $T_a=0°C$ Photoperiod= 10L:14D		111		Kendeigh et al. (1977)	8.08	0.527		1.14
Birds	R_e	$T_a=0°C$		24		Calder (1974)	8.17	0.53		
Passerines	R_s	Night; summer $T_a=0°C$		41		Kendeigh et al. (1977)	6.43	0.528		1.13
Passerines	R_s	Night, winter $T_a=0°C$		35		Kendeigh et al. (1977)	5.88	0.531		1.10
Passerines	R_e	$T_a=0°C$	0.055–1.2	71		Kendeigh et al. (1977)	7.99	0.524		1.07
Passerines	R_e	Nesting, $T_a=0°C$		70		Kendeigh et al. (1977)	8.17	0.544		1.07
Nonpasserines	R_s	Night; summer $T_a=0°C$		35		Kendeigh et al. (1977)	6.51	0.570		
Nonpasserines	R_s	Night; winter $T_a=0°C$		30		Kendeigh et al. (1977)	5.31	0.594		
Nonpasserines	R_e	$T_a=0°C$		41		Kendeigh et al. (1977)	7.77	0.529		1.18
Nonpasserines	R_e	Nesting, $T_a=0°C$		70		Kendeigh et al. (1977)	8.61	0.544		1.13
Birds	R_{sum}					Calder (1974)	20.65	0.60		
Nonpasserines	R_{sum}					Calder (1974)	20.16	0.65		

Note: W, wet body mass; T_a, ambient temperature.

249

Appendix IIId. *Allometric relations for metabolic rates during nonthermogenic activity: Maximum metabolic rate* (R_{max}), *minimum metabolic power for flight* ($R_{f(min)}$), *power during flight* (R_f), *swimming* (R_{swim}), *and activity* (R_a)

Taxon	Independent variable	Conditions	Range of W (kg)	N	No. of spp.	Reference	Standardized relation			
							Intercept at $W=1$ kg, a (Watts)	Slope $b \pm S_b$	r^2	S_{xy}
Organisms	R_{max}			24	24	Prothero (1979)	38.71	0.763 ± 0.034		
Mammals	R_{max}					Hart (1971)	23.4	0.73		
Wild mammals	R_{max}		7–263	27	22	Taylor et al. (1981)	39.0	0.79 ± 0.019	0.99	
Wild and domestic mammals	R_{max}		7–263	55	55	Taylor et al (1981)	33.6	0.845 ± 0.015	0.99	
Small wild mammals	R_{max}		0.003–3.4	12	12	Lechner (1978)	17.2	0.663		
Wild and domestic small mammals	R_{max}		0.003–3.4	23	17	Lechner (1978)	18.1	0.678		
Small mammals	R_{max}			22	4	Pasquis et al. (1970)	22.6	0.73		
Large mammals	R_{max}			7	3	Pasquis et al. (1970)	51.3	0.79		
Birds	$R_{f(min)}$	Wingspan (m) $= 1.1W^{0.33}$				Tucker (1973a)	84.7	1.0		
Birds	$R_{f(min)}$	Wingspan (m) $= 0.9W^{0.33}$				Tucker (1973a)	113	1.0		
Birds	$R_{f(min)}$	Wingspan (m) $= 0.13W^{0.33}$				Tucker (1973a)	67.3	1.0		
Passerines	$R_{f(min)}$					Greenewalt (1975)	41.6	0.989		
Shorebirds	$R_{f(min)}$					Greenewalt (1975)	43.7	1.00		
Ducks	$R_{f(min)}$					Greenewalt (1975)	56.0	0.997		
Birds	R_f		0.003–1.0	11		Berger & Hart (1974)	48.9	0.72		
Birds	R_f		0.003–1.0	17	16	Hails (1979)	43.5	0.649	0.91	
Birds	R_f					King (1974)	60.7	0.73		
Birds, bats	R_f				7	King (1974)	64.8	0.78		

Birds	R_f					108	1.02	Berger & Hart (1974)	
Birds	R_f					54.6	0.67	Kendeigh et al. (1977)	
Passerines	R_f					39.6	0.67	Kendeigh et al. (1977)	
Nonpasserines	R_f								
Birds	R_f	Hovering		3		264	1.019	Calder (1974)	
Reptiles	R_{max}	$T_b=30°C$		14		2.26	0.82±0.04	Bennett & Dawson (1976)	0.72
Reptile eggs	R_{max}	$T_b=30°C$	$0.002–0.068$	13	13	0.228	0.737	Seymour (1979)	0.83
Salmon	R_{max}	$T_a=15°C$	$0.0034–1.4$	1		2.85	0.97±0.026	Brett (1965)	
Salmon	$R_{0.75(max)}$	$T_a=15°C$	$0.0034–1.4$	1		1.42	0.926	Brett (1965)	
Salmon	$R_{0.5(max)}$	$T_a=15°C$	0.0034 ± 1.4	1		0.81	0.890	Brett (1965)	
Salmon	$R_{0.25(max)}$	$T_a=15°C$	$0.0034–1.4$	1		0.41	0.846	Brett (1965)	
Salmon	R_s	$T_a=15°C$	$0.0034–1.4$	1		0.19	0.778	Brett (1965)	
Moths	R_s	Hovering		63	23	94.4	0.818	Bartholomew & Casey (1978)	0.82
Moths	R_a	Warming up		67	23	3.22	0.896	Bartholomew & Casey (1978)	0.82
Beetles	R_a	Running	$1\times10^{-4}–6.3\times10^{-3}$	37	24	4.14	1.17	Bartholomew & Casey (1977)	0.72
Vert. poikilotherms[a]	R_a	$T_b=20°C$	$\exp(0.036\pm0.0068)T_b$			1.12	0.87	Robinson et al. (1983)	
Vert. poikilotherms[a]	R_a		$1\times10^{-4}–1$	109		0.55	0.87±0.039	Robinson et al. (1983)	0.23

Note: R_s, used in some comparisons is the standard metabolic rate, T_b, body temperatures; T_a, ambient temperature.

[a] Multiple regression to the model $\ln Y = \ln a + b \ln W + cT_b$ or $Y = aW^b \exp(cT_b)$.

Appendix IIIe. Allometric relations for the average energy expenditure of free-ranging animals (R_{AEE})

Taxon	Independent variable	Conditions	Range of W (kg)	N	No. of spp.	Reference	Standardized relation Intercept at $W=1$ kg a (Watts)	Slope $b \pm S_b$	r^2	S_{xy}
Rodents	R_{AEE}		0.009–0.61	19	19	King (1974)	8.74	0.669±0.215	0.98	0.104
Rodents	R_{AEE}		0.003–0.40	47	36	French et al. (1976)	3.51	0.5		
Insectivores	R_{AEE}			15	8	French et al. (1976)	4.74	0.5		
Birds	R_{AEE}		0.005–0.41	18	12	King (1974)	15.4	0.705±0.186	0.94	0.110
Passerines	R_{AEE}	Winter, $T_a=30°C$				Kendeigh et al. (1977)	7.88	0.67		
Passerines	R_{AEE}	Winter, $T_a=0°C$				Kendeigh et al. (1977)	7.97	0.50		
Passerines	R_{AEE}	Breeding, $T_a=30°C$				Kendeigh et al. (1977)	8.29	0.67		
Passerines	R_{AEE}	Breeding, $T_a=0°C$				Kendeigh et al. (1977)	10.66	0.50		
Nonpasserines	R_{AEE}	Winter, $T_a=30°C$				Kendeigh et al. (1977)	5.33	0.67		
Nonpasserines	R_{AEE}	Winter, $T_a=0°C$				Kendeigh et al. (1977)	9.04	0.50		
Nonpasserines	R_{AEE}	Breeding, $T_a=30°C$				Kendeigh et al. (1977)	5.26	0.67		
Nonpasserines	R_{AEE}	$T_a=0°C$				Kendeigh et al. (1977)	12.33	0.50		

Note: T_a, ambient temperature.

Appendix IIIf. *Allometric relations for other measures of energy expenditure: Specific respiration rate of a population* ($R/\bar{B}$), *the total additional respiratory cost of reproduction* ($C_{reprod.}$), *reproductive effort of feeding the litter at weaning* (R_w/R_f), *and energy expended during development of birds' eggs* (C_{egg})

Taxon	Independent variable	Conditions	Range of W (kg)	N	No. of spp.	Reference	Standardized relation Intercept at $W=1$ kg, a	Slope $b \pm S_b$	r^2
Invertebrates	$R/\bar{B}$	$W=$mature body mass		13		Banse (1979)	0.0046 Watt kg^{-1}	-0.35 ± 0.06	0.63
Mammals	C_{reprod}	$W=$total litter mass				Brody (1945)	1.32×10^7 J	1.2	
Mammals	R_w/R_f	$W=$maternal mass		100	100	Millar (1977), Kleiber (1961)	0.961 Watt	-0.17 ± 0.03	0.27
Altricial birds	C_{egg}	$W=$egg mass	0.009–0.051	13	12	Vleck, Vleck, & Hoyt, (1980)	0.54×10^6 J	0.78	0.94
Precocial birds	C_{egg}	$W=$egg mass	0.012–0.145	15	15	Vleck, Vleck, & Hoyt, (1980)	1.63×10^6 J	0.95	0.98
Procellariiforms	C_{egg}	$W=$egg mass	0.0093–0.0609	3	3	Vleck, Vleck & Hoyt, (1980)	1.12×10^6 J	0.70	1.0

Note: W, wet mass; R_f, maternal metabolic rate.

253

IV. Physiology and morphology

Appendix IVa. *Allometric relations describing aspects of respiratory physiology: volumes (V), frequencies (F), alveolar surface areas (ASA)*

Taxon	Independent variable	Range of W (kg)	N	No. of spp.	Reference	Standardized relation		
						Intercept at $W=1\,kg$, a	Slope $b \pm S_b$	r^2
Mammals	$V_{lung\ air}$		688		Stahl (1962)	11.57 ml	0.99±0.005	0.98
Mammals	$V_{lung\ vol.}$				Tenney & Remers (1963); Schmidt-Nielsen (1979)	56.7 ml	1.02	
Mammals	$V_{lung\ vol.}$				Weibel (1973)	45.2 ml	1.05	0.99
Mammals	$V_{lung\ vol.}$				Stahl (1967)	53.5 ml	1.06±0.01	0.96
Wild mammals	$V_{lung\ vol.}$	0.4–230	27	15	Gehr et al. (1981)	66.1 ml	0.986±0.017	0.99
Mammals	$V_{lung\ vol.}$	0.002–700	32		Gehr et al. (1981)	46.0 ml	1.059±0.014	0.99
Small wild mammals	$V_{lung\ vol.}$	0.002–3.7	7	7	Lechner (1978)	30.4 ml	0.933±0.029	0.99
Small mammals	$V_{lung\ vol.}$	0.002–3.7	17	13	Lechner (1978)	40.0 ml	1.021±0.055	0.98
Birds	$V_{lung\ vol.}$	0.006–88	4		Lasiewski & Calder (1971)	29.6 ml	0.94	0.98
Birds	$V_{lung\ vol.}$	0.318–2.28	7		Lasiewski & Calder (1971)	9.9 ml	0.76	1.0
Reptiles	$V_{lung\ vol.}$	0.020–100			Tenney & Tenney (1970)		1.05	
Amphibians	$V_{lung\ vol.}$	0.002–0.3			Tenney & Tenney (1970)		0.75	
Mammals	$V_{tracheal\ vol.}$				Lasiewski & Calder (1971)	0.920 ml	1.15	
Birds	$V_{tracheal\ vol.}$	0.005–67.5			Lasiewski & Calder (1971)	3.72 ml	1.06	
Mammals	$V_{total\ resp.\ syst.}$				Lasiewski & Calder (1971)	54.4 ml	1.06	
Birds	$V_{total\ resp.\ syst.}$	0.006–88	5		Lasiewski & Calder (1971)	160.8 ml	0.91	0.98
Mammals	$V_{dead\ space}$		52		Stahl (1967)	2.76 ml	0.96±0.045	0.90
Mammals	V_{frc}^{a}		261		Stahl (1967)	24.1 ml	1.13±0.015	0.94
Mammals	V_{frc}^{a}				Günther & Léon de la Barra (1966)	11.1 ml	1.19	
Mammals	V_{vc}^{b}		315		Stahl (1967)	56.7 ml	1.03±0.005	0.98
Mammals	V_{tidal}				Adolph (1949)	6.64 ml	1.01	
Mammals	V_{tidal}		300		Guyton (1947a)	7.4 ml	1.0	

Group	Variable	Range	N	N	Reference	Value	Coefficient	r
Mammals	V_{tidal}				Günther (1975)	6.3 ml	1.0	0.98
Mammals	V_{tidal}		688		Stahl (1967)	7.69 ml	1.04±0.005	
Birds	V_{tidal}	0.005–1	6		Berger & Hart (1974)	50.1 ml	0.9	
Birds	V_{tidal}		6		Lasiewski & Calder (1971)	13.2 ml	1.08	0.98
Frogs	V_{tidal}				Hutchinson et al. (1968)	65.5 ml	0.73	
Mammals	$F_{resp.}$		692		Günther & Léon de la Barra (1966a)	0.795 s^{-1}	−0.28	
Mammals	$F_{resp.}$		300		Stahl (1967)	0.891 s^{-1}	−0.26±0.005	0.83
Mammals	$F_{resp.}$				Guyton (1947a)	0.886 s^{-1}	−0.25	
Mammals	$F_{resp.}$				Adolph (1949)	0.885 s^{-1}	−0.28	
Birds	$F_{resp.}$	0.043–100	26		Lasiewski & Calder (1971)	0.286 s^{-1}	−0.31	
Birds	$F_{resp.}$	0.005–2	6		Berger & Hart (1974)	1.51 s^{-1}	−0.20	
Passerines	$F_{resp.}$				Günther (1975)	0.31 s^{-1}	−0.33	
Nonpasserines	$F_{resp.}$				Günther (1975)	0.381 s^{-1}	−0.28	
Birds	$F_{resp.(max)}$	0.005–2	13		Berger & Hart (1974)	4.32 s^{-1}	−0.083	
Mammals	Air flow				Adolph (1949)	5.53 ml s^{-1}	0.74	
Mammals	Air flow				Guyton (1947b)	6.22 ml s^{-1}	0.75	
Mammals	Air flow		691		Stahl (1967)	6.32 ml s^{-1}	0.80±0.005	0.96
Mammals	Air flow	0.043–100			Günther & Léon de la Barra (1966a)	5.34 ml s^{-1}	0.76	
Birds	Air flow		6		Lasiewski & Calder (1971)	4.73 ml s^{-1}	0.77	0.96
Birds	Air flow in flight	0.05–1	6		Berger & Hart (1974)	108 ml s^{-1}	0.73	
Mammals	ASA		29		Weibel (1973)	3.31 m^2	0.98	
Mammals	ASA	0.002–650	32	32	Gehr et al. (1981)	3.34 m^2	0.949±0.031	0.99
Wild mammals	ASA		19		Weibel (1973)	4.11 m^2	0.95	
Wild mammals	ASA	0.4–230	27	15	Gehr et al. (1981)	3.58 m^2	0.918±0.015	0.98
Small mammals	ASA	0.002–3.7	17	13	Lechner (1978)	1.87 m^2	0.888±0.051	0.96
Small wild mammals	ASA	0.002–3.7	7	7	Lechner (1978)	0.97 m^2	0.693±0.030	0.99
Reptiles	ASA	0.010–100			Tenney & Tenney (1970)	0.0018 m^2	0.75	
Amphibians	Respiratory area	0.002–0.10			Tenney & Tenney (1970)	0.0007 m^2	0.98	
Fish	Gill surface area				Muir (1969)		0.85	
Mammals	Air–blood barrier thickness	0.002–650	32	32	Gehr et al. (1981)	0.416 μm	0.05±0.007	0.63
Mammals	Air–blood barrier thickness				Weibel (1973)	0.42 μm	0.05	
Wild mammals	Air–blood barrier thickness	0.4–230	27	15	Gehr et al. (1981)	0.444 μm	0.028±0.014	
Mammals	O_2 diffusing capacity		58		Stahl (1967)	2.0×10^{-5} ml (Pa·s)$^{-1}$	1.18±0.025	0.96
Mammals	O_2 diffusing capacity				Weibel (1973)	7.9×10^{-5} ml (Pa·s)$^{-1}$	0.96	0.98
Mammals	O_2 diffusing capacity	0.002–650	32	32	Gehr et al. (1981)	4.9×10^{-5} ml (Pa·s)$^{-1}$	0.99±0.010	0.99

Appendix IVa. (*cont.*)

Taxon	Independent variable	Range of W (kg)	N	No. of spp.	Reference	Intercept at $W=1$ kg, a	Slope $b\pm S_b$	r^2
Wild mammals	O$_2$ diffusing capacity	0.4–230	27	15	Gehr et al. (1981)	5.5×10^{-5} ml (Pa·s)$^{-1}$	0.95 ± 0.030	0.98
Mammals	CO diffusing capacity	0.02–0.250	258		Stahl (1967)	2.8×10^{-5} ml (Pa·s)$^{-1}$	1.14 ± 0.02	0.92
Reptiles	Alveolar diameter	0.25–100			Tenney & Tenney (1970)	1.6×10^{-3} m	0.2	
Reptiles	Alveolar diameter				Tenney & Tenney (1970)	4.0×10^{-3} m	0.2	
Amphibians	Alveolar diameter	0.002–0.3			Tenney & Tenney (1970)	2×10^{-3} m	0	
Amphibians	Total capillary length				Tenney & Tenney (1970)	0.15 m^2	0.8	
Wild mammals	Lung capillary area	0.4–230	27	15	Gehr et al. (1981)	2.99 m^2	0.924 ± 0.025	0.98
Mammals	Lung capillary area	0.002–600	32	32	Gehr et al. (1981)	2.73 m^2	0.952 ± 0.030	0.99
Wild mammals	Lung capillary volume	0.4–230	27	15	Gehr et al. (1981)	4.19 ml	0.971 ± 0.035	0.97
Mammals	Lung capillary volume	0.002–600	32	32	Gehr et al. (1981)	3.20 ml	1.00 ± 0.040	0.94
Mammals	Lung compliance		286		Stahl (1967)	21.4×10^{-9} ml Pa^{-1}	1.08 ± 0.01	0.96
Mammals	Thoracic compliance		60		Stahl (1967)	46.1×10^{-9} ml Pa^{-1}	0.86 ± 0.025	0.96
Mammals	Thoracic compliance				Günther & Léon de la Barra (1966a)	47.9×10^{-9} ml Pa^{-1}	0.82	
Mammals	Lung and thoracic compliance		114		Stahl (1967)	1.59×10^{-9} ml Pa^{-1}	1.04 ± 0.015	0.98
Mammals	Lung and thoracic compliance				Günther & Léon de la Barra (1966a)	1.2×10^{-9} ml Pa^{-1}	1.0	
Mammals	Total airway resistance		135		Stahl (1967)	2.36×10^{6} Pa(m^3·s)$^{-1}$	-0.70 ± 0.02	0.92
Mammals	Total airway resistance				Günther & Léon de la Barra (1966a)	3.2×10^{6} Pa(m^3·s)$^{-1}$	-0.87	
Mammals	Intrapleural pressure				Günther & Léon de la Barra (1966a)	31.2 Pa	0.004	
Mammals	Breathing power		89		Stahl (1967)	1.57×10^{-3} Watts	0.78 ± 0.03	0.96
Mammals	Breathing power				Günther & Léon de la Barra (1966a)	1.30×10^{-3} Watts	0.70	
Mammals	Breathing power	0.02–70	30	5	Drorbaugh (1960)	1.30×10^{-4} Watts	0.70	
Mammals	Work/breath	0.02–70	30	5	Drorbaugh (1960)	1.77×10^{-4} J	0.965	

Note: Other variables as indicated. Masses associated with respiration appear in Appendix IVe.

[a] frc = functional residual capacity.

Appendix IVb. *Allometric relations describing aspects of blood flow and cellular respiration*

Taxon	Independent variable	Range of W (kg)	N	No. of spp.	Reference	Intercept at $W=1$ kg, a	Slope $b \pm S_b$	r^2
Mammals	V_{heart}				Stahl (1964)	5.72 ml	0.98	
Mammals	$V_{end\ systole}$ (lv)				Günther (1975)	0.59 ml	0.99	
Mammals	$V_{end\ systole}$ (rv)				Günther (1975)	0.62 ml	0.99	
Mammals	$V_{end\ diastole}$ (lv)				Günther (1975)	1.76 ml	1.02	
Mammals	$V_{end\ diastole}$ (rv)				Günther (1975)	2.02 ml	1.02	
Mammals	V_{stroke}				Günther (1975)	0.74 ml	1.03	
Mammals	V_{stroke} (lv)				Günther (1975)	0.66 ml	1.05	
Mammals	V_{stroke} (rv)				Günther (1975)	0.59 ml	1.04	
Mammals	V_{blood}	0.01–1,000	97	34	Prothero (1980)	76 ml	1.0 ± 0.001	0.99
Mammals	V_{blood}				Stahl (1962)	51.3 ml	0.99	
Mammals	V_{blood}				Stahl (1967)	65.6 ml	1.02 ± 0.05	
Mammals	V_{blood}				Brody (1945)	50.7 ml	0.98 ± 0.017	
Mammals	V_{plasma}	0.01–400	78	25	Prothero (1980)	44 ml	1.0 ± 0.01	0.99
Mammals	$V_{red\ cells}$	0.01–800	73	25	Prothero (1980)	36 ml	1.0 ± 0.01	0.98
Mammals	F_{pulse}		447		Stahl (1981)	$4.03\ s^{-1}$	-0.25 ± 0.01	0.77
Mammals	F_{pulse}				Brody (1945)	$3.62\ s^{-1}$	-0.27	
Mammals	F_{pulse}				Adolph (1943)	$3.61\ s^{-1}$	-0.27	
Mammals	F_{pulse}				Günther & Léon de la Barra (1966a)	$3.60\ s^{-1}$	-0.27	
Mammals	F_{pulse} (lv)				Günther (1975)	$3.93\ s^{-1}$	-0.25	
Mammals	F_{pulse} (rv)				Günther (1975)	$4.0\ s^{-1}$	-0.25	
Birds	F_{pulse}	0.004–7.5	27		Lasiewski & Calder (1971)	$2.60\ s^{-1}$	-0.23	
Birds	F_{pulse}	0.007–10		13	Berger & Hart (1974)	$2.93\ s^{-1}$	-0.209	

Appendix IVb. (*cont.*)

Taxon	Independent variable	Range of W (kg)	N	No. of spp.	Reference	Standardized relation		
						Intercept at $W=1$ kg, a	Slope $b \pm S_b$	r^2
Varanids (30°C)	F_{pulse}		10		Bartholomew & Tucker (1964)	$0.50\,s^{-1}$	-0.155 ± 0.032	
Birds	$F_{pulse\ (flight)}$	0.003–1.5	26	12	Berger & Hart (1974)	$8.79\,s^{-1}$	-0.146	
Birds	$F_{pulse\ (max)}$	0.003–10	28	28	Berger & Hart (1974)	$8.48\,s^{-1}$	-0.157	
Mammals	$F_{pulse\ (max)}$	0.032–67	10	10	Baudinette (1978)	$6.25\,s^{-1}$	-0.19	0.94
Mammals	Cardiac output		568		Stahl (1967)	$3.12\,ml\,s^{-1}$	0.81 ± 0.005	0.96
Mammals	Cardiac output				Günther (1975)	$1.69\,ml\,s^{-1}$	0.999	
Mammals	Cardiac output				Günther (1975)	$3.4\,ml\,s^{-1}$	0.74	
Mammals	Cardiac output				Günther & Léon de la Barra (1966a)	$5.48\,ml\,s^{-1}$	0.74	
Mammals	Cardiac output (lv)				Günther (1975)	$2.78\,ml\,s^{-1}$	0.79	
Mammals	Cardiac output (rv)				Günther (1975)	$2.98\,ml\,s^{-1}$	0.78	
Small mammals	Cardiac output	0.17–2.3	120	4	White et al. (1968)	$2.70\,ml\,s^{-1}$	0.776	0.83
Mammals	Hemoglobin mass				Adolph (1949)	$0.012\,kg$	0.99	
Mammals	Hemoglobin mass				Munro (1969)	$0.013\,kg$	1.00	
Mammals	P_{50}			17	Lutz et al. (1974)	$0.378\,Pa$	-0.054 ± 0.01	
Birds	P_{50}			7	Lutz et al. (1974)	$0.416\,Pa$	-0.079 ± 0.026	
Lizards[a]	P_{50}				Pough (1977a)	$0.517\,Pa$	-0.040	
Snakes[a]	P_{50}		72	15	Pough (1977b)	$0.162\,Pa$	0.166	
Mammals	Carbonic anhydrase Activity				Magid (1967), Munro (1969)		0.89	
Birds	O_2 uptake by blood	0.01–100	6		Girard & Grima (1980)	$127\,M\,O_2$ liter red cells^{-1}	-0.10 ± 0.016	
Birds	O_2 uptake by hemoglobin	0.01–100	6		Girard & Grima (1980)	$6.3\,M\,O_2\,mM\,H_b^{-1}$	-0.10 ± 0.014	

258

Group	Property	Body mass range	n	n	Reference	Value		
Mammals	$W_{myoglobin}$				Adolph (1949)	0.33×10^{-3} kg		1.31
Mammals	Capillary density (semitendinousus)	0.4–250	20	13	Hoppeler et al. (1981)	7.47 mm⁻²	−0.138±0.025	0.31
Mammals	Capillary density (longissimus dorsi)	0.4–250	21	13	Hoppeler et al. (1981)	7.60 mm⁻²	−0.10±0.050	0.24
Mammals	Capillary density (vastus medialis)	0.4–250	20	13	Hoppeler et al. (1981)	9.65 mm⁻²	−0.097±0.050	0.37
Mammals	Capillary density (diaphragm)	0.4–250	21	13	Hoppeler et al. (1981)	10.60 mm⁻²	−0.045±0.07	0.11
Mammals	Blood sugar	0.006–550	73	73	Umminger (1975)	1.16 g liter⁻¹	−0.100±0.008	
Nonruminants	Blood sugar				Umminger (1975)	1.19 g liter⁻¹	−0.0777±0.008	
Mammals	Mitochondria (semitendinousus muscle)	0.4–250	20	13	Mathieu et al. (1981)	0.065 %vol	−0.231±0.025	0.85
Mammals	Mitochondria (longissimus dorsi)	0.4–250	18	13	Mathieu et al. (1981)	0.066 %vol	−0.163±0.024	0.72
Mammals	Mitochondria (vastus medialis)	0.4–250	20	13	Mathieu et al. (1981)	0.071 %vol	−0.139±0.070	0.59
Mammals	Mitochondria (diaphragm)	0.4–250	21	13	Mathieu et al. (1981)	0.131 %vol	−0.055±0.028	0.32
Homeotherms	Cell diameter		3		Maldonado et al. (1974)	16.1 μm	0.028	
Poikilotherms	Cell diameter		3		Maldonado et al. (1974)	23.6 μm	0.049	
Mammals	$W_{cytochromes}$				Adolph (1949)	33×10^{-6} kg	0.84	
Mammals	$W_{cytochrome c}$				Munro (1969)	34×10^{-6} kg	0.70	
Mammals	Cytochrome oxidase				Munro (1969)		0.67	
Mammals	Aorta cross section				Günther & Léon de la Barra (1966a)	1.8×10^{-5} m²	0.67	
Mammals	Aorta length				Günther & Léon de la Barra (1966a)	0.164 m	0.32	
Mammals	Aortic blood velocity				Günther & Léon de la Barra (1966a)	0.298 m s⁻¹	0.07	

Appendix IVb. (cont.)

Taxon	Independent variable	Range of W (kg)	N	No. of spp.	Reference	Standardized relation Intercept at $W=1$ kg, a	Slope $b \pm S_b$	r^2
Mammals	Tension (aortic wall)				Günther (1975)	$27.3\,\mathrm{N\,m^{-1}}$	0.35	
Mammals	Aortic elasticity				Günther & Léon de la Barra (1966a)	$31\times10^6\,\mathrm{Pa\,m^{-3}}$	1.19	
Mammals	Systemic arterial pressure				Günther & Léon de la Barra (1966a)	$15\times10^3\,\mathrm{Pa}$	0.032	
Mammals	Total peripheral resistance				Günther & Léon de la Barra (1966a)	$0.255\,\mathrm{Pa\cdot s\,m^{-3}}$	−0.68	
Mammals	Total peripheral resistance				Günther (1975)	$0.487\,\mathrm{Pa\cdot s\,m^{-3}}$	−0.76	
Homeotherms	Cardiac force				Günther (1975)	$8.32\times10^{-3}\,\mathrm{N}$	0.999	
Homeotherms	Cardiac work				Günther (1975)	$8.91\times10^{-3}\,\mathrm{J}$	1.06	
Homeotherms	Cardiac power				Günther (1975)	$0.038\,\mathrm{Watt}$	0.77	
Mammals	Work/stroke left ventricle				Günther (1975)	$5.78\times10^{-3}\,\mathrm{J}$	1.15	
Mammals	Work/stroke (right ventricle)				Günther (1975)	$1.17\times10^{-3}\,\mathrm{J}$	1.14	
Mammals	Internal surface area (lv)				Günther (1975)	$0.617\times10^{-3}\,\mathrm{m^2}$	0.68	
Mammals	Plasma colloid pressure				Günther (1975)	$2{,}300\,\mathrm{Pa}$	0.075	

Note: V, volume; F, frequency; lv, left ventricle; rv, right ventricle other variables as indicated. Associated masses are given in Appendix IVe.
[a]These equations describe a semilogrithmic relationship $Y = a + b\log W$.

Appendix IVc. *Allometric relations describing aspects of renal physiology*

Taxon	Independent variable	Range of W (kg)	N	No. of spp.	Reference	Standardized relation		
						Intercept at $W = 1$ kg, a	Slope $b \pm S_b$	r^2
Mammals	Inulin clearance	0.15–600	119	26	Edwards (1975)	$0.089 \, \mathrm{ml\,s^{-1}}$	0.72 ± 0.02	0.98
Mammals	Inulin clearance				Adolph (1949)	$0.099 \, \mathrm{ml\,s^{-1}}$	0.77	
Birds	Inulin clearance	0.11–10	8	4	Edwards (1975)	$0.035 \, \mathrm{ml\,s^{-1}}$	0.78 ± 0.05	0.92
Reptiles	Inulin clearance	0.005–4		6	Edwards (1975)	$0.0083 \, \mathrm{ml\,s^{-1}}$	0.75	
Mammals	PAH clearance	0.15–600	287	19	Edwards (1975)	$0.363 \, \mathrm{ml\,s^{-1}}$	0.77 ± 0.04	0.98
Mammals	PAH clearance				Adolph (1949)	$0.377 \, \mathrm{ml\,s^{-1}}$	0.80	
Mammals	Creatinine clearance	0.15–600	351	14	Edwards (1975)	$0.122 \, \mathrm{ml\,s^{-1}}$	0.68 ± 0.045	0.98
Mammals	Creatinine clearance				Adolph (1949)	$0.137 \, \mathrm{ml\,s^{-1}}$	0.69	
Mammals	Diodrast clearance				Adolph (1949)	$0.278 \, \mathrm{ml\,s^{-1}}$	0.89	
Mammals	Urine clearance		35	30	Edwards (1975)	$70 \times 10^{-9} \, \mathrm{ml\,s^{-1}}$	0.75 ± 0.05	0.90
Mammals	Nephron number				Adolph (1949)	1.9×10^5 No./kidney	0.62	

Note: Equations describing related masses appear in Appendix IVe.

261

Appendix IVd. *Allometric relations for a variety of other physiological traits*

Taxon	Independent variable	Range of W (kg)	N	No. of spp.	Reference	Standardized relation			
						Intercept at $W = 1$ kg, a	Slope $b \pm S_b$	r^2	S_{xy}
Mammals	p-Hydroxylphenyl pyruvate oxidase				Munro (1969)		0.71		
Mammals	Homogentisate oxidase				Munro (1969)		0.73		
Mammals	Phenylalanine oxoglutarate transaminase				Munro (1969)		0.54		
Mammals	Phenylalanine pyruvate transaminase				Munro (1969)		0.25		
Mammals	Tyrosine oxoglutarate transaminase				Munro (1969)		0.50		
Mammals	Tyrosine pyruvate transaminase				Munro (1969)		0.29		
Mammals	Glutamate pyruvate transaminase				Munro (1969)		0.90		
Mammals	Glutamate oxaloacetate transaminase				Munro (1969)		0.83		
Mammals	Thyroxine turnover				Munro (1969)		0.75		
Mammals	Erythrocyte life span	0.026–500	34	34	Vacha & Znojil (1981)	5.88×10^6 s	0.132	0.75	
Mammals	Erythrocyte turnover rate				Munro (1969)	0.16×10^{-6} s^{-1}	−0.138		
Mammals	Albumin turnover rate				Munro (1969)	4.7×10^{-6} s^{-1}	−0.34		
Mammals	Ceruplasmin turnover rate				Munro (1969)	2.2×10^{-6} s^{-1}	−0.24		
Mammals	Protein synthesis				Munro (1969)	$4.6\ \mu$g s^{-1}	0.76		
Mammals	β-OH-Butyryl-CoA dehydrogenase		20	10	Emmett & Hochachka (1981)	246 μM (kg·s)$^{-1}$	−0.21±0.009	0.52	
Mammals	Citrate synthase		20	10	Emmett & Hochachka (1981)	410 μM (kg·s)$^{-1}$	−0.106±0.032	0.37	
Mammals	Malate dehydrogenase		20	10	Emmett & Hochachka (1981)	10,500 μM (kg·s)$^{-1}$	−0.065±0.036	0.42	

Group	Parameter	Range	n_1	n_2	Reference	Value	Coefficient	r
Mammals	Pyruvate kinase		20	10	Emmett & Hochachka (1981)	$10{,}500\ \mu M\ (kg \cdot s)^{-1}$	0.14 ± 0.026	0.41
Mammals	Lactate dehydrogenase		20	10	Emmett & Hochachka (1981)	$16{,}300\ \mu M\ (kg \cdot s)^{-1}$	0.15 ± 0.027	0.59
Mammals	Phosphorylase		20	10	Emmett & Hochachka (1981)	$1{,}100\ \mu M\ (kg \cdot s)^{-1}$	0.09 ± 0.021	0.50
Mammals	Gravitational tolerance	0.010–30	3	3	Economos (1981, 1979b)	39.2 N	-0.14	
Mammals	Gut beat duration				Adolph (1949)	2.8 s	0.31	
Mammals	Stride frequency (trot-gallop transition)	0.03–680	19		Heglund et al. (1974)	$4.48\ s^{-1}$	-0.14	0.98
Sphingid moths	$F_{wing\ beat}$		29		Bartholomew & Casey (1978)	$7.38\ s^{-1}$	-0.244	0.299
Sphingid moths	$F_{wing\ beat}$				Bartholomew & Casey (1978)	$8.75\ s^{-1}$	-0.115	0.079
Mammals	Total sleep time		49	49	From Zepelin & Rechtschaffen (1974)	44,300 s	-0.12 ± 0.019	0.460
Herbivorous mammals	Total sleep time		19	19	From Zepelin & Rechtschaffen (1974)	32,500 s	-0.145 ± 0.0216	0.724
Carnivorous mammals	Total sleep time		30	30	From Zepelin & Rechtschaffen (1974)	47,200 s	-0.036 ± 0.0215	0.092 (NS)
Mammals	Sleep cycle		24	24	From Zepelin & Rechtschaffen (1974)	1,200 s	0.194 ± 0.028	0.690
Herbivorous mammals	Sleep cycle		12	12	From Zepelin & Rechtschaffen (1974)	955 s	0.190 ± 0.034	0.758
Carnivorous mammals	Sleep cycle		12	12	From Zepelin & Rechtschaffen (1974)	1,530 s	0.244 ± 0.037	0.859
Birds	Survival time (−13 to −18°C)				Calder (1974)	9.08 d	0.59 ± 0.14	0.91
Birds	Survival time (−1 to −9°C)				Calder (1974)	12.4 d	0.58 ± 0.16	0.92
Birds	Survival time (2 to 6°C)				Calder (1974)	15.0 d	0.39 ± 0.279	0.63

Appendix IVe. *Allometric relations describing sizes of various organs and tissues*

Taxon	Independent variable	Range of W (kg)	N	No. of spp.	Reference	Intercept at $W=1$ kg, a (kg)	Slope $b \pm S_b$	r^2
Mammals	Skeleton mass	0.006–6,600	49	29	Prange et al. (1979)	0.061	1.06	0.984
Mammals	Skeleton mass		10		Reynolds & Karlotski (1977)	0.073	1.123±0.021	0.998
Mammals	Skeleton mass	0.02–10,000	7	7	Kayser & Heusner (1964)	0.093	1.14±0.07	
Whales	Skeleton mass	8×10^3–130×10^6	170		Anderson et al, (1979)	0.024	1.11	0.951
Birds	Skeleton mass	0.0031–81	311	209	Prange et al. (1979)	0.065	1.071	0.986
Birds	Skeleton mass		16		Reynolds & Karlotski (1977)	0.068	1.119±0.033	0.986
Rattlesnakes	Skeleton mass	0.049–3.5	12		Anderson et al, (1979)	0.065	1.17	0.99
Fish	Skeleton mass		11		Reynolds & Karlotski (1977)	0.041	1.030	0.992
Spiders	Exoskeleton mass	2.5×10^{-5}–1.2×10^{-3}	61	3	Anderson et al, (1979)	0.184	1.135	0.94
Mollusks	Shell mass	8.5×10^{-4}–0.055	106	4	Anderson et al, (1979)	0.448	1.096	0.99
Birds' eggs	Shell mass	0.5×10^{-3}–8.5	368		Anderson et al, (1979)	0.118	1.13	0.99
Birds' eggs	Shell thickness		47	367	Ar et al. (1979)	1.20[a]	0.456±0.77	0.94
Birds' eggs	Shell thickness		47	47	Ar et al. (1974)	1.34[a]	0.458	0.951
Birds' eggs	Shell thickness		10		Ricklefs (1974)	1.26[a]	0.291±0.036	
Mammals	Skin mass	0.01–10	60	5	Pace et al. (1979)	0.139	0.942±0.015	0.99
Birds	Plumage mass	0.002–4	249	91	Turček (1966)	0.064	0.95	0.63
Mammals	Mass of skinned, eviscerated carcass	0.01–10	60	5	Pace et al. (1979)	0.503	1.055±0.0065	0.998
Mammals	Muscle mass				Munro (1969)	0.45	1.0	
Mammals	Visceral mass	0.01–10	60	5	Pace et al. (1979)	0.158	0.871±0.012	0.99
Mammals	Liver mass		100		Brody (1945), Stahl (1965)	0.033	0.867±0.009	0.98
Birds	Liver mass				Brody (1945)	0.033	0.877	
Mammals	Lung mass		7,100		Brody (1945), Stahl (1965)	0.0113	0.986±0.012	0.92

Group	Variable	Mass range (kg)	N		Reference	a	b	r
Birds	Lung mass				Brody (1945)	0.0149	0.937±0.029	
Birds	Lung mass	0.005–125	51		Lasiewski & Calder (1971)	0.0126	0.95	0.85
Highland birds	Lung mass		220		Carey & Morton (1976)	0.0083	0.94	0.90
Lowland birds	Lung mass		186		Carey & Morton (1976)	0.0103	1.06	0.90
Reptiles	Lung mass				Tenney & Tenney (1970)	0.03	1.0	
Amphibians	Lung mass				Tenney & Tenney (1970)	0.03	1.0	
Mammals	Gut mass				Brody (1945)	0.0746	0.941±0.044	
Herbivorous mammals	Total gut capacity	0.68–1,100			Parra (1978)	0.0936	1.077	0.98
Ruminants	Total gut capacity				Parra (1978)	0.0896	1.048	
Nonruminants	Total gut capacity				Parra (1978)	0.102	1.080	
Herbivorous mammals	Fermentation contents of gut				Parra (1978)	0.076	1.096	
Ruminants	Fermentation contents of gut				Parra (1978)	0.105	1.045	
Nonruminants	Fermentation contents of guts				Parra (1978)	0.0633	1.092	
Heteromyid rodents	Cheek pouch content	0.01–0.2	62	12	Morton et al. (1980)	0.030	0.887	0.73
Heteromyid rodents	Cheek pouch content	0.01–0.2	32	11	Morton et al. (1980)	0.059	1.043±0.059	0.94
Birds	Gut mass				Brody (1945)	0.0899	0.985±0.047	
Birds	Gut mass				Calder (1974)	0.098	1.003	
Mammals	Kidney mass	0.0025–100	>100		Brody (1945), Stahl (1965)	0.00732	0.846±0.010	
Mammals	Kidney mass				Fujita et al. (1966)	0.0096	0.84	
Birds	Kidney mass		334		Johnson (1968)	0.0087	0.913	
Birds	Kidney mass				Brody (1945)	0.0087	0.852±0.032	
Birds with salt glands	Kidney mass		51		Hughes (1970)	0.0112	0.879	
Birds with no salt glands	Kidney mass		52		Hughes (1970)	0.0074	0.928	
Mammals	Brain mass				Brody (1945)	0.00996	0.697±0.009	
Mammals	Brain mass		108		Stahl (1965)	0.0093	0.73	0.94

Appendix IVe. (cont.)

Taxon	Independent variable	Range of W (kg)	N	No. of spp.	Reference	Standardized relation		
						Intercept at $W=1$ kg, a (kg)	Slope $b \pm S_b$	r^2
Recent mammals	Brain mass				Jerison (1961)	0.00956	0.64	
Mammals	Brain mass	2–1,000	40		Jerison (1973)	0.012	0.67	
Placental mammals	Brain mass	0.005–1×10^5	309		Martin (1981)	0.011	0.76 ± 0.012	0.92
Precocial mammals	Brain mass		159		Martin (1981)	0.015	0.72	0.86
Altricial mammals	Brain mass		87		Martin (1981)	0.0098	0.79	0.92
Oligocene mammals	Brain mass				Jerison (1961)	0.00507	0.655	
Eocene mammals	Brain mass				Jerison (1961)	0.00243	0.657	
Fossil ungulates	Brain mass				Jerison (1973)	0.0069	0.67	
Primates	Brain mass		50		Stahl (1965)	0.0203	0.66	0.96
Birds	Brain mass				Brody (1945)	0.0066	0.498 ± 0.022	
Birds	Brain mass	0.010–100	180		Martin (1981)	0.0071	0.58 ± 0.018	0.86
Higher vertebrates	Brain mass	0.004–4.0×10^4	15		Jerison (1973)	0.007	0.67	
Lower vertebrates	Brain mass	0.01–200	4		Jerison (1973)	0.0007	0.67	
Reptiles	Brain mass	0.002–100	59		Martin (1981)	0.00069	0.54 ± 0.02	0.92
Fish	Brain mass	0.0001–1	110		Bauchot et al. (1977)	0.0010	0.657	0.78
Mammals	Heart mass		>100		Brody (1945), Stahl (1965)	0.00588	0.983 ± 0.009	0.98
Mammals	Heart mass		416		Berger & Hart (1974)	0.0057	0.935	
Mammals	Heart mass				Günther (1975)	0.0057	0.98	
Birds	Heart mass	0.004–7.5	76		Lasiewski & Calder (1971)	0.0082	0.91	0.94
Birds	Heart mass		78		Berger & Hart (1974)	0.0097	0.899 ± 0.045	
Birds	Heart mass		416		Berger & Hart (1974)	0.0076	0.834 ± 0.0085	
Hummingbirds	Heart mass		36		Berger & Hart (1974)	0.0155	0.937 ± 0.026	

					Berger & Hart (1974)	0.0067	0.874±0.009	
Birds (no hummingbirds)	Heart mass		380					
Highland birds	Heart mass		272		Carey & Morton (1976)	0.0070	0.84±0.021	0.96
Lowland birds	Heart mass		206		Carey & Morton (1976)	0.0081	0.91±0.023	0.98
Turtles	Heart mass				Brody (1945)	0.0048	0.99	
Frogs	Heart mass				Brody (1945)	0.0019	0.89	
Fish	Heart mass				Brody (1945)	0.0013	1.06	
Teleosts	Heart mass		21		Hughes (1977)	0.0024	1.07	
Elasmobranchs	Heart mass		23		Hughes (1977)	0.0014	0.90	
Mammals	Left-ventricle mass				Günther (1975)	0.00165	1.11	
Mammals	Right-ventricle mass				Günther (1975)	0.00074	1.06	
Mammals	Spleen mass		75		Stahl (1965)	0.0025	1.02±0.05	0.96
Mammals	Mammary mass	18	18		Linzell (1972)	0.04	0.87±0.03	0.98
Mammals	Mammary mass	14	14		Hanwell & Peaker (1977)	0.045	0.82±0.026	
Mammals	Adrenal mass		>100		Brody (1945), Stahl (1965)	273×10^{-6}	0.792±0.014	0.94
Birds	Adrenal mass				Brody (1945)	176×10^{-6}	0.89±0.029	
Mammals	Thyroid mass		>100		Brody (1945), Stahl (1945)	129×10^{-6}	0.924±0.017	0.94
Birds	Thyroid mass				Brody (1945)	117×10^{-6}	0.86±0.037	
Birds	Pituitary mass				Brody (1945)	24×10^{-6}	0.76±0.018	
Mammals	Pituitary mass		60		Stahl (1965)	30×10^{-6}	0.56±0.02	0.98
Primates	Pancreas mass		201		Stahl (1965)	2.0×10^{-3}	0.91±0.06	0.85

[a]Shell thickness in mm.

Appendix IVf. *Allometric relations describing tissue composition*

Taxon	Independent variable	Range of W (kg)	N	No. of spp.	Reference	Standardized relation		
						Intercept at $W = 1$ kg, a (kg)	Slope $b \pm S_b$	r
Mammals	Fat content	0.003–86,000	328	49	Pitts & Bullard (1968)	0.060	0.2	0.74
Mammals	Liver phospholipid				Munro (1969)	1.01×10^{-3}	0.84	
Mammals	Liver protein				Munro (1969)	5.9×10^{-3}	0.835	
Mammals	Thyroid protein				Munro (1969)	14.1×10^{-6}	1.05	
Mammals	Body albumin				Munro (1969)	4.13×10^{-3}	1.02	
Mammals	Skeletal DNA				Munro (1969)	203×10^{-6}	0.91	
Mammals	Liver DNA				Munro (1969)	91×10^{-6}	0.884	
Mammals	Thyroid DNA				Munro (1969)	321×10^{-9}	0.895	
Mammals	Skeletal DNA				Munro (1969)	1.66×10^{-3}	0.91	
Mammals	Liver RNA				Munro (1969)	256×10^{-6}	0.755	
Mammals	Thyroid RNA				Munro (1969)	324×10^{-9}	0.824	
Crustaceans	Total RNA				Båmstedt & Skjoldal (1980)	700×10^{-6}	0.683	
Mammals	Potassium				Fujita et al. (1966)	2.4×10^{-6}	1.0	

Appendix Va. The effects of size and ambient temperature on heat flux: Conductance, coefficients of heat stress, rate constants of heating or cooling (λ), and the temperature coefficient of existence metabolism (K)

Taxon	Independent variable	Range of W (kg)	N	No. of spp.	Reference	Intercept at W=1kg, a	Slope b±S_b	r²	S_xy
Mammals	Conductance				Morrison & Ryser (1951)	0.177 Watt°C⁻¹	0.50		
Mammals	Conductance	0.003–0.6		24	Herreid & Kessel (1967)	0.174 Watt°C⁻¹	0.495±0.053		
Mammals	Conductance				Hart (1971)	0.168 Watt°C⁻¹	0.50±0.014		
Mammals	Conductance	0.0035–150	192		Bradley & Deavers (1980)	0.224 Watt°C⁻¹	0.574	0.88	
Arctic mammals	Conductance (summer)	0.01–10		6	Casey et al. (1979)	0.193 Watt°C⁻¹	0.47	0.94	
Arctic mammals	Conductance (winter)	0.01–10		6	Casey et al. (1979)	0.109 Watt°C⁻¹	0.49	0.92	
Eutheria	Conductance (active phase)	0.08–6.6	27		Aschoff (1981)	0.161 Watt°C⁻¹	0.483		
Eutheria	Conductance (resting phase)	0.004–4.4	59		Aschoff (1981)	0.158 Watt°C⁻¹	0.481		
Eutheria	Conductance	0.004–6.6	86		Aschoff (1981)	0.203 Watt°C⁻¹	0.538		
Dasyurid marsupials	Conductance	0.0072–5.05		12	MacMillen & Nelson (1969)	0.208 Watt°C⁻¹	0.537±0.025	0.94	
Chiroptera	Conductance		30		Bradley & Deavers (1980)	0.0995 Watt°C⁻¹	0.43		
Sciurids	Conductance		14		Bradley & Deavers (1980)	0.239 Watt°C⁻¹	0.62	0.67	
Heteromyids	Conductance		22		Bradley & Deavers (1980)	0.267 Watt°C⁻¹	0.56	0.61	
Cricetids	Conductance		59		Bradley & Deavers (1980)	0.134 Watt°C⁻¹	0.46	0.94	
Murids	Conductance		19		Bradley & Deavers (1980)	0.055 Watt°C⁻¹	0.33	0.90	
Birds	Conductance	0.01–2.8		31	Herreid & Kessel (1967)	0.132 Watt°C⁻¹	0.464±0.016		
Birds	Conductance				Lasiewski et al. (1967)	0.142 Watt°C⁻¹	0.492		
Birds	Conductance	0.01–100	9		Calder & King (1974)	0.110 Watt°C⁻¹	0.49		
Birds	Conductance (0–11°C)	0.01–100	9		Calder & King (1974)	0.091 Watt°C⁻¹	0.374±0.029		
Dead birds	Conductance	0.01–2.8	15		Herreid & Kessel (1967)	0.193 Watt°C⁻¹	0.48±0.135		
Featherless dead birds	Conductance	0.01–2.8	17		Herreid & Kessel (1967)	0.176 Watt°C⁻¹	0.56±0.023		
Skinned dead birds	Conductance	0.01–2.8	15		Herreid & Kessel (1967)	0.608 Watt°C⁻¹	0.52±0.028		
Passerines	Conductance				Calder & King (1974)	0.113 Watt°C⁻¹	0.46±0.018		
Passerines	Conductance (summer)		41		Kendeigh et al. (1977)	0.162 Watt°C⁻¹	0.532		

Appendix Va. (cont.)

Taxon	Independent variable	Range of W (kg)	N	No. of spp.	Reference	Intercept at $W=1$ kg, a	Slope $b \pm S_b$	r^2	S_{xy}
Passerines	Conductance (winter)		35		Kendeigh et al. (1977)	0.150 Watt°C⁻¹	0.543		
Passerines	Conductance (active phase)	0.006–1.1	28		Aschoff (1981)	0.195 Watt°C⁻¹	0.537		
Passerines	Conductance (resting phase)	0.10–0.36	26		Aschoff (1981)	0.133 Watt°C⁻¹	0.539		
Nonpasserines	Conductance (active phase)	0.0027–2	39		Aschoff (1981)	0.186 Watt°C⁻¹	0.516		
Nonpasserines	Conductance (resting phase)	0.040–2	11		Aschoff (1981)	0.129 Watt°C⁻¹	0.462		
Nonpasserines	Conductance				Calder & King (1974)	0.113 Watt°C⁻¹	0.46±0.036		
Birds' eggs	Conductance	0.001–0.18	10	10	Ricklefs (1974)	0.133 Watt°C⁻¹	0.582±0.12		
Lizards	Conductance	0.37–2.8			Ellis & Ross (1978)	0.85 Watt°C⁻¹	0.571		
Lizards	Conductance				Ellis & Ross (1978); Bartholomew & Lasiewski (1965)	1.54 Watts°C⁻¹	0.379		
Varanids	Conductance	0.01–4	9	4	Bartholomew & Tucker (1964)	2.04 Watts°C⁻	0.63		
Fish	Conductance	5×10⁻⁴–0.2	164		Stevens & Fry (1970)	3.98 Watts°C⁻¹	0.50		
Sphinx moths	Conductance	1×10⁻⁴–5×10⁻³	41	2	Bartholomew & Epting (1975)	0.32 Watt°C⁻¹	0.531	0.72	
Birds' eggs	$\lambda_{cooling}$	0.001–0.8	10	10	Ricklefs (1974)	0.00011 s⁻¹	−0.412±0.124		
Birds' eggs	$\lambda_{cooling}$				Kendeigh et al. (1977)	0.00011 s⁻¹	−0.397		
Homeotherms	$\lambda_{heating}$	0.003–0.3	23		Heinrich & Bartholomew (1971)	0.0021 s⁻¹	−0.40		
Homeotherms & insects	$\lambda_{heating}$	0.1×10⁻³–0.3	33		Heinrich & Bartholomew (1971)	0.0016 s⁻¹	−0.51		
Lizards	$\lambda_{cooling}$				Ellis & Ross (1978)	0.00025 s⁻¹	−0.429		
Lizards	$\lambda_{cooling}$				Bartholomew & Lasiewski (1965)	0.00046 s⁻¹	−0.621		
Reptiles	$\lambda_{cooling}$	0.2–4	32	17	Spray & May (1977)	0.00127 s⁻¹	−0.418		

Terrestrial reptiles	$\lambda_{cooling}$		22		Spray & May (1977)	0.00167 s^{-1}	−0.384
Aquatic reptiles	$\lambda_{cooling}$		10		Spray & May (1977)	0.000523 s^{-1}	−0.591
Reptiles	$\lambda_{heating}$	0.2–4	31	17	Spray & May (1977)	0.00134 s^{-1}	−0.421
Terrestrial reptiles	$\lambda_{heating}$		22		Spray & May (1977)	0.00178 s^{-1}	−0.382
Aquatic reptiles	$\lambda_{heating}$		7		Spray & May (1977)	0.00204 s^{-1}	−0.329
Varanids	$\lambda_{cooling}$	0.01–4	9	4	Bartholomew & Tucker (1964)	0.00054 s^{-1}	−0.385±0.025
Lizards	$\lambda_{cooling}$	0.002–0.5	34		Claussen & Art (1981)	0.0024 s^{-1}	−0.38	0.90
Lizards	$\lambda_{cooling}$	0.01–5	26		Claussen & Art (1981)	0.0053 s^{-1}	−0.40	0.74
Lizards	$\lambda_{cooling}$	0.002–0.1	32		Claussen & Art (1981)	0.0034 s^{-1}	−0.35	0.61
Lizards	$\lambda_{heating}$	0.01–5	27		Claussen & Art (1981)	0.0069 s^{-1}	−0.34	0.83
Fish	$\lambda_{cooling}$	0.01–8.0	83		Spigarelli et al. (1977)	0.0015 s^{-1}	−0.36	0.94
Fish	$\lambda_{cooling}$	0.0006–1.2	229	1	Stevens & Fry (1974)	0.0012 s^{-1}	−0.57
Fish	$\lambda_{cooling}$	0.0005–0.2	160	2	Stevens & Fry (1970)	0.0016 s^{-1}	−0.5
Bumblebees	$\lambda_{cooling}$	0.00024–0.0006	6	4	Pyke (1980)	0.000060 s^{-1}	−0.55	0.89
Sphinx moths	$\lambda_{cooling}$	0.0001–0.005	41		Bartholomew & Epting (1975)	0.00010 s^{-1}	−0.469	0.85
Tropical dragonflies	$\lambda_{cooling}$ (25°)	0.0001–0.001	79	16	May (1976)	0.00012 s^{-1}	−0.481	0.943
Temperate dragonflies	$\lambda_{cooling}$ (15°C)	0.0001–0.001	60	9	May (1976)	0.00015 s^{-1}	−0.432	0.924
Temperate dragonflies	$\lambda_{cooling}$ (25°C)	0.0001–0.001	60	9	May (1976)	0.00027 s^{-1}	−0.378	0.863
Temperate dragonflies	$\lambda_{cooling}$ (30°C)	0.0001–0.001	60	9	May (1976)	0.00037 s^{-1}	−0.341	0.843
Dead tropical dragonflies	$\lambda_{cooling}$	0.0001–0.001	79	16	May (1976)	0.000077 s^{-1}	−0.524	0.975
Dead temperate dragonflies	$\lambda_{cooling}$		44	9	May (1976)	0.000096 s^{-1}	−0.488	0.966
Birds	Heat stress coefficient	0.0064–3.9	26		Weathers (1981)	0.14 Watt s^{-1}	0.35±0.174	0.86
Passerines	K_{winter} (10L:14D)		71		Kendeigh et al. (1977)	0.041 Watt°C^{-1}	0.243	1.26
Passerines	K_{summer} (15L:9D)		70		Kendeigh et al. (1977)	0.41 Watt°C^{-1}	0.216	1.16
Nonpasserines	K_{winter} (10L:14D)		40		Kendeigh et al. (1977)	0.081 Watt°C^{-1}	0.326	1.57
Nonpasserines	K_{summer} (15L:9D)		70		Kendeigh et al. (1977)	0.091 Watt°C^{-1}	0.282	1.50
Anseriformes	K_{winter} (10L:14D)		9		Kendeigh et al. (1977)	0.119 Watt°C^{-1}	0.290	1.34
Psittaciformes	K		5		Kendeigh et al. (1977)	0.139 Watt°C^{-1}	0.303	1.13

Note Standardized equations for $t_{1/2}$ give the rate constant (Table 5.2).

Appendix Vb. Equations relating body temperature (T_b), lower critical temperature (T_{lc}), and upper critical temperature (T_{uc}) to body mass of homeotherms

Taxon	Independent variable	Range of W (kg)	N	No. of spp.	Intercept at $W = 1$ kg, a (°C)	Slope $b \pm S_b$	r^2	S_{xy}	Reference
Small birds	T_b				$32.2 + 5.42W^b$	-0.09			McNab (1966b)
Birds	T_b	0.002–100			$32.2 + 7.7W^b$	-0.09			McNab (1966b)
Birds	T_b			34	40.08	0.04 ± 0.021		0.075	Calder & King (1974)
Mammals	$T_b - T_{lc}$				19.6	0.23			Gordon (1972)
Mammals	$T_b - T_{lc}$	0.003–150	28		20.1	0.26			McNab (1970)
Mammals	$38 - T_{lc}$				22.5	0.25			Morrison (1960)
Wintering arctic mammals	$T_b - T_{lc}$	0.01–10		6	42.7	0.20	0.90		Casey et al. (1979)
Summering arctic mammals	$T_b - T_{lc}$	0.01–10		6	25.1	0.20	0.90		Casey et al. (1979)
Birds	$T_b - T_{lc}$	0.006–100	12		27.8	0.17			McNab (1970)
Passerines	$T_b - T_{lc}$				43.8	0.266			Calder & King (1974)
Nonpasserines	$T_b - T_{lc}$				25.5	0.274			Calder & King (1974)
Nonpasserines	T_{lc}	0.003–4	74		13.5	-0.181	0.50	1.38	Kendeigh et al. (1977)
Summering passerines	T_{lc}	0.009–1.2	43		11.4	-0.184	0.67	1.13	Kendeigh et al. (1977)
Wintering passerines	T_{lc}	0.009–1.2	35		7.08	-0.250	0.67	1.26	Kendeigh et al. (1977)
Homeotherms[a]	T_{uc}		10		29.1	-5.4			Chin (unpubl.)

Note: Not all equations have the same form as Equation 1.1.

[a] This regression uses the semilogarithmic model $Y = a - b \ln W$.

VI. Locomotion

Appendix VIa. *Transport costs* (T_c) *as a function of body size*

Taxon	Independent variable	Range of W (kg)	N	No. of spp.	Reference	Standardized relation Intercept at $W=1$kg, a ($\mathrm{J\,m^{-1}}$)	Slope $b \pm S_b$	r^2	S_{xy}
Walking homeotherms[a]	$T_{c\,(total)}$	0.01–1,000	20		Tucker (1970)	$0.419 \times 10^{1.67W^{-0.126}}$			
Walking homeotherms[a]	$T_{c\,(total)}$	0.01–1,000	11		Taylor et al. (1970)	$10.8W^{0.60}+6.0W^{0.75}/V$			
Walking birds[a]	$T_{c\,(total)}$				Taylor (1977)	$11.5W^{0.76}+5.6W^{0.72}/V$			
Walking mammals[a]	$T_{c\,(total)}$				Taylor (1977)	$10.7W^{0.76}+6W^{0.75}/V$			
Walking mammals[a]	$T_{c\,(total)}$	0.03–1,000	20		Tucker (1973a)	$18.7W^{0.473}W^{0.0775\,\log W}$			
Running animals	$T_{c\,(total)}$				Tucker (1970)	10.73	0.60		
Running animals	$T_{c\,(net)}$	0.01–100	69	69	Fedak & Seeherman (1979)	11.3	0.72 ± 0.015	0.84	0.14
Running animals	$T_{c\,(net)}$	0.01–100	52		Paladino & King (1979)	11.1	0.68 ± 0.01	0.96	0.76
Running animals	$T_{c\,(net)}$				Schmidt-Nielsen (1972)	13.2	0.60		
Running bipeds	$T_{c\,(net)}$	0.01–100	7		Fedak et al. (1974)	11.1	0.76 ± 0.04		
Running lizards	$T_{c\,(net)}$				Cragg (1975) in Hughes (1977)	4.16	0.52		0.19
Flying birds	T_c	0.01–1	16		Tucker (1970)	5.23	0.773		
Flying birds and insects	T_c	0.03–0.2	16		Tucker (1973a)	5.23	0.773		
Flying birds	T_c				Schmidt-Nielsen (1972)	5.95	0.75		
Flying birds	T_c				Berger & Hart (1974)	4.19	0.773		
Flying birds	T_c				Tucker (1973a)	5.76	0.80		
Flying passerines	$T_{c\,(min)}$	0.003–10			Greenewalt (1975)	3.56	0.861 ± 0.0031	0.94	0.067
Flying shorebirds	$T_{c\,(min)}$	0.003–10			Greenewalt (1975)	3.20	0.847 ± 0.0073	0.92	0.071
Flying ducks	$T_{c\,(min)}$	0.1–1			Greenewalt (1975)	3.20	0.841 ± 0.018	0.81	0.042
Swimming fish	T_c	0.001–1	12		Tucker (1973a)	1.24	0.705		
Swimming fish	T_c				Schmidt-Nielsen (1972)	0.66	0.60		

Note: Terrestrial transport costs are treated as total ($T_{c\,(total)}$) or net ($T_{(net)}$). Some relations require complex equations. Greenewalt's (1975) values for birds are calculated assuming 25% efficiency (muscular power/metabolic power).
[a] These equations use complex models listed as "standardized relations."

273

Appendix VIb. Allometric relations for velocity of locomotion

Taxon	Independent variable	Range of W (kg)	N	No. of spp.	Reference	Standardized relation		
						Intercept at $W=1$ kg, a (m s^{-1})	Slope $b \pm S_b$	r^2
Running animals	V_{max}		24		From Bonner (1965)	10.4	0.38	0.95
African mammals	V_{max}	0.0072–263	27	27	Taylor et al. (1981)	2.34	0.122	0.45
Mammals	$V_{trot\text{-}gallop\ transition}$	0.03–680			Heglund et al. (1974)	1.53	0.24	0.86
Mammals	$V_{walking}$				From Buddenbrock (1934)	0.33	0.21	0.94
Beetles	$V_{walking}$		16		From Buddenbrock (1934)	0.30	0.29	0.60
Lamellicorn beetles	$V_{walking}$	120×10^{-9}–370×10^{-6}			From Hempel (1954)	0.21	0.36	
Carabid beetles	$V_{walking}$	120×10^{-9}–370×10^{-6}			From Hempel (1954)	1.96	0.35	
Flies	$V_{walking}$	7.3×10^{-9}–84×10^{-6}	59	7	From Hempel (1954)	1.84	0.36	
Birds	V_{max}		12		From Bonner (1965)	22.2	0.144	0.73
Birds	V				Berger & Hart (1974)	28.0	0.288	
Birds $(W_s = 1.1W^{0.33})$	V				Tucker (1973b)	14.6	0.20	
Birds $(W_s = 0.9W^{0.33})$	V				Tucker (1973b)	13.1	0.21	
Birds $(W_s = 1.3W^{0.33})$	V				Tucker (1973b)	16.7	0.20	
Passerines	$V_{min\ power}$				Greenewalt (1975)	9.46	0.12 ± 0.0057	0.76
Shorebirds	$V_{min\ power}$				Greenewalt (1975)	16.1	0.17 ± 0.0083	0.90
Ducks	$V_{min\ power}$				Greenewalt (1975)	14.6	0.156 ± 0.029	0.59
Fish	$V_{min\ (sprint)}$				Webb (1975)	4.4	0.35	
Fish	V_{max}		17		From Bonner (1965)	2.37	0.35	
Salmon, 2°C	$V_{max\ (sustained)}$	0.002–2		1	Brett & Glass (1973)	0.89	0.201	0.93

Salmon, 5°C	$V_{\text{max (sustained)}}$	0.002–2	1	Brett & Glass (1973)	1.00	0.201
Salmon, 10°C	$V_{\text{max (sustained)}}$	0.002–2	1	Brett & Glass (1973)	1.20	0.201
Salmon, 15°C	$V_{\text{max (sustained)}}$	0.002–2	1	Brett & Glass (1973)	1.52	0.203
Salmon, 15°C	$V_{\text{max (sustained)}}$	0.003–1.5	1	Brett (1965)	1.29	0.175
Salmon, 20°C	$V_{\text{max (sustained)}}$	0.002–2	1	Brett & Glass (1973)	1.41	0.201
Salmon	$V_{\text{min power}}$	0.01–1.5	1	Ware (1978)	0.39	0.136
Trout, 10°C	V_{cruising}	0.004–0.10	128	Fry & Cox (1970)	0.98	0.13±0.014
Fish	V			Trump & Leggett (1980)	0.53	0.35

Note: Some velocities are qualified as those requiring minimum power expenditure or maximal speeds. W_s, wingspan (m)

275

VII. Feeding

Appendix VIIa. Allometric relations describing ingestion rates (I)

Taxon	Independent variable	Range of W (kg)	N	Reference	Standardized relation Intercept at $W=1$ kg, a	Slope $b \pm S_b$	r^2	S_{xy}
Homeotherms	I	0.003–5,000	238	From Farlow (1976)	10.7 Watts	0.703	0.947	
Herbivorous homeotherms	I		119	Farlow (1976)	11.26 Watts	0.716±0.015	0.949	
Carnivorous homeotherms	I		149	Farlow (1976)	10.60 Watts	0.692±0.012	0.958	
Herbivorous mammals	I		183	Farlow (1976)	11.26 Watts	0.728±0.020	0.942	
Carnivorous mammals	I		100	Farlow (1976)	11.31 Watts	0.697±0.013	0.968	
Mammals	I			Harestad & Bunnell (1979)	15.1 Watts	0.68±0.02		
Zoo mammals	I	0.01–3,000	18	Evans & Miller (1968)	7.08 Watts	0.75		
Whales	I			Hinga (1979)	7.62 Watts	0.75		
Raptorial birds	I		19	Schoener (1968)	15.1 Watts	0.63±0.11		
Raptorial birds	I		19	Calder (1974)	15.7 Watts	0.63±0.09		0.03
Finches	I_{max}		8	Calder (1974)	946 Watts	1.02±0.293	0.81	0.216
Carnivorous poikilotherms	I		48	Farlow (1976)	0.779 Watts d^{-1}	0.820±0.031	0.935	
Benthic detritivores	I_{total}		16	Cammen (1980)	8.2 kg d^{-1}	1.115	0.86	
Benthic detritivores	$I_{organic}$		19	Cammen (1980)	1.07 Watts	0.742	0.95	
Benthic detritivores	$I_{organic}$		13	Cammen (1980)	1.98 Watts	0.799	0.97	
Benthic detritivores[a]	I_{total}		19	Cammen (1980)	$1.77\,OM^{-0.92}\,W^{0.771}$ Watts		0.98	
Forest floor arthropods	I		11	Reichle (1968)	23 Watts	0.68±0.129		
Crustacea	I			Sushchenya & Khmeleva (1967) in Conover (1978)	1.52 Watts	0.80		
Copepods	I			Ikeda (1977)	0.13 Watts	0.623±0.097	0.65	
Ciliates	I_{max}			Fenchel (1980)	23 Watts	0.84		

Note: Dry mass = wet mass/4; OM, organic fraction in food; I_{max}, maximum ingestion rate.
[a] A multiple regression of Y on mass and organic content (OM) of the substrate.

Appendix VIIb. *Relationships between body mass of a predator* (W_{pred}) *and that of its prey* (W_{prey})

Taxon	Independent variable	Range of W (kg)	N	No. of spp.	Reference	Intercept at $W_{pred.}=1\,kg$ a (kg)	Slope $b \pm S_b$	r^2
Small-prey eaters	$W_{prey\,(max)}$	0.0001–2.1	61	53	Vézina (unpubl.)	0.0735	1.24±0.10	0.70
Small-prey eaters	$W_{prey\,(mean)}$	0.0001–2.1	61	53	Vézina (unpubl.)	0.00187	1.18±0.089	0.75
Small-prey eaters	$W_{prey\,(min)}$	0.0001–2.1	61	53	Vézina (unpubl.)	0.0000481	1.12±0.11	0.64
Large-prey eaters	$W_{prey\,(max)}$	0.08–210	49	47	Vézina (unpubl.)	4.03	1.45±0.13	0.72
Large-prey eaters	$W_{prey\,(mean)}$	0.08–210	49	47	Vézina (unpubl.)	0.109	1.16±0.098	0.74
Large-prey eaters	$W_{prey\,(min)}$	0.08–210	49	47	Vézina (unpubl.)	0.00293	0.87±0.19	0.31
Raptorial birds	W_{prey}				Schoener (1968)	0.179	0.93±0.21	
Codfish	W_{prey}				Ware (1980)	0.0035	1.26	

The "Intercept", "Slope", and "r^2" columns fall under the heading **Standardized relation**.

VIII. Reproduction

Appendix VIIIa. *Allometric relations describing masses and numbers of offspring at birth*

Taxon	Independent variable	Range of W (kg)	N	No. of spp.	Reference	Intercept at $W=1$ kg, a (kg)	Slope $b \pm S_b$	r^2
Mammals	Litter mass	0.01–5,000	205		Blueweiss et al. (1978)	0.159	0.82	0.97
Mammals	Litter mass	0.05–100	114		Leitch et al. (1959)	0.170	0.83	
Mammals	Litter mass	0.004–6,000	76		Cabana et al. (1982)	0.112	0.84±0.027	0.93
Mammals	Litter mass		250	250	Millar (1981)	0.112	0.767±0.013	0.92
Mammals[a]	Litter mass				Millar (1981)	$0.0032e^{0.101N}$	0.797	0.93
Marsupials	Litter mass (pouch exit)	0.01–27	16	16	From Russell (1982)	0.111	1.03	0.94
Ungulates	Litter mass	0.01–3,000	42		Robbins & Robbins (1979)	0.214	0.793	0.92
Halophrine primates	Litter mass		25		Leutenegger (1979)		0.69	0.98
Strepsichine primates	Litter mass		9		Leutenegger (1979)		0.63	0.98
Birds	Clutch mass	0.008–100	200		Blueweiss et al. (1978)	0.206	0.74	0.85
Nonpasserine birds	Clutch mass		114		Cabana et al. (1982)	0.286	0.63±0.043	0.661
Nonpasserine birds[a]	Clutch mass		114		Cabana et al. (1982)	$0.266W_m^{1.64}$	−1.05±0.26	0.752
Reptiles	Clutch mass	0.003–100	27		Blueweiss et al. (1978)	0.153	0.88	0.96
Lizards[b]	Clutch mass	0.001–0.1	74		Vitt & Congdon (1978)	−0.34	0.43	0.835
Salamanders	Clutch mass	1×10^{-5}–100	200		Kaplan & Salthe (1979)	0.0202	0.64	0.810
Aquatic poikilotherms	Clutch mass		95		Blueweiss et al. (1978)	0.253	0.92	0.86
Mammals	Neonate mass				Millar (1977)	0.0270	0.71±0.03	0.88
Mammals	Neonate mass	0.004–6,000	250	250	Millar (1981)	0.0452	0.888±0.015	0.94

Mammals[a]	Neonate mass	0.004–6,000	250	Millar (1981)	250	$0.0034e^{-0.206N}$	0.826	0.95
Mammals	Neonate mass		76	Cabana et al. (1982)		0.040	0.94±0.036	0.94
Mammals	Neonate mass	0.008–100,000	200	Blueweiss et al. (1978)		0.0558	0.92	0.94
Marsupials	Individual mass (pouch exit)	0.01–27	16	From Russell (1982)	16	0.319	0.60	0.45
Felids	Neonate mass	0.003–10	15	Hemmer (1979)		0.0333	0.72	0.92
Birds	Hatchling mass		71	Blueweiss et al. (1978)		0.0329	0.69	0.86
Birds	Egg mass	0.01–100	200	Blueweiss et al. (1978)		0.0531	0.77	0.83
Birds	Egg mass			Brody (1945)		0.040	0.73	
Nonpasserine birds	Egg mass		114	Cabana et al. (1982)		0.075	0.68±0.024	0.88
Nonpasserine birds[a]	Egg mass		114	Cabana et al. (1982)		$0.065D^{0.46}$	0.67±0.023	0.89
Passerines	Egg mass		25	Ricklefs (1974)		0.038	0.733±0.047	0.96
Galliformes	Egg mass		11	Ricklefs (1974)		0.027	0.401±0.024	0.38
Raptors	Egg mass		15	Ricklefs (1974)		0.049	0.588±0.034	0.86
Ducks	Egg mass		5	Ricklefs (1974)		0.053	0.64±0.052	0.96
Shorebirds	Egg mass		13	Ricklefs (1974)		0.064	0.618±0.047	0.85
Shorebirds	Egg mass	0.02–0.7	60	Ross (1979)		0.085	0.71±0.03	
Shorebirds	Egg mass	0.02–0.7	60	Ross (1979)		0.089	0.71±0.03	
Suboscines	Egg mass		21	Ricklefs (1974)		0.090	0.707±0.021	0.49
Oscines	Egg mass		17	Ricklefs (1974)		0.078	0.889±0.014	0.16
Temperate passerines	Egg mass		19	Ricklefs (1974)		0.044	0.758±0.017	0.59
Gulls	Egg mass		11	Ricklefs (1974)		0.095	0.666±0.042	0.98
Reptiles	Egg mass	0.005–100	36	Blueweiss et al. (1978)		0.0075	0.42	0.70
Salamanders	Egg mass		25	Kaplan & Salthe (1979)		20×10^{-6}	0.26	0.19
Fish	Egg mass	0.01–100	61	Blueweiss et al. (1978)		11.5×10^{-6}	0.43	0.26
Crustacea	Egg mass	1×10^{-1}–1×10^{-2}	23	Blueweiss et al. (1978)		52×10^{-9}	0.24	0.35
Fish and crustaceans	Egg mass	1×10^{-6}–100	84	Blueweiss et al. (1978)		12×10^{-6}	0.77	0.82
Mammals[c]	Litter size	0.004–6,000	250	Millar (1981)	250	3.03[d]	−0.298	0.25
Mammals	Litter size	0.008–2,750	91	Sacher & Staffeldt (1974)		2.45[d]	−0.136±0.076	
Mammals	Litter size	0.005–2,750		Blueweiss et al. (1978)		2.71[d]	0	

Appendix VIIIa. (cont.)

Taxon	Independent variable	Range of W (kg)	N	No. of spp.	Reference	Intercept at $W=1$kg, a (kg)	Slope $b\pm S_b$	r^2
Birds	Clutch size	0.005–100	114		Blueweiss et al. (1978)	4.85[d]	0	0.10
Birds[a]	Clutch size	0.003–100			Cabana et al. (1982)	$4.15W_m^{1.18}$[d]	-1.26 ± 0.325	0.78
Reptiles	Fecundity		54		Blueweiss et al. (1978)	27.5[d]	0.33	0.25
Salamanders	Fecundity	50×10^{-6}–100	25		Kaplan & Salthe (1979)	789[d]	0.47	0.58
Aquatic poikilotherms	Fecundity		166		Blueweiss et al. (1978)	$8{,}900$[d]	0.64	0.78
Nonsalmonid fish	Fecundity	0.1×10^{-3}–100			Blueweiss et al. (1978)	$34{,}000$[d]	0.70	0.69
Salmonids	Fecundity	0.05–10			Blueweiss et al. (1978)	$1{,}900$[d]	1.22	0.50
Fish (within species)	Fecundity				Wooton (1979)			
Spiders (within species)	Fecundity		94		Peterson (1950)	4.30×10^{-8}[d]	1.03	
Marsupials	Litter mass at weaning	0.01–27	18	18	From Russell (1982)	0.97	0.73	0.72
Mammals	Individual mass at weaning		100		Millar (1977)	0.160	0.73 ± 0.02	0.90
Marsupials	Individual mass at weaning	0.01–27	18	18	From Russell (1982)	0.74	1.06	0.90

[a] Multiple regression of Y on body mass and number of offspring per litter (N), male body mass, W_m, female body mass, W_f, or sexual size dimorphism, $D = \log W_m/W_f$.

[b] Regression uses linear model $Y = a + bW$.

[c] Regression uses the semilogarithmic model $Y = a + b \ln W$.

[d] Units in numbers of animals or eggs.

280

Appendix VIIIb. *Allometric relations describing time periods associated with phases of reproduction*

Taxon	Independent variable	Range of W (kg)	N	No. of spp.	Reference	Intercept at $W=1$ kg, a (d)	Slope $b \pm S_b$	r^2
						Standardized relation		
Eutheria	Gestation time	0.017–2,750	84		Sacher & Staffeldt (1974), Blueweiss et al. (1978)	65.3	0.258±0.032	0.72
Eutheria	Gestation time	0.004–6,000	250	250	Millar (1981)	63	0.238±0.010	0.69
Artiodactyla	Gestation time	6–2,600	13		Western (1979)	120	0.16	0.77
Artiodactyla	Gestation time		49		Kihlström (1972)	94	0.177	0.61
Perissodactyla	Gestation time		5		Kihlström (1972)	121	0.195	0.94
Carnivora	Gestation time	0.06–210	18		Western (1979)	48	0.12±0.039	0.64
Carnivora	Gestation time		44		Kihlström (1972)	48	0.158	0.69
Felids	Gestation time		6		Hemmer (1979)	101	0.08	0.90
Pantherines	Gestation time		6		Hemmer (1979)	69	0.09	0.86
Cetacea	Gestation time		9		Kihlström (1972)	458	0.037	0.41
Primates	Gestation time		16		Kihlström (1972)	95	0.161	0.66
Primates	Gestation time	0.15–210	15		Western (1979)	135	0.14	0.53
Rodents	Gestation time		48		Kihlström (1972)	39	0.19	0.35
Chiroptera	Gestation time		13		Kihlström (1972)	133	0.18	0.28
Insectivora	Gestation time		16		Kihlström (1972)	45	0.14	0.59
Birds	Incubation time	0.005–100	100		Blueweiss et al. (1978)	28	0.16	0.47
Birds[a]	Incubation time	0.0004–2	475		Rahn & Ar (1974)	54	0.22±0.092	0.74
Birds[a]	Incubation time	0.004–2	194		Heinroth (1921) in Rahn & Ar (1974)	50	0.20±0.086	0.71
Birds[a]	Incubation time				From Needham (1931) in Rahn & Ar[a] (1974)	52	0.24	

Appendix VIIIb. (*cont.*)

Taxon	Independent variable	Range of W (kg)	N	No. of spp.	Reference	Standardized relation — Intercept at $W=1$ kg, a (d)	Standardized relation — Slope $b \pm S_b$	Standardized relation — r^2
Birds[a]	Incubation time			104	From Worth in Rahn & Ar (1974)	58	0.23	
Mammals	Age at weaning	0.01–1,000	11		Blaxter (1971)	34	0.15	
Mammals	Age at weaning		98		Millar (1977)	28	0.05±0.02	
Mammals	Time from conception to implantation				Taylor (1965)	3.5	0	
Mammals	Age (from implantation) to end of exponential growth				Taylor (1965)	50	0.27	
Mammals	$T_{0.50}$ (from implantation)				Taylor (1965)	120	0.27	
Mammals	$T_{0.50}$ (from birth)				Stahl (1962)	129	0.25	
Mammals	$T_{0.50}$ (from implantation)				Taylor (1965)	190	0.27	
Mammals	$T_{0.98}$ (from implantation)				Taylor (1965)	440	0.27	
Mammals	$T_{0.98}$ (from birth)				Brody (1945)	440	0.26	
Mammals[b]	T_p (from birth,), $p > 0.37$				Taylor (1968)	$40-36[\ln(-\ln p)]$	0.27	
Homeotherms	$T_{maturity}$ (from implantation)				Taylor (1968)		0.286±0.026	
Mammals	$T_{maturity}$ (from implantation)				Taylor (1965)	150	0.27	
Mammals	$T_{maturity}$ (from birth)				Taylor (1968)		0.271±0.028	
Carnivores	$T_{maturity}$ (from birth)	0.06–120	15		Western (1979)	591	0.32±0.039	0.84
Artiodactyls	$T_{maturity}$ (from birth)	6–2,600	11		Western (1979)	229	0.27±0.030	0.90
Primates	$T_{maturity}$ (from birth)	0.16–210	15		Western (1979)	871	0.32±0.052	0.74
Felids	Age at family dissolution		5		Hemmer (1979)	64	0.49	0.98
Birds[b]	T_p, $p > 0.37$				Taylor (1968)	$35-25[\ln(-\ln p)]$	0.27	
Birds	$T_{0.10-0.90}$	0.007–3	53		Ricklefs (1968)	26.2	0.278	0.89

Group	Variable	Range	N	Source		a	b
Birds	$T_{maturity}$ (from birth)			Taylor (1968)		0.284±0.074	
Altricial land birds	Age at first flight	0.1–5	16	Ricklefs (1973)	45	0.25	
Precocial birds	Age at first flight	0.04–5	27	Ricklefs (1973)	45	0.37	
Fish	$T_{maturity}$ (from birth)		4	Ware (1980)	507	0.20	
Organisms	$T_{maturity}$ (from birth)	10^{-15}–10^3		Blueweiss et al. (1978)	342	0.27	
Animals	Maximum life span	10^{-7}–10^4		Blueweiss et al. (1978)	4,425	0.15	
Mammals	Maximum life span			Sacher (1959)	4,240	0.20±0.021	0.56
Mammals	Longevity	0.01–3,000	63	Sacher (1959)	5,730	0.20±0.021	0.60
Mammals	Longevity			Boddington (1978)	3,550	0.23±0.030	0.62
Mammals	Average life span			Stahl (1962)	8,850	0.29	
Mammals	Average life span	0.005–5,000	67	Blueweiss et al. (1978)	2,040	0.17	0.56
Primates	Average life span	0.16–210	14	Western (1979)	4,700	0.24±0.038	0.77
Carnivores	Average life span	1.0–120	17	Western (1979)	3,200	0.17±0.026	0.74
Artiodactyls	Average life span	6–2,600	14	Western (1979)	2,100	0.22±0.027	0.85
Artiodactyls	Life expectancy at birth	30–2,600	8	Western (1979)	810	0.20±0.092	0.45
Wild birds	Maximum life span		152	Lindstedt & Calder (1976)	6,400	0.20±0.01	
Wild passerines	Maximum life span		71	Lindstedt & Calder (1976)	7,900	0.26±0.02	
Wild nonpasserines	Maximum life span		81	Lindstedt & Calder (1976)	6,100	0.18±0.02	
Captive birds	Maximum life span		58	Lindstedt & Calder (1976)	10,300	0.19±0.02	

[a] Egg mass is the independent variable.
[b] Equation is not in the standard form.

Appendix VIIIc. Allometric relations describing population production

Taxon	Independent variable	Range of W (kg)	N	No. of spp.	References	Intercept at $W=1\,\mathrm{kg}$, a (Watts kg^{-1})	Slope $b \pm S_b$	r^2
Organisms	r_{max}	10^{-15}–10^3	42		Blueweiss et al. (1978)	0.336	−0.26	
Homeotherms	r_{max}	0.01–1,000			Fenchel (1974)	0.627	−0.27	
Mammals	$P/\overline{B}$	0.005–4,400	7		Banse & Mosher (1980)	0.230	−0.33±0.04	0.97
Mammals	$P/\overline{B}$		61		Farlow (1976)	0.200	−0.266±0.013	0.87
Terrestrial verts.	$P/\overline{B}$		87		Farlow (1976)	0.163	−0.23	0.72
Animals	$P/\overline{B}$		67		From Banse & Mosher (1980)	0.0698	−0.17	0.56
Poikilotherms	$P/\overline{B}$		60		From Banse & Mosher (1980)	0.0613	−0.179	0.38
Aquatic animals	Turnover rate				Parsons (1980)	0.470	−0.217	
Marine poikilotherms	Turnover rate				Sheldon et al. (1972)	0.291	−0.213	
Fish	$P/\overline{B}$	0.0005–0.28	11		Banse & Mosher (1980)	0.230	−0.26	0.60
Invertebrates	$P/\overline{B}$		49		From Banse & Mosher (1980)	0.033	−0.23	0.37
Invertebrates, 10–14°C	$P/\overline{B}$	1×10^{-8}–0.01	16		Banse & Mosher (1980)	0.0086	−0.35±0.045	0.85
Invertebrates, 5–20°C	$P/\overline{B}$		31		Banse & Mosher (1980)	0.0067	−0.37±0.032	0.82
Noninsect inverts.	$P/\overline{B}$		19		Banse & Mosher (1980)	0.0055	−0.39±0.025	0.93
Insects	$P/\overline{B}$		12		Banse & Mosher (1980)	0.764	0.06±0.13	0.017
Planktonic crustaceans	r	2×10^{-9}–3.9×10^{-3}	25	21	Båmstedt & Skjoldal (1980)	0.0254	−0.32	
Poikilotherms	r_{max}	1×10^{-10}–1,000			Fenchel (1974)	0.281	−0.27	
Unicells	r_{max}	1×10^{-16}–1×10^{-8}			Fenchel (1974)	0.136	−0.28	
Ciliates, 20°C	r_{max}	1×10^{-12}–1×10^{-9}	35	35	Taylor & Shuter (1981)	1.41	−0.24	0.42
Ciliates, 20°C	r	1×10^{-12}–1×10^{-8}	10		Finlay (1977)	0.095	−0.303	

Ciliates, 15°C	r	1×10^{-12}–1×10^{-8}		10	Finlay (1977)	0.108	−0.276	
Ciliates, 8.5°C	r	1×10^{-12}–1×10^{-8}		10	Finlay (1977)	0.0089	−0.332	
Amoebae, 25°C	r	0.2×10^{-12}–1×10^{-8}		5	Baldock et al. (1980)	0.073	−0.306±0.06	0.90
Amoebae, 20°C	r	0.2×10^{-12}–1×10^{-8}		6	Baldock et al. (1980)	0.192	−0.271±0.12	0.59
Amoebae, 15°C	r	0.2×10^{-12}–1×10^{-8}		6	Baldock et al. (1980)	0.0288	−0.323±0.14	0.63
Amoebae, 10°C	r	0.2×10^{-12}–1×10^{-8}		6	Baldock et al. (1980)	0.131	−0.243±0.085	0.80
Amoebae	r	0.2×10^{-12}–1×10^{-8}		6	Baldock et al. (1980)	0.0049	−0.385±0.052	0.72
Phytoplankton, 21°C	r	21×10^{-15}–2.1×10^{-9}	23	11	Eppley & Sloan (1966)	6.8	−0.11	0.643
Algae	r_{max}	0.7×10^{-15}–2.1×10^{-10}		26	Schlesinger et al. (1981)	0.059	−0.32±0.044	0.70
Algae 20°C	r	30×10^{-15}–1.3×10^{-12}	11	8	Banse (1976)	15.6	−0.06±0.11	0.90
Mammals[a]	Birth rate				Western (1979)	1.26	−0.33	

[a] Units expressed in $year^{-1}$.

285

Appendix VIIId. *Allometric equations describing aspects of individual productivity*

Taxon	Independent variable	Range of W (kg)	N	No. of spp.	Reference	Intercept at $W=1$ kg, a	Slope $b \pm S_b$	r^2
Animals	Individual growth rate	1×10^{-7}–1			Blueweiss et al. (1978)	0.336 Watt	0.74	0.92
Mammals	G_{10-90}		167		Case (1978)	0.445 Watt	0.72	0.92
Eutheria	G_{10-90}		162		Case (1978)	0.456 Watt	0.72	0.97
Marsupials	G_{10-90}		4		Case (1978)	0.0748 Watt	0.82	0.76
Insectivores	G_{10-90}		8		Case (1978)	0.435 Watt	0.67	0.82
Primates	G_{10-90}		16		Case (1978)	0.204 Watt	0.37	0.86
Prosimians	G_{10-90}		5		Case (1978)	0.275 Watt	0.62	0.79
Anthropoids	G_{10-90}		11		Case (1978)	0.204 Watt	0.35	0.64
Chiroptera	G_{10-90}		10		Case (1978)	0.370 Watt	0.65	0.88
Rodents	G_{10-90}		60		Case (1978)	0.397 Watt	0.67	0.94
Hystricomorphs	G_{10-90}		9		Case (1978)	0.416 Watt	0.67	0.84
Myomorphs	G_{10-90}		32		Case (1978)	0.574 Watt	0.72	0.70
Sciuromorphs	G_{10-90}		17		Case (1978)	0.456 Watt	0.82	0.82
Lagomorphs	G_{10-90}		9		Case (1978)	0.690 Watt	0.61	0.86
Fissipeds	G_{10-90}		23		Case (1978)	0.548 Watt	0.70	0.38
Pinnipeds	G_{10-90}		11		Case (1978)	0.909 Watt	0.74	0.99
Cetaceans	G_{10-90}		3		Case (1978)	0.644 Watt	0.75	0.62
Ruminants	G_{10-90}		17		Case (1978)	2.08 Watts	0.43	0.69
Ungulates	G_{10-90}		20		Case (1978)	1.65 Watts	0.52	0.89
Birds	G_{10-90}	0.007–3	50		Ricklefs (1968)	2.41 Watts	0.722	0.86
Reptiles	G_{10-90}		43		Case (1978)	0.0388 Watt	0.67	0.62
Fish	G_{10-90}		10		Case (1978)	0.0064 Watt	0.61	

Group	Variable	W	n	Value	Reference		r^2
Birds	K_g	0.003–10	150	$0.105\,\mathrm{d}^{-1}$	Ricklefs (1979)	-0.34 ± 0.026	
Altricial land birds	K_g	0.1–5	16	$0.12\,\mathrm{d}^{-1}$	Ricklefs (1973)	-0.26	
Precocial fowl	K_g	0.1–6	14	$0.035\,\mathrm{d}^{-1}$	Ricklefs (1973)	-0.36	
Gulls	K_g	0.1–2	21	$0.08\,\mathrm{d}^{-1}$	Ricklefs (1973)	-0.42	
Felids	Growth rate 0–1 month		15	$0.494\,\mathrm{Watt}$	Hemmer (1979)	0.51	0.94
Mammals	Neonate production	0.01–5,000		$0.154\,\mathrm{Watt}$	Blueweiss et al. (1978)	0.57	
Nonprimate mammals	Neonate production			$0.288\,\mathrm{Watt}$	Payne & Wheeler (1968)	0.60	
Primates	Neonate production			$0.068\,\mathrm{Watt}$	Payne & Wheeler (1968)	0.56	
Nonprimate mammals	Neonate production	0.025–3,600		$0.381\,\mathrm{Watt}$	Payne & Wheeler (1968)	0.54	
Primates	Neonate production	0.12–100		$0.070\,\mathrm{Watt}$	Payne & Wheeler (1967)	0.54	
Mammals	Weanling metabolic rate		100	$3.15\,\mathrm{Watts}$	Millar (1977); Kleiber (1961)	0.586 ± 0.03	0.27
Mammals	Peak milk production		22	$1.16\,\mathrm{mg\,s}^{-1}$	Linzell (1972)	0.79 ± 0.04	0.94
Mammals	Peak milk production		12	$1.39\,\mathrm{mg\,s}^{-1}$	Linzell (1972)	0.73 ± 0.03	0.98
Mammals	Peak milk production		22	$1.0\,\mathrm{mg\,s}^{-1}$	Hanwell & Peaker (1977)	0.765 ± 0.022	
Mammals	Peak milk production		22	$1.32\,\mathrm{mg\,s}^{-1}$	Linzell (1972)	0.83 ± 0.05	0.92
Mammals	Peak milk production		14	$7.03\,\mathrm{Watts}$	Linzell (1972)	0.69 ± 0.04	0.92
Mammals	Peak milk production		22	$6.11\,\mathrm{Watts}$	Linzell (1972)	0.73 ± 0.045	0.92
Mammals	Peak milk production		14	$6.01\,\mathrm{Watts}$	Payne & Wheeler (1968)	0.75	
Mammals	Peak milk production			$6.16\,\mathrm{Watts}$	Hanwell & Peaker (1977)	0.69 ± 0.041	
Mammals	Peak milk production		20	$8.33\,\mathrm{Watts}$	Linzell (1972)	0.74	
Mammals[a]	Milk protein concentration				Payne & Wheeler (1968)	-0.28	
Deer	Total milk production			$2.71\,\mathrm{kg}$	Sadleir (1980)	0.75	
Deer	Total milk production			$17\times10^{6}\,\mathrm{J}$	Sadleir (1980)	0.75	

Note: W in these equations refers to adult body size unless otherwise indicated. K_g is the growth rate constant and represents the maximum proportional rate.

[a] Independent variable refers to neonate body mass.

287

IX. Mass flow

Appendix IXa. *Allometric relations for defecation rates* (D)

Taxon	Independent variable	Range of W (kg)	Reference	Standardized relation	
				Intercept at $W=1\,kg$, a	Slope b
Mammals	D		Blueweiss et·al. (1978)	3.82 Watts	0.63
Benthic detritivores	D	320×10^{-9}–26×10^{-3}	From Hargrave (1972)	$0.60\,kg\,d^{-1}$	0.905

Appendix IXb. *Allometric relations describing nutritional requirements*

Taxon	Independent variable	Range of W (kg)	N	Reference	Standardized relation		
					Intercept at $W=1$ kg, a	Slope $b \pm S_b$	r^2
Mammals	Threonine requirement			Munro (1969)	220 ng s^{-1}	0.76	
Mammals	Methionine requirement			Munro (1969)	450 ng s^{-1}	0.78	
Mammals	Nitrogen requirement			Miller & Payne (1964)	2,900 ng s^{-1}	0.75	
Mammals	Vitamin B requirement			Brody (1945)		0.75	
Mammals	Vitamin A requirement			Brody (1945)		1.00	
Unicells	P cell quota	$0.4 \times 10^{-15} - 4 \times 10^{-12}$	27	Shuter (1978)	760 ng P	0.554 ± 0.050	0.82
Eucaryotic algae	P cell quota	$31 \times 10^{-15} - 4 \times 10^{-12}$	20	Shuter (1978)	41 µg P	0.69 ± 0.11	0.68
Eucaryotic algae	P cell quota	$31 \times 10^{-15} - 4 \times 10^{-12}$	45	Shuter (1978)	71 µg P	0.709 ± 0.33	0.90
Algae	P cell quota	$60 \times 10^{-15} - 18 \times 10^{-12}$	7	Smith (1981)	186 µg P	0.70 ± 0.16	0.90
Unicells	N cell quota	$0.9 \times 10^{-15} - 17 \times 10^{-9}$	28	Shuter (1978)	1.4 mg N	0.723 ± 0.028	0.96
Eucaryotic algae	N cell quota	$0.9 \times 10^{-15} - 17 \times 10^{-9}$	25	Shuter (1978)	1.1 mg N	0.711 ± 0.032	0.96
Eucaryotic algae	N cell quota	$0.9 \times 10^{-15} - 17 \times 10^{-9}$	45	Shuter (1978)	1.0 mg N	0.709 ± 0.033	0.90

end os setting Table page 359 only
Edited by Meow Ee   14.3.83   P773/125

289

Appendix IXc. Allometric relations describing the flux of water

Taxon	Independent variable	Range of W (kg)	N	No. of spp.	Reference	Standardized relation Intercept at $W=1\,\mathrm{kg}$ a	Slope $b \pm S_b$	r^2
Mammals	Water intake	0.01–2,000	14		Adolf (1943)	$1.21\times10^{-3}\,\mathrm{ml\,s^{-1}}$	0.88	
Mammals	TWT	0.01–1,000	42		Altman & Dittmer (1968)	$1.36\times10^{-3}\,\mathrm{ml\,s^{-1}}$	0.86	
Mammals	TWT				Eberhardt (1969)	$1.42\times10^{-3}\,\mathrm{ml\,s^{-1}}$	0.80	
Mammals	Urine production	0.15–590	35	30	Edwards (1975)	$7.0\times10^{-4}\,\mathrm{ml\,s^{-1}}$	0.75±0.10	
Mammals	Urine production	0.01–2,000	9		Adolf (1943)	$5.1\times10^{-4}\,\mathrm{ml\,s^{-1}}$	0.82	
Frogs	Urine production	0.02–0.2	5		Adolf (1943)	$6.4\times10^{-3}\,\mathrm{ml\,s^{-1}}$	1.03	
Mammals	EWL	0.01–1,000	49	49	Altman & Dittmer (1968)	$4.53\times10^{-4}\,\mathrm{ml\,s^{-1}}$	0.86	
Mammals	EWL	0.02–4,000			Crawford & Lasiewski (1968)	$4.4\times10^{-4}\,\mathrm{ml\,s^{-1}}$	0.88	
Birds	EWL_{max}	0.012–88	13	13	Calder & King (1974)	$0.0048\,\mathrm{ml\,s^{-1}}$	0.80	
Birds	EWL	0.005–100	53	53	Altman & Dittmer (1968)	$2.84\times10^{-4}\,\mathrm{ml\,s^{-1}}$	0.58	
Birds	EWL	0.010–100	31		Gordon (1977)	$2.80\times10^{-4}\,\mathrm{ml\,s^{-1}}$	0.61	
Birds	EWL				Calder & King (1974)	$2.85\times10^{-4}\,\mathrm{ml\,s^{-1}}$	0.585	
Birds	EWL	0.010–6		13	Walter & Hughes (1978)	$8.2\times10^{-4}\,\mathrm{ml\,s^{-1}}$	0.75	
Birds	EWL	0.003–100	53	53	Crawford & Lasiewski (1968)	$2.84\times10^{-4}\,\mathrm{ml\,s^{-1}}$	0.585±0.18	
Passerines	EWL	0.005–0.3		18	Altman & Dittmer (1968)	$8.10\times10^{-5}\,\mathrm{ml\,s^{-1}}$	0.22	
Passerines	EWL	0.010–0.10		18	Crawford & Lasiewski (1968)	$8.45\times10^{-5}\,\mathrm{ml\,s^{-1}}$	0.217±0.18	
Nonpasserines	EWL	0.003–100	35		Altman & Dittmer (1968)	$2.80\times10^{-4}\,\mathrm{ml\,s^{-1}}$	0.613±0.15	
Lizards	EWL	0.001–0.5	35		Altman & Dittmer (1968)	$8.23\times10^{-4}\,\mathrm{ml\,s^{-1}}$	0.72	
Other reptiles	EWL	0.01–1	15		Altman & Dittmer (1968)	$3.0\times10^{-3}\,\mathrm{ml\,s^{-1}}$	0.84	
Frogs, 20°C	EWL_{wind}	0.001–0.1	11		Altman & Dittmer (1968)	$1.0\times10^{-3}\,\mathrm{ml\,s^{-1}}$	0.55	
Frogs, 20°C	$EWL_{still\ air}$	0.001–0.02	15		Altman & Dittmer (1968)	$3.6\times10^{-5}\,\mathrm{ml\,s^{-1}}$	0.61	
Frogs, 25°C	$EWL_{still\ air}$	0.005–0.055	36	6	Claussen (1969)	$4.06\times10^{-5}\,\mathrm{ml\,s^{-1}}$	0.25	

Frogs, 25°C	$EWL_{\text{moderate wind}}$	0.005–0.055	6	2	Claussen (1969)	$8.8 \times 10^{-6}\,\text{ml s}^{-1}$	0.18	
Frogs, 25°C	$EWL_{\text{strong wind}}$	0.005–0.055	6	2	Claussen (1969)	$4.3 \times 10^{-4}\,\text{ml s}^{-1}$	0.36	
Birds' eggs	EWL	0.0015–0.50	93	80	Ar & Rahn (1980)	$2.8 \times 10^{-5}\,\text{ml s}^{-1}$	0.754 ± 0.04	0.93
Birds' eggs	EWL	0.001–1	57	46	Drent (1970)	$2.92 \times 10^{-5}\,\text{ml s}^{-1}$	0.742	
Birds' eggs	EWL		77		Hoyt (1980)	$2.76 \times 10^{-5}\,\text{ml s}^{-1}$	0.747 ± 0.021	0.98
Birds' eggs	IWL		93		Ar & Rahn (1980)	144 ml	0.992	0.98
Birds' eggs	IWL		93		Ar & Rahn (1980)	154 ml	1	
Altricial birds eggs	IWL		32		Ar & Rahn (1980)	162 ml	1	
Semialtricial birds' eggs	IWL				Ar & Rahn (1980)	152 ml	1	
Semiprecocial birds' eggs	IWL				Ar & Rahn (1980)	150 ml	1	
Precocial birds' eggs	IWL				Ar & Rahn (1980)	140 ml	1	
Birds' eggs	Water vapor conductance		143		Hoyt (1980)	$10 \times 10^{-9}\,\text{ml(s·Pa)}^{-1}$	0.829 ± 0.014	0.96
Birds' eggs	Water vapor conductance		29	29	Ar et al. (1974)	$8 \times 10^{-9}\,\text{ml(s·Pa)}^{-1}$	0.78 ± 0.104	0.92

Note: TWT, total water turnover; EWL, evaporative water loss; IWL, water loss during incubation.

Appendix IXd. *Allometric relations describing flux rates of essential nutrients*

Taxon	Independent variable	Range of W (kg)	N	No. of spp.	Reference	Standardized relation Intercept $W=1\,kg$ a $(g\,s^{-1})$	Slope b	r^2
Mammals	Voluntary N intake	0.03–3,000	16		Evans & Miller (1968)	18.5×10^{-6}	0.75	
Mammals	Total N release				Brody (1945)	3.28×10^{-6}	0.735	
Mammals	Total N release				Miller & Payne (1964)	2.89×10^{-6}	0.75	
Mammals	Urinary N release				Brody (1945)	2.3×10^{-6}	0.72	
Mammals	N excretion				Stahl (1962)	3.41×10^{-6}	0.74	
Mammals	Endogenous N release				Brody (1945)	1.69×10^{-6}	0.72	
Fish	Endogenous N release	0.001–0.4	19	10	From Brett & Groves (1979)	9.5×10^{-9}	0.85	0.14
Mammals	Endogenous S release				Brody (1945)	78×10^{-9}	0.741	
Marine invertebrates	P excretion				Johannes (1964)	10×10^{-9}	0.33	0.92
Marines invertebrates	P excretion	5×10^{-5}–1×10^{-4}	24		Johannes (1964)	243×10^{-9}	0.67	0.97
Well-fed zooplankton	P excretion				Peters & Rigler (1973)	88×10^{-9}	0.62	0.54
Poorly fed zooplankton	P excretion				Peters & Rigler (1973)	36×10^{-9}	0.62	0.54

Appendix IXe. *Allometric relations describing the rate constants of flux of contaminants into (K_{in}) or out of (K_{out}) animals and the equilibrium levels of some contaminants*

Taxon	Independent variable	Range of W (kg)	N	No. of spp.	Reference	Standardized relation Intercept $W=1$ kg a	Slope b	r^2
Vertebrates	K_{out} Cs				DiGregorio et al. (1978)	$0.197\,\mathrm{d^{-1}}$	-0.24	0.93
Vertebrates	K_{out} Cs				DiGregorio et al. (1978)	$0.097\,\mathrm{d^{-1}}$	-0.15	0.32
Animals	K_{out} ^{60}Co				DiGregorio et al. (1978)	$0.157\,\mathrm{d^{-1}}$	-0.13	0.28
Vertebrates	K_{out} ^{60}Co				DiGregorio et al. (1978)	$0.265\,\mathrm{d^{-1}}$	-0.67	0.67
Animals	K_{out} ^{131}I				DiGregorio et al. (1978)	$0.087\,\mathrm{d^{-1}}$	-0.37	0.57
Animals	K_{out} I				DiGregorio et al. (1978)	$0.101\,\mathrm{d^{-1}}$	-0.13	
Animals	K_{out} ^{90}Sr				DiGregorio et al. (1978)	$0.168\,\mathrm{d^{-1}}$	-0.20	0.79
Animals	K_{out} ^{90}Sr				DiGregorio et al. (1978)	$0.0064\,\mathrm{d^{-1}}$	-0.26	
Vertebrates	K_{out} ^{3}H				DiGregorio et al. (1978)	$0.863\,\mathrm{d^{-1}}$	-1.55	0.90
Animals	K_{out} ^{65}Zn				DiGregorio et al. (1978)	$0.0324\,\mathrm{d^{-1}}$	-0.16	0.30
Mammals	K_{out} ^{65}Zn	0.01–0.2	8	8	Golley et al. (1965)		-0.204	0.62
Animals	K_{out} ^{59}Fe				DiGregorio et al. (1978)	$0.434\,\mathrm{d^{-1}}$	-0.10	0.31
Plaice	K_{in} ^{134}Cs	2×10^{-4}–1	5	1	Morgan (1964)	$0.0198\,\mathrm{d^{-1}}$	-0.25	
Lobster	K_{in} ^{134}Cs	1×10^{-5}–1.1	4	1	Morgan (1964)	$0.0462\,\mathrm{d^{-1}}$	-0.29	
Eel	K_{in} ^{134}Cs	3×10^{-4}–0.5	5	1	Morgan (1964)	$0.0087\,\mathrm{d^{-1}}$	-0.25	
Fish	K_{out} methyl mercury	0.001–0.2	27	2	Norstrom et al. (1976)	$0.0005\,\mathrm{d^{-1}}$	-0.58	
Mammals	K equil. level	0.02–100	25		Fujita et al (1966)	10 multiples	0.45	
Mammals	Cs equil. level	0.02–100	38		Fujita et al. (1966)	19 multiples	0.45	
Mammals	Cs equil. level				Eberhardt (1969)	19.2 multiples	0.46	
Mammals	^{131}I equil. level	0.02–90	5		Furcher & Richmond (1963)	1.67 multiples	0.24	
Mammals	^{131}I equil. level				Eberhardt (1969)	4.64 multiples	0.37	
Mammals	^{65}Zn equil. level	0.02–100	4	4	Richmond et al. (1962)	21.8 multiples	0.38	

Note: Equations describing the equilibrium level of contamination are expressed in multiples of daily dose.

X. Animal abundance

Appendix Xa. *Allometric relations describing the abundance or density individual animals species (D)*

Taxon	Independent variable	Range of W (kg)	N	W̄ (kg)	Reference	Standardized relation			
						Intercept $W=1$ kg a (No. km^{-2})	Slope $b \pm S_b$	r^2	S_{xy}
Temperate animals	D	1×10^{-11}–1×10^3	291	8.4×10^{-5}	Peters & Wassenberg (unpubl.)	32	-0.98 ± 0.029	0.81	1.44
Mammals	D	0.03–1×10^3	337	0.10	Peters & Raelson (unpubl.)	55	-0.90 ± 0.036	0.66	0.93
Temperate mammals	D	0.01–100	92	0.35	From Mohr (1947)	204	-0.91 ± 0.10	0.47	0.87
Temperate mammals	D	0.01–200	57	0.37	Peters & Wassenberg (unpubl.)	98	-0.77 ± 0.09	0.57	0.90
Tropical mammals	D	0.05–5×10^3	245	34	Peters & Raelson (unpubl.)	11	-0.58 ± 0.04	0.44	0.76
Herbivorous mammals	D	0.01–5×10^3	250	16	Peters & Raelson (unpubl.)	103	-0.93 ± 0.04	0.71	0.87
Herbivorous mammals	D	0.01–1,000	307		Damuth (1981)	96	-0.75 ± 0.026	0.74	
Temperate herb. mammals	D	0.008–30	68	0.33	From Mohr (1947)	468	-0.66 ± 0.12	0.30	0.77
Temperate herb. mammals	D	0.01–200	45	0.26	Peters & Wassenberg (unpubl.)	214	-0.61 ± 0.073	0.62	0.64
Tropical herb. mammals	D	0.1–5,000	182	34	Peters & Raelson (unpubl.)	16	-0.60 ± 0.049	0.45	0.72
Carnivorous mammals	D	0.008–150	87	0.38	Peters & Raelson (unpubl.)	15	-1.16 ± 0.063	0.80	0.67
Temperate carn. mammals	D	0.008–10	24	0.42	From Mohr (1947)	36	-1.14 ± 0.12	0.81	0.70
Temperate carn. mammals	D	0.01–10	12	1.7	Peters & Wassenberg (unpubl.)	13	-0.94 ± 0.260	0.56	1.08
Tropical carn. mammals	D	0.01–150	63	7.5	Peters & Raelson (unpubl.)	8.5	-1.01 ± 0.078	0.73	0.60
North American birds	D	0.01–4	60	0.055	Peters & Wassenberg (unpubl.)	6.2	-0.19 ± 0.14	0.031 (NS)	0.71

Category		Range	n		Source		Slope ± SE		
North American carn. birds	D	0.01–4	38	0.043	Peters & Wassenberg (unpubl.)	1.8	-0.52 ± 0.18	0.18	0.70
North American herb. birds	D	0.01–4	22	0.092	Peters & Wassenberg (unpubl.)	25	0.21 ± 0.20	0.054 (NS)	0.63
Vert. poikilotherms	D	0.0001–1	11	0.0015	Peters & Wassenberg (unpubl.)	64	-0.77 ± 0.30	0.42	1.7
Invertebrates	D	1×10^{-11}–0.1	162	0.37×10^{-6}	Peters & Wassenberg (unpubl.)	85×10^3	-0.54 ± 0.038	0.55	1.1
Aquatic invertebrates	D	1×10^{-7}–0.1	56	7.9×10^{-6}	Peters & Wassenberg (unpubl.)	237×10^3	-0.58 ± 0.07	0.54	1.2
Terrestrial invertebrates	D	1×10^{-11}–0.01	106	60×10^{-9}	Peters & Wassenberg (unpubl.)	3.0×10^3	-0.69 ± 0.04	0.72	0.88
Soil animals	D				Ghilarov (1967)	14×10^3	-0.12		
Animals	D_{max}	1.0×10^{-9}–1,000	143	0.6×10^{-3}	Peters & Wassenberg (unpubl.)	15	-1.17 ± 0.039	0.86	1.39
Animals	D_{min}	1×10^{-9}–1,000	147	0.6×10^{-3}	Peters & Wassenberg (unpubl.)	0.87	-1.16 ± 0.041	0.85	1.45
Temperate mammals	D_{max}	0.01–1,000	37	0.4	Peters & Wassenberg (unpubl.)	129	-0.96 ± 0.12	0.66	0.88
Temperate mammals	D_{min}	0.01–1,000	38	0.4	Peters & Wassenberg (unpubl.)	15	-0.74 ± 0.13	0.46	1.03
Temperate herb. mammals	D_{max}	0.01–500	30	0.25	Peters & Wassenberg (unpubl.)	320	0.73 ± 0.78	0.75	0.64
Temperate herb. mammals	D_{min}	0.01–500	31	0.25	Peters & Wassenberg (unpubl.)	36	-0.52 ± 0.12	0.39	0.84
Temperate carn. mammals	D_{max}	0.06–16	7	4.1	Peters & Wassenberg (unpubl.)	17	-1.34 ± 0.56	0.53 (NS)	1.10
Temperate carn. mammals	D_{min}	0.06–16	7	4.1	Peters & Wassenberg (unpubl.)	2	-0.99 ± 0.55	0.39 (NS)	1.07
Invertebrates	D_{max}	1×10^{-9}–0.01	59	0.4×10^{-6}	Peters & Wassenberg (unpubl.)	53×10^3	-0.69 ± 0.064	0.67	0.86
Invertebrates	D_{min}	1×10^{-9}–0.01	55	0.4×10^{-6}	Peters & Wassenberg (unpubl.)	3.3×10^3	-0.73 ± 0.080	0.61	1.04
Aquatic invertebrates	D_{max}	1×10^{-9}–0.01	32	0.2×10^{-6}	Peters & Wassenberg (unpubl.)	46×10^3	-0.72 ± 0.094	0.65	0.81
Aquatic invertebrates	D_{min}	1×10^{-9}–0.01	31	0.2×10^{-6}	Peters & Wassenberg (unpubl.)	1.6×10^3	-0.82 ± 0.11	0.63	0.98
Terrestrial invertebrates	D_{max}	1×10^{-9}–0.01	26	1.1×10^{-6}	Peters & Wassenberg (unpubl.)	74×10^3	-0.63 ± 0.09	0.66	0.92
Terrestrial invertebrates	D_{min}	1×10^{-9}–0.01	23	1.1×10^{-6}	Peters & Wassenberg (unpubl.)	14×10^3	-0.55 ± 0.10	0.58	0.93

Note: D_{max} and D_{min} are maximum and minimum densities reported in different studies.

Appendix Xb. Allometric relations for animal home range (H)

Taxon	Independent variable	Range of W (kg)	N	$\bar{W}$ (kg)	Reference	Intercept at $W=1\,\mathrm{kg}$ a (km^2)	Slope $b \pm S_b$	r^2	S_{xy}
Mammals	H		38		McNab (1963)	0.031	0.63		
Mammals	H	0.01–200	53	2.0	From Harestad & Bunnell (1979)	0.154	1.06±0.12	0.60	1.12
Primates	H				Milton & May (1976)	0.010	0.78	0.44	
Mammalian hunters	H				McNab (1963)	0.057	0.71		
Carnivorous mammals	H	0.008–100	19	0.27	From Harestad & Bunnell (1979)	1.39	1.37±0.16	0.80	0.90
Hunting primates	H				Milton & May (1976)	0.21	0.83	0.66	
Omnivorous mammals	H	0.02–100	7		From Harestad & Bunnell (1979)	0.33	0.92±0.13	0.90	0.44
Omnivorous and carnivorous mammals	H	0.008–100	26	1.8	From Harestad & Bunnell (1979)	0.83	1.17±0.14	0.75	0.92
Herbivorous mammals	H	0.01–100	28	2.2	From Harestad & Bunnell (1979)	0.032	0.998±0.1	0.74	0.77
Cropping mammals	H				McNab (1963)	0.014	0.69		
Cropping primates	H				Milton & May (1976)	0.0030	1.06	0.49	
Birds	H	0.01–10	46		Armstrong (1965)	1.64	1.23		
Birds	H	0.01–2	77	0.045	From Schoener (1968)	1.00	1.14±0.11	0.57	0.71
Carnivorous birds	H	0.01–2	47	0.046	From Schoener (1968)	8.26	1.37±0.73	0.86	0.46
Herbivorous birds	H	0.05–1.05	3	0.31	From Schoener (1968)	0.026	0.70±0.042	0.89	0.41
Secondary consumers	H		8		Buskirk in Calder (1974)	0.67	1.27	0.91	
Herbivores	H		32		Buskirk in Clader (1974)	0.047	1.03	0.87	
Lizards	H		29		Turner et al. (1969)	0.121	0.95±0.15	0.58	
Primates	Group home range				Milton & May (1976)	0.068	1.23		

References

Ackerman, R. A., G. C. Whittow, C. V. Paganelli, & T. N. Pettit. 1980. Oxygen consumption, gas exchange and growth of embryonic wedged-tailed shearwaters (*Puffinus pacificus chlororhynchus*). *Physiological Zoology* 53: 210–211.

Adolf, E. F. 1943. *Physiological Regulations*. Lancaster, Pa.: Cattell Press.

– 1949. Quantitative relations in the physiological constitutions of animals. *Science* 109: 579–585.

Allison, T., & D. V. Cicchetti. 1976. Sleep in mammals: Ecological and constitutional correlates. *Science* 194: 732–734.

Altman, P. L., & D. S. Dittmer. 1968. *Metabolism*. Bethesda, Md.: Federation of the American Society of Experimental Biologists.

Ames, D. 1980. Thermal efficiency affects production efficiency of livestock. *Bioscience* 30: 457–460.

Anderson, J. F., H. Rahn, & H. D. Prange. 1979. Scaling of supportive tissue mass. *The Quarterly Review of Biology* 54: 139–148.

Andrewartha, H. G. 1961. *Introduction to the Study of Animal Populations*. Chicago: University of Chicago Press.

Andrewartha, H. G., & L. C. Birch. 1954. *The Distribution and Abundance of Animals*. Chicago: University of Chicago Press.

Apple, M. S., & M. A. Korostyshevskiy. 1980. Why many biological parameters are connected by power dependence. *Journal of Theoretical Biology* 85: 569–573.

Ar, A., C. V. Paganelli, R. B. Reeves, D. G. Greene, & H. Rahn. 1974. The avian egg: Water vapour conductance, shell thickness and functional pore area. *Condor* 76: 153–158.

Ar, A., & H. Rahn. 1980. Water in the avian egg: Overall budget of incubation. *American Zoologist* 20: 373–384.

Ar, A., H. Rahn, & C. V. Paganelli. 1979. The avian egg: Mass and strength. *Condor* 81: 331–337.

Armstrong, J. T. 1965. Breeding home range in the nighthawk and other birds: Its evolutionary and ecological significance. *Ecology* 46: 619–629.

Aschoff, J. 1981. Thermal conductance in mammals and birds: Its dependence on body size and circadian phase. *Comparative Biochemistry and Physiology* 69: 611–619.

Aschoff, J., & H. Pohl. 1970. Der Ruheumsatz von Vögeln als Funktion der Tageszeit und der Körpergrösse. *Journal für Ornithologie* 111: 38–47.

Bainbridge, R. 1958. The speed of swimming of fish as related to size and to the frequency and amplitude of the tail beat. *Journal of Experimental Biology* 35: 109–133.

Bakker, R. T. 1972. Locomotor energetics of lizards and mammals compared. *The Physiologist* 15:76 (abstract).

– 1975. Experimental and fossil evidence for the evolution of tetrapod bio-energetics. In *Perspectives in Biophysical Ecology*, D. M. Gates, & R. B. Schmerl, eds., pp. 365–399. New York: Springer-Verlag.

Baldock, B. M., J. H. Baker, & M. A. Sleigh. 1980. Laboratory growth rates of six species of freshwater Gymnamoebia. *Oecologia* (Berlin) 47:156–159.

Båmstedt, U., & H. R. Skjoldal. 1980. RNA concentration of zooplankton: Relationship with size and growth. *Limnology and Oceanography* 25:304–316.

Banse, K. 1976. Rates of growth, respiration and photosynthesis of unicellular algae as related to cell size – a review. *Journal of Phycology* 12:135–140.

– 1979. On weight dependence of net growth efficiency and specific respiration rates among field populations of invertebrates. *Oecologia* (Berlin) 38:111–126.

Banse, K., & S. Mosher. 1980. Adult body mass and annual production/biomass relationships of field populations. *Ecological Monographs* 50:355–379.

Bartels, H. 1970. *Prenatal Respiration*. Amsterdam: North Holland.

Bartholomew, G. A., & T. C. Cade. 1963. The water economy of land birds. *The Auk* 80:504–539.

Bartholomew, G. A., & T. M. Casey. 1977. Body temperature and oxygen consumption during rest and activity in relation to body size in some tropical beetles. *Journal of Thermal Biology* 2:173–176.

– 1978. Oxygen consumption of moths during rest, pre-flight warm-up, and flight in relation to body size and wing morphology. *Journal of Experimental Biology* 76:11–25.

Bartholomew, G. A., & W. R. Dawson. 1953. Respiratory water loss in some birds of the southwestern United States. *Physiological Zoology* 26:162–166.

Bartholomew, G. A., & R. J. Epting. 1975. Rates of post-flight cooling in sphinx moths. In *Perspectives of Biophysical Ecology*, D. M. Gates, & R. B. Schmerl, eds., pp. 405–415. New York: Springer-Verlag.

Bartholomew, G. A., & R. C. Lasiewski. 1965. Heating and cooling rates, heart rate and simulated diving in the Galapagos marine iguana. *Comparative Biochemistry Physiology* 16:573–582.

Bartholomew, G. A., P. Leitner, & J. E. Nelson. 1964. Body temperature, oxygen consumption, and heart rate in three species of Australian flying fox. *Physiological Zoology* 37:179–198.

Bartholomew, G. A., & V. A. Tucker. 1964. Size, body temperature, thermal conductance, oxygen consumption, and heart rate in Australian varanid lizards. *Physiological Zoology* 27:341–354.

Batholomew, G. A., D. Vleck, & C. M. Vleck. 1981. Instantaneous measurements of oxygen consumption during pre-flight warm-up and post-flight cooling in sphingid and saturniid moths. *Journal of Experimental Biology* 90:17–32.

Bauchot, R., M. L. Bauchot, R. Platel, & J. M. Ridet. 1977. Brains of Hawaiian tropical fishes: Brain size and evolution. *Copeia* 1977:42–46.

Baudinette, R. U. 1978. Scaling of heart rate during locomotion. *Journal of Comparative Physiology* 127:337–342.

Baumann, F. H., & R. Baumann. 1977. A comparative study of the respiratory properties of bird blood. *Respiration Physiology* 31:333–343.

Beamish, F. W. H. 1978. Swimming capacity. In *Fish Physiology*, Vol. VIII, W. S. Hoar, & D. J. Randall, eds., pp. 101–187. New York: Academic Press.

Benedict, F. G. 1938. *Vital Energetics. A Study in comparative basal Metabolism.* Washington D.C. Carnegie: Institute.

Bennett, A. F. 1972. A comparison of activities of metabolic enzymes in lizards and rats. *Comparative Biochemistry and Physiology B* 42:637–647.

– 1978. Activity metabolism of the lower vertebrates. *Annual Review of Physiology* 40:447–469.

– 1980. The metabolic foundations of vertebrate behavior. *Bioscience* 30:452–456.

Bennett, A. F., & W. R. Dawson. 1976. Metabolism. In *Biology of the Reptilia*, Vol. V, C. Gans, & W. R. Dawson, eds., pp. 127–223 New York: Academic Press.

Bennett, A. F., & K. A. Nagy. 1977. Energy expenditure in free ranging lizards. *Ecology* 58:697–700.

Bennett, A. F., & J. A. Ruben. 1979. Endothermy and activity in vertebrates. *Science* 206:649–654.

Berger, M., & J. S. Hart. 1974. Physiology and energetics of flight. In *Avian Biology*, Vol. IV, D. S. Farner, & J. R. King, eds., pp. 415–477. New York: Academic Press.

Blaxter, K. L. 1971. The comparative biology of lactation. In *Lactation*, I. R. Falconer, ed., pp. 51–69. London: Butterworth.

Blueweiss, L., H. Fox, V. Kudzma, D. Nakashima, R. Peters, & S. Sams. 1978. Relationships between body size and some life history parameters. *Oecologia* (Berlin) 37:257–272.

Blum, J. J. 1977. On the geometry of four-dimensions and the relationship between metabolism and body mass. *Journal of Theoretical Biology* 64:599–601.

Boddington, M. J. 1978. An absolute metabolic scope for activity. *Journal of Theoretical Biology* 75:443–449.

Bonner, J. T. 1965. *Size and Cycle: An Essay on the Structure of Biology.* Princeton, N.J.: Princeton University Press.

Bottrell, H. H. 1975. The relationship between temperature and duration of egg development in some epiphytic Cladocera and Copepoda from the river Thames, Reading, with a discussion of temperature functions. *Oecologia* (Berlin) 18:63–84.

Bowen, H. J. M. 1979. *Environmental Chemistry of the Elements.* New York: Academic Press.

Bowen, W. D. 1981. Variation in coyote social organization: The influence of prey size. *Canadian Journal of Zoology* 59:639–652.

Bradley, S. R., & D. R. Deavers. 1980. A reexamination of the relationship between thermal conductance and body weight in mammals. *Comparative Biochemistry and Physiology A* 65:465–476.

Braithwaite, R. W., & A. K. Lee. 1979. A mammalian example of semelparity. *The American Naturalist* 113:151–155.

Brett, J. R. 1965. The relation of size to rate of oxygen consumption and sustained swimming speed of sockeye salmon (*Onchorhynchus nerka*). *Journal of the Fisheries Research Board of Canada* 22:1491–1501.

Brett, J. R., & N. R. Glass. 1973. Metabolic rates and critical swimming speeds of sockeye salmon (*Onchorhynchus nerka*) in relation to size and temperature. *Journal of the Fisheries Research Board of Canada.* 30:379–387.

Brett, J. R., & T. D. D. Groves. 1979. Physiological energetics. In *Fish Physiology*, Vol. VIII, W. S. Hoar, D. J. Randall, & J. R. Brett, eds., pp. 272–352. New York: Academic Press.

Brody, S. 1945. *Bioenergetics and Growth*. Baltimore, Md.: Reinhold.

Brown, J. H. 1971. Mammals on mountain tops: Nonequilibrium insular biogeography. *The American Naturalist* 105:467–478.

– 1981. Two decades of homage to Santa Rosalia: Toward a general theory of diversity. *American Zoologist* 21:877–888.

Buddenbrock, W. V. 1934. Uber die kinetische and statische Leistung grosser und kleiner Tiere und ihre bedeutung für dem Gesamtstoffwechsel. *Naturwissenschaft* 22:675–680.

Burnison, B. K. 1975. Microbial ATP studies. *International Association of Theoretical and Applied Limnology Proceedings* 19:286–290.

Cabana, G., A. Frewin, R. H. Peters, & L. Randall. 1982. The effect of sexual size dimorphism on variations in reproductive effort of birds and mammals. *The American Naturalist* 120:17–25.

Calder, W. A. 1968. Respiratory and heart rates of birds at rest. *Condor* 70:358–365.

– 1969. Temperature relations and underwater endurance of the smallest homeothermic diver, the water shrew. *Comparative Biochemistry and Physiology* 30:1075–1082.

– 1974. Consequences of body size for avian energetics. In *Avian Energetics*, R. A. Paynter, ed., pp. 86–151. Cambridge, Mass.: Nuttall Ornithological Club.

– 1981. Scaling of physiological processes in homeothermic animals. *Annual Review of Physiology*. 43:301–322.

Calder, W. A. & T. J. Dawson. Resting metabolic rates of rattite birds: The kiwis and the emu. *Comparative Biochemistry and Physiology A* 60:479–481.

Calder, W. A. & J. R. King. 1974. Thermal and caloric relations of birds. In *Avian Biology*, Vol. IV, D. S. Farner, & J. R. King, eds., pp 259–413. New York: Academic Press.

Cammen, L. M. 1980. Ingestion rate: An empirical model for aquatic deposit feeders and detritivores. *Oecologia (Berlin)* 44:303–310.

Carey, C., & M. L. Morton. 1976. Aspects of circulatory physiology of mountain and lowland birds. *Comparative Biochemistry and Physiology A* 54:61–74.

Carlander, K. D. 1977. *Handbook of Freshwater Fishery Biology*, Vol. II. Ames: Iowa State University Press.

Case, T. J. 1978. On the evolution and adaptive significance of postnatal growth rates in the terrestrial vertebrates. The Quarterly Review of Biology 53:243–282.

– 1979. Optimal body size and an animal's diet. *Acta Biotheoretica* 28:54–69.

Casey, T. M., P. C. Withers, N. & K. K. Casey. 1979. Metabolic and respiratory responses of arctic mammals to ambient temperature during the summer. *Comparative Biochemistry and Physiology A* 64:331–341.

Claussen, D. L. 1969. Studies on water loss and rehydration in anurans. *Physiological Zoology* 42:1–14.

Claussen, D. L., & G. R. Art. 1981. Heating and cooling rates in *Anolis carolinensis* and comparisons with other lizards. *Comparative Biochemistry and Physiology A* 62:23–29.

Clutton-Brock, T. H., S. D. Albon, & P. H. Harvey. 1980. Antlers, body size, and breeding group size in the Cervidae. *Nature* (London) 285:565–567.

Clutton-Brock, T. H., & P. H. Harvey. 1977a. Species differences in feeding and ranging behavior in primates. In *Primate Ecology*, T. H. Clutton-Brock, ed., pp. 557–584. New York: Academic Press.

– 1977b. Primate ecology and social organization. *Journal of Zoology* 183:1–39.

Coe, M. J., D. H. Cumming, & J. Phillipson. 1976. Biomass and production of large African herbivores in relation to rainfall and primary production. *Oecologia* (Berlin) 22:341–354.

Colinvaux, P. A. 1973. *Introduction to Ecology*. New York: Wiley.

Collins, S. 1979. This fall's kill. *MacLean's Magazine* October 29:46.

Conover, R. J. 1978. Transformaton of organic matter. In *Marine Ecology*, Vol. IV, O. Kinne, ed., pp. 221–499. New York: Wiley.

Cousins, S. H. 1980. A trophic continuum derived from plant structure, animal size and a detritus cascade. *Journal of Theoretical Biology* 82:607–618.

Covo, C. 1965. *Ornitologia Practica*. Milano: Editore Ulrico Hoepli.

Crawford, E. C., Jr., & R. C. Lasiewski. 1968. Oxygen consumption and respiratory evaporation of the emu and rhea. *Condor* 70:333–339.

Cummins, K.W., & J. C. Wuychek. 1971. Caloric equivalents for investigations in ecological energetics. *Mitteilungen Internationale Vereinigung für Theoretische und Angewandte Limnologie* 18:1–158.

Curio, E. 1976. *The Ethology of Predation*. Berlin: Springer.

Damuth, J. 1981. Population density and body size in mammals. *Nature* (London) 290:699–700.

Davidson, J. 1955. Body weight, cell surface, and metabolic rate in anuran amphibia. *The Biological Bulletin* 109:407–419.

Davies, I. J. 1980. Relationships between dipteran emergence and phytoplankton production in the experimental lakes area, northwestern Ontario. *Canadian Journal of Fisheries and Aquatic Sciences* 37:523–533.

Dawson, T. J., & W. R. Dawson. 1982. Metabolic scope and conductance in response to cold of some dasyurid mammals and Australian rodents. *Comparative Biochemistry and Physiology A* 71:59–64.

Dawson, T. J., & A. J. Hurlbert. 1970. Standard metabolism, body temperature, and surface areas of Australian marsupials. *American Journal of Physiology* 218:1233–1238.

Dejours, P. 1975. *Principles of Comparative Respiratory Physiology*. Amsterdam: North Holland.

Derome, J. R. 1977. Biological similarity and group theory. *Journal of Theoretical Biology* 65: 369–378.

DiGregorio, D., T. Kitchings, & P. Van Voris. 1978. Radionuclide transfer in terrestrial animals. *Health Physics* 34: 3–31.

Dillon, P. J., & F. H. Rigler. 1974. The phosphorus chlorophyll relationship in lakes. *Limnology and Oceanography* 19: 767–773.

– 1975. A simple method for predicting the capacity of a lake for development based on lake trophic status. *Journal of the Fisheries Researach Board of Canada* 32: 1519–1531.

Dmi'el, R. 1972. Relation of metabolism to body weight in snakes. *Copeia* 1972: 179–181.

Downing, J. A. 1979. Aggregation, transformation and the design of benthos sampling programs. *Journal of the Fisheries Research Board of Canada* 36: 1454–1463.

– 1980. Precision vs generality: A reply. *Canadian Journal of Fisheries and Aquatic Sciences* 37: 1329–1330.

– 1981. In situ foraging responses of three species of littoral cladocerans. *Ecological Monographs* 51: 85–103.

Drent, R. H. 1970. Functional aspects of incubation in the herring gull. *Behavior Supplement* 17: 1–132.

Drent, R .H., & B. Stonehouse. 1971. Thermoregulatory responses of the Peruvian penguin, *Spheniscus humboldti*. *Comparative Biochemistry and Physiology* 40: 689–710.

Drorbaugh, J. E. 1960. Pulmonary function in different animals. *Journal of Applied Physiology* 15: 1069–1072.

Durnin, J. V. G. A., & R. Passmore. 1967. *Energy, Work, and Leisure*. London: Heinemann.

Eberhard, W. G. 1979. Rates of egg production by tropical spiders in the field. *Biotropica* 11: 292–300.

Eberhardt, L. L. 1969. Similarity, allometry and food chains. *Journal of Theoretical Biology* 24: 43–55.

Economos, A. C. 1979a. On structural theories of basal metabolic rate. *Journal of Theoretical Biology* 80: 445–450.

– 1979b. Gravity, metabolic rate, and body size of mammals. *The Physiologist* 22: S-71–S-72.

– 1981. The largest land mammal. *Journal of Theoretical Biology* 89: 211–215.

Edwards, N. A. 1975. Scaling of renal functions in mammals. *Comparative Biochemistry and Physiology A* 52: 63–66.

Eisenberg, J. F. 1980. The density and biomass of tropical mammals. In *Conservation Biology: an Evolutionary-Ecological Perspective*, M. E. Soulé & B. A. Wilcox, eds., pp 35–55. Sunderland, Mass.: Sinauer Association.

Ellis, H. I., & J. P. Ross. 1978. Field observations of cooling rates of Galapagos land iguanas (*Conolophus subcristatus*). *Comparative Biochemistry and Physiology A* 59: 203–209.

Elton, C. S. 1966. *The Pattern of Animal Communities*. London: Methuen & Co.
– 1973. The structure of invertebrate populations inside neotropical rain forest. *Journal of Animal Ecology* 42: 55–104.
Emlen, J. T. 1972. Size and structure of a wintering avian community in southern Texas. *Ecology* 53: 317–329.
Emmett, B., & P. W. Hochachka. 1981. Scaling of oxidative and glycolytic enzymes in mammals. *Respiration Physiology* 45: 261–272
Enders, F. 1975. The influence of hunting manner on prey size, particularly in spiders with long attack distances (Araneidae, Linphiidae, and Salticidae). *The American Naturalist* 109: 737–763.
Eppley, R. W., J. N. Rogers, & J. J. McCarthy. 1969. Half saturation constants for uptake of nitrate and ammonium by marine phytoplankton. *Limnology and Oceanography* 14: 912–920.
Eppley, R. W. & P. R. Sloan. 1966. Growth rates of marine phytoplankton: Correlation with light absorption by cell chlorophyll a. *Physiologica Plantarum* 19: 47–59.
Evans, E., & D. S. Miller, 1968. Comparative nutrition, growth and longevity. *Proceedings of the Nutrition Society* 27: 121–129.
Eviator, N. 1973 Adaptive variation in size of cricket frogs. *Ecology* 54: 1271–1281.
Fagerström, T., B. Asell, & A. Jernelov. 1974. Modes for accumulation of methyl mercury in northern pike, *Esox lucius*. *Oikos* 25: 14–20.
Farlow, J. O. 1976. A consideration of the trophic dynamics of a late Cretaceous large-dinosaur community (Oldman Formation). *Ecology* 57: 841–857.
Fedak, M. A., B. Pinshaw, & K. Schmidt–Nielsen. 1974. Energy cost of bipedal running. *American Journal of Physiology* 227: 1038–1044.
Fedak, M. A., & H. J. Seeherman. 1979. Reappraisal of energetics of locomotion shows identical costs in bipeds and quadrupeds including ostrich and horse. *Nature* (London) 282: 713–716.
Feder, M. E. 1976. Oxygen consumption and body temperature in neotropical and temperature zone lungless salamonders (Amphibia: Plethodontidae). *Journal of Comparative Physiology* 110: 197–208.
Fenchel, T. 1974. Intrinsic rate of natural increase: The relationship with body size. *Oecologia* (Berlin) 14: 317–326.
– 1980. Suspension feeding in ciliated Protozoa: Feeding rates and their ecological significance. *Microbial Ecology* 6: 13–26.
Finlay, B. J. 1977. The dependence of reproductive rate on cell size and temperature in freshwater ciliated Protozoa. *Oecologia* (Berlin) 30: 75–81.
– 1978. Community production and respiration by ciliated Protozoa in the benthos of a small eutrophic loch. *Freshwater Biology* 8: 327–341.
Fleming, T. H. 1973. Numbers of mammal species in north and central American forest communities. *Ecology* 54: 555–563.
Forsyth, D. J. 1976. A field study of growth and development of nestling masked shrews (*Sorex cinereus*). *Journal of Mammalogy* 57: 708–721.
Foster, W. L., & J. Tate. 1966. The activities and coactions of animals at sapsucker trees. *The Living Bird* 5: 87–113.

French, N. R., W. E. Grant, W. Grodzinski, & D. N. Swift. 1976. Small mammal energetics in grassland ecosystems. *Ecological Monographs* 46:201–220.

Friebele, E. S., D. L. Correll, & M. A. Faust. 1978. Relationship between phytoplankton cell size and the rate of orthophosphate uptake: In situ observations of an estuarine population. *Marine Biology* 45:39–52.

Fry, F. E. J. 1947. Effects of the environment on animal activity. University of Toronto, Studies in Biology 55. *Publications of the Ontario Fisheries Research Laboratory* 68:1–62.

Fry, F. E. J., & E. T. Cox. 1970. A relation of size to swimming speed in rainbow trout. *Journal of the Fisheries Research Board of Canada* 27:976–978.

Fujita, M., J. Iwamoto, & M. Kondo. 1966. Comparative metabolism of cesium and potassium in mammals. Interspecies correlation between body weight and equilibrium level. *Health Physics* 12:1237–1247.

Furchner, J. E. & C. R. Richmond. 1963. Comparative metabolism of radioisotopes in mammals. II. Retention of iodine-131 by four mammalian species. *Health Physics* 9:277–282.

Gates, D. M. 1980. *Biophysical Ecology.* New York: Springer-Verlag.

Gates, D. M., & R. B. Schmerl (eds.). 1975. *Perspectives of Biophysical Ecology.* New York: Springer-Verlag.

Gehr, P., D. K. Mwangi, A. Ammann, G. M. O. Maloiy, C. R. Taylor, & E. R. Weibel. 1981. Design of the mammalian respiratory system. V. Scaling morphometric pulmonary diffusing capacity to body mass: Wild and domestic mammals. *Respiration Physiology* 44:61–86.

Geller, W. 1975. Die Nahrungsaufnahme von *Daphnia pulex* in Abhangigkeit von der Futterkonzentration, der Temperatur, der Körpergrösse und dem Hungerzustand der Tiere. *Archiv für Hydrobiologie Supplementband* 48:47–107.

Gerking, S. D. 1955a. Influence of rate of feeding on body composition and protein metabolism of bluegill sunfish. *Physiological Zoology* 28:267–282.

– 1955b. Endogeneous nitrogen excretion of bluegill sunfish. *Physiological Zoology* 28:283–289.

Gettinger, R. D. 1975. Metabolism and thermoregulation of a fossorial rodent, the northern pocket gopher (*Thomomys talpoides*). *Physiological Zoology* 48:311–322.

Ghilarov, M.S. 1967. Abundance, biomass and vertical distribution of soil animals in different zones. In *Secondary Productivity of Terrestrial Ecosystems; Principles and Methods*, K. Petrusewicz, ed., pp. 611–630. Warsaw: Państwowe Wydawnictwo Naukowe.

Gibb, J. 1954. Feeding ecology of tits, with notes on treecreeper and gold crest. *Ibis* 96:513–543.

Girard, H., & M. Grima. 1980. Allometric relation between blood oxygen uptake and body mass in birds. *Comparative Biochemistry and Physiology A* 66:485–491.

Glebe, B. D., & W. C. Leggett. 1981. Latitudinal differences in energy allocation and use during the freshwater migrations of American shad (*Alosa sapidissima*) and their life history consequences. *Canadian Journal of Fisheries and Aquatic Sciences* 38:806–820.

Gleeson, T. T. 1979. Foraging and transport costs in the Galapagos marine iguana, *Amblyrhyncus cristatus. Physiological Zoology* 52:549–557.

Gliwicz, Z. M. 1969a. The share of algae, bacteria, and trypton in the food of the pelagic zooplankton in lakes with various trophic characteristics. *Bulletin de l' Academie Polonaises des Sciences. II. Series des Sciences Biologiques* 17:159–165.

– 1969b. Studies on the feeding of pelagic zooplankton in lakes with varying trophy. *Ekologia Polska A* 17:663–708.

– 1980. Filtering rates, food size selection and feeding rates in cladocerans– another aspect of interspecific competition in filter-feeding zooplankton. In *Evolution and Ecology of Zooplankton Communities*, W. C. Kerfoot, ed., pp. 282–291. Hanover, N. H.: University Press of New England.

Godley, J. S. 1980. Foraging ecology of the striped swamp snake, *Regina alleni*, in southern Florida. *Ecological Monographs* 50:411–436.

Golley, F. B., R. G. Wiegart, & R. W. Walter. 1965. Excretion of orally administered zinc-65 by wild small mammals. *Health Physics* 11:719–722.

Gordon, M. S. 1977. *Animal Physiology: Principles and Adaptations*, 3rd ed. New York: Macmillan.

Gould, S. J. 1966. Allometry and size in ontogeny and phylogeny. *Biological Reviews of the Cambridge Philosophical Society* 41:587–640.

– 1979. An allometric interpretation of species-area curves. The meaning of the coefficient. *The American Naturalist* 114:335–343.

Goulden, C. E., & L. L. Hornig. 1980. Population oscillations and energy reserves in planktonic Cladocera and their consequences to competition. *Proceedings of the National Academy of Sciences of the United States of America* 77:1716–1720.

Gray, B. F. 1981. On the "surface law" and basal metabolic rate. *Journal of Theoretical Biology* 93:757–767.

Greenewalt, C. H. 1975. Flight of birds. The significant dimensions, their departure from the requirements for dimensional similarity, and the effect on flight aerodynamics of that departure. *Transactions of the American Philosophical Society, New Series* 65:1–67.

Greenstone, M. H., & A. F. Bennett. 1980. Foraging strategy and metabolic rate in spiders. *Ecology* 61:1255–1259.

Griesbach, S., R. H. Peters, & S. Youakim. 1982. An allometric model for pesticide bioaccumulation. *Canadian Journal of Fisheries and Aquatic Sciences* 39:727–735.

Günther, B. 1975. Dimensional analysis and theory of biological similarity. *Physiological Reviews* 55:659–699.

Günther, B., & E. Guerra. 1955. Biological similarities. *Acta Physiologica Latino Americana* 5:169–186.

Günther, B., & B. Léon de la Barra. 1966a. Physiometry of the mammalian circulatory system. *Acta Physiologica Latino Americana* 16:32–42.

– 1966b. A unified theory of biological similarities. *Journal of Theoretical Biology* 13:48–59.

Guyton, A. C. 1947a. Analysis of respiratory patterns in laboratory animals. *American Journal of Physiology* 150: 78–83.

– 1947b. Measurement of the respiratory volumes of laboratory animals. *American Journal of Physiology* 150: 70–77.

Hails, C. J. 1979. A comparison of flight energetics in hirundines and other birds. *Comparative Biochemistry and Physiology A* 63: 581–585.

Hall, D. J., S. T. Threlkeld, C. W. Burns, & P. H. Crowley. 1976. The size efficiency hypothesis and the size structure of zooplankton communities. *Annual Review of Ecology and Systematics* 7: 177–208.

Hall, T. 1978. *Nickel Uptake, Retention and Loss in Daphnia magna*, M.Sc. Thesis, University of Toronto.

Haney, J. F. 1973. An in situ examination of the grazing activities of natural zooplankton communities. *Archiv für Hydrobiologie Supplementband* 72: 87–132.

Hanson, J. M., & W. C. Leggett. 1982. Empirical prediction of fish biomass and yield. *Canadian Journal of Fisheries and Aquatic Sciences* 39: 257–263.

Hanwell, A., & M. Peaker. 1977. Physiological effects of lactation on the mother. *Symposia of the Zoological Society of London* 41: 297–312.

Harestad, A. S., & F. L. Bunnell. 1979. Home range and body weight–A reevaluation. *Ecology* 60: 389–402.

Hargrave, B. T. 1972. Prediction of egestion by the deposit feeding amphipod. *Oikos* 23: 116–124.

Hargrave, B. T., & G. H. Geen. 1968. Phosphorus excretion by zooplankton. *Limnology and Oceanography* 13: 332–343.

Harper, J. L. 1969. The role of predation in vegetational diversity. *Brookhaven Symposia in Biology* No. 22: 48–62.

Hart, J. S. 1971. Rodents. In *Comparative Physiology of Thermorgulation*, Vol. II, *Mammals*, G. C. Whittow, ed., pp 2–149. New York: Academic Press.

Hayward, R. S., & D. N, Gallup. 1976. Feeding, filtering and assimilation in *Daphnia schoedleri* as affected by environmental conditions. *Archiv für Hydrobiologie Supplementband* 77: 139–163.

Hecky, R. E., & P. Kilham. 1974. Environmental control of phytoplankton cell size. *Limnology and Oceanography* 19: 361–365.

Heglund, N. C., G. A. Cavagna, M. A. Fedak, & C. R. Taylor. 1979. Muscle efficiency during locomotion: How does it vary with body size and speed? *Federation Proceedings* 38: 1443.

Heglund, N. C., C. R. Taylor, & T. A. McMahon. 1974. Scaling stride frequency and gait to animal size: Mice to horses. *Science* 186: 1112–1113.

Heldmaier, G. 1971. Zitterfrei Wärmebildung und der Körpergrösse bei Saügetieren. *Zeitschrift fur Vergleichende Physiologie* 73: 222–248.

Heinrich, B, & G. A. Bartholomew. 1971. An analysis of pre-flight warm-up in the sphinx moth, *Manduca sexta*. *Journal of Experimental Biology* 55: 223–239.

Hemmer, H. 1979. Gestation period and postnatal development in felids. *Carnivore* 2: 90–100.

Hemmingsen, A. M. 1960. Energy metabolism as related to body size and respiratory surfaces, and its evolution. *Reports of the Steno Memorial Hospital and Nordinsk Insulin Laboratorium* 9: 6–110.

Hempel, C. G. 1965. *Aspects of Scientific Explanation*. New York: Free Press.

Hempel, G. 1954. Laufgeschwindigkeit und Körpergrösse bei Insekten. *Zeitschrift für Vergleichende Physiologie* 36:261–265.

Hensel, H., K. Brück, & P. Raths. 1973. Homeothermic organisms. In *Temperature and Life*, H. Precht, J. Christophersen, H. Hensel, & W. Larcher, eds., pp. 505–762. New York: Springer-Verlag.

Heron, A. C. 1972. Population ecology of a colonizing species: The pelagic tunicate *Thalia democratica*. II. Population growth rate. *Oecologia* (Berlin) 10:294–312.

Herreid, C. F. 1969. Water loss of crabs from different habitats. *Comparative Biochemistry and Physiology* 28:829–839.

Herreid, C. F., & B. Kessel. 1967. Thermal conductance in birds and mammals. *Comparative Biochemistry and Physiology* 21:405–414.

Hespenheide, H. A. 1973. Ecological inferences from morphological data. *Annual Review of Ecology and Systematics* 4:213–219.

Hinds, D. S., & W. A. Calder, 1971. Tracheal dead space in the respiration of birds. *Evolution* 25:429–440.

Hinga, K. 1979. The food requirements of whales in the southen hemisphere. *Deep Sea Research* 26A:569–577.

Holeton, G. F. 1974. Metabolic cold adaptation of polar fish: Fact or artefact? *Physiological Zoology* 47:137–152.

Holm–Hansen, D. 1969. Algae: Amounts of DNA and organic carbon in single cells. *Science* 163:87–88.

Holt, J. P., E. A. Rhode, & H. Kines. 1968. Ventricular volumes and body weight in mammals. *American Journal of Physiology* 215:704–715.

Hoppeler, H., O. Mathieu, E. R. Weibel, R. Krower, S. Lindstedt, & C. R. Taylor. 1981. Design of the mammalian respiratory system. VIII. Capillaries in skeletal muscles. *Respiration Physiology* 44:129–150.

Hoyt, D. F. 1980. Adaptation of avian eggs to incubation period: Variability around allometric regressions is correlated with time. *American Zoologist* 20:417–425.

Hoyt, D. F., R. G. Board, H. Rahn, & C. V. Paganelli. 1979. The eggs of the Anatidae: Conductance, pore structure and metabolism. *Physiological Zoology* 52:438–450.

Hoyt, D. F., & Rahn, H. 1980. Respiration of avian embroys–A comparative analysis. *Respiration Physiology* 39:255–264.

Huggett, A. St. G., & W. F. Widdas. 1951. The relation between mammalian foetal weight and conception age. *Journal of Physiology* 114:306–317.

Hughes, G. M. 1977. Dimensions and the respiration of lower vertebrates. In *Scale Effects in Animal Locomotion*, T. J. Pedley, ed., pp. 57–81. New York: Academic Press.

Hughes, M. R. 1970. Relative kidney size in non passerine birds with functional salt glands. *Condor* 72:164–168.

Hull, D. L. 1968. The operational imperative: Sense and nonsense in operationism. *Systematic Zoology* 17:438–457.

Hume, D. 1739, 1740. *A Treatise of Human Nature*, 2 vols. London: Dent.

– 1748. An enquiry concerning human understanding. In *Theory of Knowledge*, D. C. Yalden–Thomson, ed., Edinburgh: Nelson, 1951.

Humphreys, W. F. 1979.. Production and respiration in animal populations. *Journal of Animal Ecology* 48: 427–453.

Hutchinson, G. E. 1959. Homage to Santa Rosalia or why are there so many kinds of animals. *The American Naturalist* 93: 145–159.

Hutchinson, G. E., & R. H. MacArthur. 1959. A theoretical ecological model of size distributions among species of animals. *The American Naturalist* 93: 117–125.

Hutchinson, V. H., W. G. Whitford, & M. Kohl. 1968. Relation of body size and surface area to gas exchange in anurans. *Physiological Zoology* 41: 65–85.

Huxley, J. S. 1972. *Problems of Relative Growth*, 2nd ed. New York: Dover.

Ikeda, T. 1977. Feeding rates of planktonic copepods from a tropical sea. *Journal of Experimental Marine Biology and Ecology* 29: 263–277.

Isaacs, J. D. 1973. Potential trophic biomasses and trace-substance concentrations in unstructured marine food webs. *Marine Biology* 22: 97–103.

Ivlea, I. V. 1980. The dependence of crustacean respiration rate on body mass and habitat temporative. *Internationale Revue der Gesamten Hydrobiologie* 65: 1–47.

Jankowsky, H. D. 1973. Body temperature and external temperatures. In *Temperature and Life*, H. Precht, J. Christophersen, H. Hensel, & W. Larcher, eds., pp. 293–301. New York: Springer-Verlag.

Janzen, D. H. 1973. Sweep samples of tropical foliage insects: Description of study sites, with data on species abundance and size distribution. *Ecology* 54: 659–686.

Janzen, D. H., & T. W. Schoener. 1968. Differences in insect abundance and diversity between wetter and drier sites during a tropical dry season. *Ecology* 49: 96–110.

Jarman, P. J. 1974. The social organization of antelope in relation to their ecology. *Behavior* 48: 215–267.

Jerison, H. J. 1961. Quantitative analysis of evolution of the brain in mammals. *Science* 133: 1012–1014.

– 1973. *Evolution of the Brain and Intelligence*. New York: Academic Press.

Johannes, R. E. 1964. Phosphorus excretion and body size in marine animals: Microzooplankton and nutrient regeneration. *Science* 146: 923–924.

Johnson, O. W. 1968. Some morphological features of avian kidneys. *The Auk* 85: 216–228.

Jørgensen, S. E. 1979. Modelling the distribution and effect of heavy metals in aquatic ecosystems. *Ecological Modelling* 6: 199–223.

Kaplan, R. H. & S. N. Salthe. 1979. The allometry of reproduction: An empirical view of salamanders. *The American Naturalist* 113: 671–689.

Kayser, C. 1950. Le problème de la loi des tailles et de la loi des surfaces tel qu' il apparait dans l' étude de la calorification des batraciens et reptiles et des mammifères hibernants. *Archives des Sciences Physiologiques* 4: 361–378.

– 1964. La dépense d'énergie des mammiféres en hiberation. *Archives des Sciences Physiologiques* 18: 137–150.

Kayser, C., & A. Heusner. 1964. Etude comparative du métabolisme énergétique dans la série animale. *Journal de Physiologie (Paris)*. 56: 489–524.

Kendeigh, S. C., V. R. Dol' nik & V. M. Govrilov. 1977. Avian energetics. In

Granivorous Birds in Ecosystems, J. Pinowski, & S. C. Kendeigh, eds., pp. 127–204. Cambridge: Cambridge University Press.

Kerr, S. R. 1974. Theory of size distributions in ecological communities. *Journal of the Fisheries Research Board of Canada* 31:1859–1862.

Kihlström, J. E. 1972. Period of gestation and body weight in some placental mammals. *Comparative Biochemistry and Physiology A* 43:673–679.

Kilham, P., & S. S. Kilham. 1980. The evolutionary ecology of phytoplankton. *The Physiological Ecology of Phytoplankton*, L. Morris, ed., pp. 571–597. London: Blackwell Publisher.

King, F. D., & T. T. Packard 1975. Respiration and the activity of the respiratory electron transport system in marine zooplankton. *Limnology and Oceanography* 20:849–854.

King, J. R. 1974. Seasonal allocation of time and energy resources in birds. In *Avian Energetics*, R. A. Paynter, ed., pp. 4–85. Cambridge, Mass.: Nuttall Ornithological Club.

King, J. R., & D. S. Farner. 1961. Energy metabolism, thermoregulation and body temperature. In *Biology and Comparative Physiology of Birds*, A. J. Marshall, ed., pp. 215–288. New York: Academic Press.

Kleiber, M. 1961. *The Fire of Life: An Introduction to Animal Energetics*. New York: Wiley.

Klekowski, R. Z. 1981. Size dependence of metabolism in protozoans. *Verhandlungen Internationale Vereinigung für Theoretische und Angewandte Limnologie* 21:1498–1502

Klekowski, R. Z., L. Wasilewsko, & E. Paplinska. 1972. Oxygen consumption by soil inhabiting nematodes. *Nematologica* 18:391–403.

Konoplev, V. A., V. E. Sokolov, & A. L. Zotin. 1978. Criterion of orderliness and some problems of taxonomy. In *Thermodynamics of Biological Processes*, I. Lamprecht, & A. L. Zotin, eds., pp. 349–359. Berlin: deGruyter.

Kozlovsky, D. 1968. A critical evaluation of the trophic level concept. I. Ecological efficiencies. *Ecology* 49:48–60.

Krausman, P. R., & J. A. Bissonette. 1978. Bone chewing behavior of desert mule deer. *Southwest Naturalist* 16:55–63.

Krebs, C. J. 1978. *Ecology: The Experimental Analysis of Distribution and Abundance*, 2nd ed. New York: Harper & Row.

Kuhn, T. S. 1961. The function of measurement in modern physical science. *Isis* 52:169–190.

– 1970. *The Structure of Scientific Revolution*, 2nd ed. Chicago: University of Chicago Press.

– 1977. *The Essential Tension*. Chicago: University of Chicago Press.

Laird, A. K., S. A. Tyler, & A. D. Barton. 1965. Dynamics of normal growth. *Growth* 29:233–248.

Lal, D. 1977. The oceanic microcosm of particles. *Science* 198:997–1009.

Lal, D., & A. Lerman. 1975. Size spectra of biogenic particles in ocean water and sediments. *Journal of Geophysical Research* 80:423.

Lampert, W. 1975. A tracer study of the carbon turnover of *Daphnia pulex*. *Verhandlungen Vereinigung für Theoretische und Angewandte Limnologie* 19:2913–2921.

Larimer, J. L. & K. Schmidt–Nielsen. 1960. A comparison of blood carbonic anhydrase of various mammals. *Comparative Biochemistry and Physiology* 1:19–23.

Lasiewski, R. C., & W. A. Calder. 1971. Preliminary allometric analysis of respiratory variables in resting birds. *Respiration Physiology* 11:152–166.

Lasiewski, R. C., & W. R. Dawson. 1967. A reexamination of the relation between standard metabolic rate and body weight in birds. *Condor* 69: 13–23.

Lasiewski, R. C., M. W. Weathers, & M. N. Bernstein. 1967. Physiological responses of the giant hummingbird. *Patagona gigas*. *Comparative Biochemistry and Physiology* 23:797–813.

Laudien, H. 1973. Changing reaction systems. In *Temperature and Life*, H. Precht, J. Christophersen, H. Hensel, & W. Larcher, eds., pp. 355–399. New York: Springer-Verlag.

Lawler, G. H. 1965. The food of the pike, *Esox lucius*, in Hemming Lake, Manitoba. *Journal of the Fisheries Research Board of Canada* 22:1357–1377.

Laws, R. M., I. S. C. Parker, & R. C. B. Johnstone. 1975. *Elephants and Their Habitats*. Oxford: Oxford University Press (Clarendon Press).

Lawton, J. H. 1970. Feeding and food assimilation in larvae of the damselfly *Pyrrhosoma nymphula* (Sulz.) (Odonata: Zygoptera). *Journal of Animal Ecology* 39:669–689.

Laybourn, J., & B. J. Finlay. 1976. Respiratory energy losses related to cell weight and temperature in ciliated Protozoa. *Oecologia (Berlin)* 24:349–355.

Lean, D. R. S., M. N. Charlton, B. K. Burnison, T. P. Murphy, S. E. Millard, & K. R. Young. 1975. Phosphorus: Changes in ecosystem metabolism from reduced loading. *Verhandlungen Internationale Vereinigung für Theoretische und Angewandte Limnologie* 19: 249–257.

Lechner, A. J. 1978. The scaling of maximal oxygen consumption and pulmonary dimensions in small mammals. *Respiration Physiology* 34: 29–44.

Leggett, W. C. 1977. The ecology of fish migrations. *Annual Review of Ecology and Systematics* 8:285–308.

Leick, V. 1968. Ratios between DNA, RNA, and protein in different microorganisms as a function of maximal growth rate. *Nature* (London) 217: 1153–1155.

Leitch, I., F. E. Hytten, & W. Z. Billewicz. 1959. The maternal and neonatal weights of some Mammalia. *Proceedings of the Zoological Society of London* 133:11–28.

Leitner, P., & J. E. Nelson. 1967. Body temperature, oxygen consumption and heart rate in the Australian false vampire bat, *Macroderma gigas*. *Comparative Biochemistry and Physiology* 21: 65–74.

Leutenegger, W. 1979. Evolution of litter size in primates. *The American Naturalist* 114:525–531.

Leuthold, W. 1977. *African Ungulates. A Comparative Review of Their Ethology and Behavioural Ecology*. Berlin: Springer.

Lindeman, R. L. 1942. The trophic-dynamic aspect of ecology. *Ecology* 23: 399–418.

Lindsey, C. C. 1966. Body sizes of poikilotherm vertebrates at different latitudes. *Evolution* 20:456–465.

Lindstedt, S. L. & W. A. Calder. 1976. Body size and longevity in birds. *Condor* 78:91–94.

– 1981. Body size, physiological time and longevity of homeothermic animals. *The Quarterly Review of Biology* 56:1–16.

Linzell, J. L. 1972. Milk yield, energy loss in milk, and mammary gland weight in different species. *Dairy Science Abstracts* 34:351–360.

Lutz, P. L., I. S. Langmuir, & K. Schmidt–Nielsen. 1974. Oxygen affinity of bird blood. *Respiration Physiology* 20:325–330.

Lynch, M. 1980. The evolution of cladoceran life histories. *The Quarterly Review of Biology* 55:23–42.

MacMillen, R. E., & J. E. Nelson. 1969. Bioenergetics and body size in dasyurid marsupials. *American Journal of Physiology* 217:1246–1251.

McCauley, E., & F. Briand. 1979. Zooplankton grazing and phytoplankton species richness: Field tests of the predation hypothesis. *Limnology and Oceanography* 24:243–252.

McCauley, E., & J. Kalff. 1981. Empirical relationships between phytoplankton and zooplankton biomass in lakes. *Canadian Journal of Fisheries and Aquatic Science* 38:458–463.

McLaren, I. A. 1963. Effects of temperature on growth of zooplankton and the adaptive value of vertical migration. *Journal of the Fisheries Research Board of Canada* 20:685–727.

McMahon, T. A. 1973. Size and shape in biology. *Science* 179:1201–1204.

– 1977. Scaling quadrupedal galloping: frequencies, stresses and joint angles. In *Scale Effects in Animal Locomotion*, T. J. Pedley, ed., pp. 143–151. New York: Academic Press.

– 1980. Scaling physiological time. *Lectures on Mathematics in the Life Sciences* 13:131–133.

McNab, B. K. 1963. Bioenergetics and the determination of home range size. *The American Naturalist* 97:133–140.

– 1966a. The metabolism of fossorial rodents: A study in convergence. *Ecology* 47:712–733.

– 1966b. An analysis of the body temperatures of birds. *Condor* 68: 47–55.

– 1969. The economics of temperature regulation in neotropical bats. *Comparative Biochemistry and Physiology* 31:227–268.

– 1970. Body weight and the energetics of temperature regulation. *Journal of Experimental Biology* 53:329–348.

– 1974. The energetics of endotherms. *The Ohio Journal of Science* 74:370–380.

– 1978a. Energetics of arboreal folivores: Physiological problems and ecological consequences of feeding on an ubiquitous food supply. In *The Ecology of Arboreal Folivores*, G. G. Montgomery, ed., pp. 153–162. Washington D.C.: Smithsonian Institute Press.

– 1978b. The evolution of endothermy in the phylogeny of mammals. *The American Naturalist* 112:1–21.

– 1979. The influence of body size on the energetics and distribution of fossorial and burrowing mammals. *Ecology* 60:1010–1021.

– 1980. Food habits, energetics, and the population biology of mammals. *The American Naturalist* 116:106–124.

McNab, B. K., & W. Auffenberg. 1976. The effects of large body size of the temperature regulation of the Komodo dragon *Varanus komodoensis*. *Comparative Biochemistry and Physiology A* 55:345–350.

McNiel, S., & J. H. Lawton. 1970. Annual production and respiration in animal populations. *Nature* (London) 225:472–474.

Magid, E. 1967. Activity of carbonic anhydrase in mammalian blood in relation to body size. *Comparative Biochemistry and Physiology* 21:357–360.

Maldonado, R., H. San José, C. Martinoya, & B. Günther. 1974. Cell size and body weight in some homeotherms and poikilotherms. *Acta Physiologica Latino Americana* 24:328–335.

Malone, T. C. 1980. Algal size. In *The Physiological Ecology of Phytoplankton*, I. Morris, ed., pp 433–463. London: Blackwell Publisher.

Margalef, R. 1963. Successions of populations. *Advancing Frontiers of Plant Science* 2:137–188.

Martin, R. D. 1981. Relative brain size and basal metabolic rate in terrestrial vertebrates. *Nature* (London) 293:57–60.

Mathieu, O., R. Krauer, H. Hoppeler, P. Gehr, S. L. Lindstedt, R. M. Alexander, C. R. Taylor, & E. R. Weibel. 1981. Design of the mammalian respiratory system. VII. Scaling mitochondrial volume in skeletal muscle to body mass. *Respiration Physiology* 44:113–128.

Mead, T. 1960. Control of respiratory frequency. *Journal of Applied Physiology* 15:325–336.

Mertens, J. A. L. 1969. The influence of brood size on the energy metabolism and water loss of nestling great tits, *Parus major major*. *Ibis* 111:11–16.

Millar, J. S. 1977. Adaptive features of mammalian reproduction. *Evolution* 31:370–386.

– 1981. Post partum reproductive characteristics of eutherian mammals. *Evolution* 35:1149–1163.

Miller, D. S., & P. R. Payne. 1964. Dietary factors influencing nitrogen balance. *Proceedings of the Nutrition Society* 23:11–19.

Milton, K., & M. L. May. 1976. Body weight, diet and home range area in primates. *Nature* (London) 259:459–462.

Mitchell, H. H. 1962. *Comparative Nutrition of Man and Domestic Animals*, Vol. I. New York: Academic Press.

Moen, A. N. 1973. *Wildlife Ecology: An Analytical Approach*. San Francisco: Freeman.

Mohr, C. O. 1940. Comparative populations of game, fur and other mammals. *The American Midland Naturalist* 24:581–584.

– 1947. Table of equivalent populations of North American small mammals. *The American Midland Naturalist* 37:223–249.

Morgan, F. 1964. The uptake of radioactivity of fish and shell fish. I. [134]Caesium by whole animals. *Journal of the Marine Biological Association of the United Kingdom* 44:259–271.

Morgan, N. C. 1980. Secondary production. In *The Functioning of Freshwater*

Ecosystems, E. D. LeCren & R. H. Lowe-McConnell, eds., pp. 247–251. Cambridge: Cambridge University Press.

Morgan, N. C., T. Backiel, G. Bretschko, A. Duncan, A. Hillbricht–Ilkowska, Z. Kajak, J. F. Kitchell, P. Larsson, C. Levêque, A. Nauwerck, F. Schiemer, & J. E. Thorpe. 1980. Secondary production. In *The Functioning of Freshwater Ecosystems,* E. D. LeCren & R. H. Lowe–McConnell, eds., pp. 247–340. Cambridge: Cambridge University Press.

Morrison, P. R. 1960. Some interrelations between weight and hibernation function. *Bulletin of the Museum of Comparative Zoology, Harvard University* 124:75–91.

Morrison, P. R., & F. A. Ryser. 1951. Temperature and metabolism in some Wisconsin mammals. *Federation Proceedings* 10:93–94.

– 1952. Weight and body temperature in mammals. *Science* 116:231–232.

Morse, D. H. 1974. Niche breadth as a function of social dominance. *The American Naturalist* 108:818–830.

Morton, S. R., D. S. Hinds, & R. E. MacMillen. 1980. Cheek pouch capacity in heteromyid rodents. *Oecologia* (Berlin) 46:143–146.

Muir, D. S. 1969. Gill dimensions as a function of fish size. *Journal of the Fisheries Research Board of Canada* 26:165–170.

Munro, H. N. 1969. Evolution of protein metabolism in mammals. In *Mammalian Protein Metabolism,* Vol. III, H. N. Munro, ed., pp. 133–182. New York: Academic Press.

Nagy, K. A., & V. H. Shoemaker. 1975. Energy and nitrogen budgets of the free-living desert lizard *Sauromalus obesus. Physiological Zoology* 48:252–262.

Nakashima, B. S., & W. C. Leggett. 1978. Daily ration of yellow perch (*Perca flavescens*) from Lake Memphremagog, Québec-Vermont, with a comparison of methods for in situ determination. *Journal of the Fisheries Research Board of Canada* 35:1597–1603.

– 1980. The role of fishes in the regulation of phosphorus availability in lakes. *Canadian Journal of Fisheries and Aquatic Science* 37:1540–1549.

Nauwerck, A. 1963. Die Beziehungen zwischen Zooplankton und Phytoplankton im See Erken. *Symbolae Botanicae Upsalensis* 17:1–163.

Newell, N. D. 1949. Phyletic size increase, an important trend illustrated by fossil invertebrates. *Evolution* 3:103–124.

Nice, M. M. 1938. The biological significance of bird weights. *Bird Banding* 9:1–11.

Nicholas, W. I. 1975. *The Biology of Free-Living Nematodes.* Oxford: Oxford University Press (Clarendon Press).

Nival, P., & S. Nival. 1976. Particle retention efficiences of an herbivorous copepod *Arcartia clausi* (adult and copepodite stages): Effects on grazing. *Limnology and Oceanography* 21:24–38.

Norstrom, R. J., A. E. McKinnon, & A. S. W. DeFreitas. 1976. A bioenergetics based model for pollutant accumulation by fish. Simulation of PCB and methylmercury residue levels in Ottawa River yellow perch (*Perca flavescens*). *Journal of the Fisheries Research Board of Canada* 33:248–267.

Odum, E. P. 1969. The strategy of ecosystem development. *Science* 164:262–270.

Owens, D., & M. Owens. 1980. Hyenas of the Kalahari. *Natural History* 89 (February):44–53.

Pace, N. D., D. F. Rahlman, & A. H. Smith. 1979. Scale effects in the musculoskeletal system, viscera and skin of small terrestrial mammals. *The Physiologist* 22:551–552.

Paladino, F. V., & J. R. King. 1979. Energy cost of terrestrial locomotion: Biped and quadruped runners compared. *Revue Canadienne de Biologie* 38:321–323.

Parra, P. 1978. Comparison of foregut and hindgut fermentation in herbivores. In *The Ecology of Arboreal Folivores*, G. Montgomery, ed., pp. 205–229. Washington, D.C.: Smithsonian Institute Press.

Parsons, T. R. 1980. Zooplanktonic production. In *Fundamentals of Aquatic Ecosystems*, R. S. K. Barnes & K. H. Mann, eds., pp. 46–66. Oxford: Blackwell Scientific Publications.

Parsons, T. R., & M. Takahashi. 1973. Environmental control of phytoplankton cell size. *Limnology and Oceanography* 18:511–515.

Pasquis, P., A. LaCaisse, & P. Dejours. 1970. Maximal oxygen uptake in four species of small mammals. *Respiration Physiology* 9:298–309.

Passingham, R. E. 1975a. The brain and intelligence. *Brain, Behavior and Evolution* 11:1–15.

– 1975b. Changes in the size and organization of the brain in man and his ancestors. *Brain, Behavior and Evolution* 11:73–90.

Pauly, D. 1979. On the interrelationships between natural mortality, growth parameters and mean environmental temperature in 175 fish stocks. *Journal de la Conseil International pour l' Exploration de la Mer* 39:175–192.

Pavoni, M. 1963. Die Bedeutung des Nannoplanktons im Vergleich zum Netzplankton. *Schweizerische Zeitschrift für Hydrologie* 25:220–231.

Payne, P. R., & E. F. Wheeler. 1967a. Growth of the foetus. *Nature* (London) 215:849–850.

– 1967b. Comparative nutrition in pregnancy. *Nature* (London) 215:1134–1136.

– 1968. Comparative nutrition in pregnancy and lactation. *Proceedings of the Nutrition Society* 27:129–138.

Payne, R. T. 1980. Food webs: Linkage, interaction strength and community infrastructure. *Journal of Animal Ecology* 49:667–686.

Pearson, T. H. 1968. The feeding biology of sea-bird species breeding on the Farne Islands. Northumberland. *Journal of Animal Ecology* 37:521–552.

Pedley, T. J. 1977. *Scale Effects in Animal Locomotion*. New York: Academic Press.

Pennycuick, C. J. 1972. *Animal Flight*. London: Arnold.

– 1979. Energy costs of locomotion and the concept of "foraging radius." In *Serengeti: Dynamics of an Ecosystem*, A. R. E. Sinclair & M. Norton-Griffiths, eds., pp. 164–184. Chicago: University Chicago Press.

Perry, M. J., & R. W. Eppley. 1981. Phosphate uptake by phytoplankton in the central north Pacific Ocean. *Deep Sea Research A* 28:39–49.

Peters, R. H. 1971. Ecology and the world view. *Limnology and Oceanography* 16:143–148.

– 1972. *Phosphorus regeneration by zooplankton*, Ph.D. Thesis, University of Toronto.

– 1975a. Phosphorus regeneration by natural populations of limnetic zooplankton. *Verhandlungen Internationale Vereinigung für Theoretische und Angewandte Limnologie* 19:273–279.

– 1975b. Orthophosphate turnover in central European lakes. *Memorie del Istituto Italiano di Idrobiologia Dottore Marco de Marchi* 32:297–311.

– 1976. Tautology in evolution and ecology. *The American Naturalist* 110:1–12.

– 1977. Unpredictable problems with trophodynamics. *Environmental Biology of Fishes* 2:97–102.

– 1978a. Predictable problems with tautology in evolution and ecology. *The American Naturalist* 112:759–762.

– 1978b. Empirical physiological models of ecosystem processes. *Verhandlungen Internationale Vereinigung für Theoretische und Angewandte Limnologie* 20:110–118.

– 1980a. From natural history to ecology. *Perspectives in Biology and Medicine* 23:191–203.

– 1980b. Useful concepts for predictive ecology. *Synthese* 43:257–269.

Peters, R. H., & D. R. S. Lean. 1973. The characterization of soluble phosphorus released by limnetic zooplankton. *Limnology and Oceanography* 18:270–279.

Peters, R. H., & F. H. Rigler. 1973. Phosphorus release by *Daphnia*. *Limnology and Oceanography* 18:821–839.

Petersen, B. 1950. The relation between size of mother and number of eggs and young in some spiders and its significance for the evolution of size. *Experimentia* 6:96–98.

Pianka, E. R. 1970. On r- and K-selection. *The American Naturalist* 104:592–597.

Pitts, G. C., & T. R. Bullard. 1968. Some interspecific aspects of body composition in mammals. In *Body Composition in Animals and Man*, pp. 45–79. Washington, D.C.: National Academy of Sciences.

Platt, T., & K. Denman. 1977. Organization in the pelagic ecosystem. *Helgoländer Wissenschaftliche Meeresuntersuchungen* 30:575–581.

– 1978. The structure of pelagic marine ecosystems. *Rapports et Procès-Verbaux des Réunions du Conseil International pour l' Exploration de la Mer* 173:60–65.

– 1980. Patchiness in phytoplankton distribution. In *The Physiological Ecology of Phytoplankton*, I. Morris, ed., pp. 413–431. London: Blackwell Publisher.

Platt, T., & W. Silvert. 1981. Ecology, physiology, allometry and dimensionality. *Journal of Theoretical Biology* 93:855–860.

Pond, C. M. 1977. The significance of lactation in the evolution of mammals. *Evolution* 31:177–199.

– 1978. Morphological aspects and the ecological and mechanical consequences of fat deposition in wild vertebrates. *Annual Review of Ecology and Systematics* 9: 519–570.

Popper, K. R. 1962. *Conjectures and Refutations: The Growth of Scientific Knowledge*. New York: Harper.

– 1972. *Objective Knowledge: An Evolutionary Approach*. London: Oxford University Press.

Porter, K. G. 1977. The plant animal interface in fresh water ecosystems. *American Scientist* 65: 159–170.

Pough, F. H. 1977a. The relationship of blood oxygen affinity to body size in lizards. *Comparative Biochemistry and Physiology A* 57: 435–442.

– 1977b. The relationship between body size and blood oxygen affinity in snakes. *Physiological Zoology* 50: 77–87.

– 1980. The advantages of ectothermy for tetrapods. *The American Naturalist* 115: 92–112.

Powell, J. A. 1978. Evidence of carnivory in manatees (*Trichechus manatus*). *Journal of Mammalogy* 59: 442.

Prange, H. D., J. F. Anderson, & H. Rahn. 1979. Scaling of skeletal mass to body mass in birds and mammals. *The American Naturalist* 113: 103–122.

Precht, H., J. Cristophersen, H. Hensel, & W. Larcher, eds. 1973. *Temperature and Life*. New York: Springer-Verlag.

Prinzinger, R., & I. Hännsler. 1980. Metabolism–weight relationship in some small nonpasserine birds. *Experientia* 36: 1299–1300.

Prosser, C. L. 1973. *Comparative Animal Physiology*. Philadelphia: Saunders.

Prothero, J. W. 1979. Maximal oxygen consumption in various animals and plants. *Comparative Biochemistry and Physiology A* 64: 463–466.

– 1980. Scaling of blood parameters in mammals. *Comparative Biochemistry and Physiology A* 67: 649–657.

Pyke, G. H. 1980. Optimal foraging in bumble bees: Calculation of net rate of energy intake and optimal path choice. *Theoretical Population Biology* 17: 232–246.

Quiring, D. P. 1941. The scale of being according to the power formula. *Growth* 5: 301–327.

Radinsky, L. 1978. Evolution of brain size in carnivores and ungulates. *The American Naturalist* 112: 815–831.

Rahn, H., & A. Ar. 1974. The avian egg: Incubation time and water loss. *Condor* 76: 147–152.

Rahn, H., C. V. Paganelli, & A. Ar. 1974. The avian egg: Air-cell tension, metabolism and incubation time. *Respiration Physiology* 22: 297–309.

Regier, H. A. 1973. Sequences of exploitation of stocks in multispecies fisheries in the Laurentian Great Lakes. *Journal of the Fisheries Research Board of Canada* 30: 1992–1999.

Regier, H. A., & K. H. Loftus. 1972. Effects of fisheries exploitation on salmonid communities in oligotrophic lakes. *Journal of the Fisheries Research Board of Canada* 29: 959–968.

Reichle, D. E. 1968. Relation of body size to food intake, oxygen consumption, and trace element metabolism in forest floor arthropods. *Ecology* 49: 538–542.

Reichman, O. J., & S. Aitchison. 1981. Mammal trails on mountain slopes. Optimal paths in relation to slope and body weight. *The American Naturalist* 117:416–420.

Remmert, H. 1980. *Ecology. A Textbook.* Berlin: Springer.

Rensch, B. 1956. Increase of learning capability with increase of brain size. *The American Naturalist* 90:81–95.

– 1960. *Evolution above the Species Level.* New York: Columbia University Press.,

Reynolds, W. W., & W. J. Karlotski. 1977. The allometric relationship of skeleton weight to body weight in teleost fishes: A preliminary comparison with birds and mammals. *Copeia* 1977:160–163.

Richmond, C. R., J. E. Furchner, G. A. Trafton, & W. H. Langham. 1962. Comparative metabolism of radionuclides in mammals. I. Uptake and retention of orally administered Zn^{65} by four mammalian species. *Health Physics* 8:481–489.

Ricker, W. E. 1973. Linear regressions in fishery research. *Journal of the Fisheries Research Board of Canada* 30:409–434.

Ricklefs, R. E. 1967. A graphical method of fitting equations to growth curves. *Ecology* 48:978–983.

– 1968. Patterns of growth in birds. *Ibis* 110:419–431.

– 1969. Preliminary models for growth rates in altricial birds. *Ecology* 50:1031–1039.

– 1973. Patterns of growth in birds. II. Growth rate and mode of development. *Ibis* 115:177–201.

– 1974. Energetics of reproduction in birds. In *Avian Energetics*, R. A. Paynter, ed., pp. 152–297. Cambridge, Mass.: Nuttall Ornithological Club.

– 1979. Adaptation, constraint and compromise in avian post natal development. *Biological Reviews* 54:269–290.

Riddell, W. I., & K. G. Corl. 1977. Comparative investigation of the relationship between cerebral indices and learning abilities. *Brain, Behavior and Evolution* 14:385–398.

Riggs, A. 1960. The nature and significance of the Bohr effect in mammalian hemoglobins. *Journal of General Physiology* 43:737–752.

Rigler, F. H. 1956. A tracer study of the phosphorus cycle in lake water. *Ecology* 37:550–562.

– 1973. A dynamic view of the phosphorus cycle in lakes. In *Environmental Phosphorus Handbook*, E. Griffith, A. Beeton, J. Spencer, & D. Mitchell, eds., New York: Wiley.

– 1975. The concept of energy flow and nutrient flow between trophic levels. In *Unifying Concepts in Ecology*, W. H. Van Dobben, & R. H. Lowe–McConnell, eds., pp. 15–26. The Hague: Junk.

Robinson, W. R., R. H. Peters & J. Zimmerman. 1983. The effects of body size and temperature on metabolic rate of organisms. *Canadian Journal of Zoology* 61:281–288.

Robbins, C. T., & B. L. Robbins. 1979. Fetal and neonatal growth patterns and maternal reproductive effort in ungulates and subungulates. *The American Naturalist* 114:101–116.

Rodbard, S. 1950. Weight and body temperature. *Science* 111:465–466.

Rogers, L. E., W. T. Hinds, & R. L. Buschborn. 1976. A general weight vs length relationship for insects. *Annals of the Entomological Society of America* 69:387–389.

Rosenzweig, M. L. 1966. Community structure in sympatric Carnivora. *Journal of Mammalogy* 47:602–612.

Ross, H. A. 1979. Multiple clutches and shorebird egg and body weight. *The American Naturalist* 111:917–938.

Russell, B. 1963. *Mysticism and Logic and Other Essays.* London: Unwin Books.

Russell, E. M. 1982. Parental investment and desertion of young in marsupials. *The American Naturalist* 119:744–743.

Sacher, G. A., 1959. Relation of lifespan to brain weight and body weight in mammals. In *The Lifespan of Animals*, G. E. Wolstenholme & M. O' Connors eds., pp. 115–141. CIBA Foundation Colloquia on Ageing 5, London: Churchill.

Sacher, G. A., & E. F. Staffeldt. 1974. Relation of gestation time to brain weight for placental mammals: Implications for the theory of vertebrate growth. *The American Naturalist* 108:593–615.

Sadleir, R. M. F. S. 1969. *The Ecology of Reproduction in Wild and Domestic Mammals.* London: Methuen & Co.

– 1980. Milk yield of black tailed deer. *Journal of Wildlife Management* 44:472–478.

Schaffer, W. M., & P. F. Elson. 1975. The adaptive significance of variations in life history among local populations of Atlantic salmon in North America. *Ecology* 56:577–590.

Schaller, G. B. 1972. *The Serengeti Lion.* Chicago: University of Chicago Press.

Schlesinger, D. A., L. A. Molot, & B. J. Shuter. 1981. Specific growth rates of freshwater algae in relation to cell size and light intensity. *Canadian Journal of Fisheries and Aquatic Sciences* 38: 1052–1058.

Schmidt-Nielsen, K. 1964. *Desert Animals: Physiological Problems of Heat and Water.* Oxford: Oxford University Press.

– 1972. Locomotion: Energy cost of swimming, flying and running. *Science* 177:222–228.

– 1979. *Animal Physiology: Adaptation and Environment*, 2nd ed. Cambridge: Cambridge University Press.

Schmidt-Nielsen, K., & P. Pennycuick. 1961. Capillary density in mammals in relation in body size and oxygen consumption. *American Journal of Physiology* 200:746–750.

Schoener, T. W. 1968. Sizes of feeding territories among birds. *Ecology* 49:123–141.

Schoener, T. W., & D. H. Janzen. 1968. Notes on environmental determinants of tropical versus temperate insect size patterns. *The American Naturalist* 102:207–224.

Schroeder, L. A. 1981. Consumer growth efficiencies: Their limits and relationships to ecological efficiencies. *Journal of Theoretical Biology* 93:805–828.

Schwinghamer, P. 1981. Characteristic size distributions of integral benthic communities. *Canadian Journal of Fisheries and Aquatic Sciences* 38:1255–1263.

Scriber, J. M., & P. Feeny. 1979. Growth of herbivorous caterpillars in relation to feeding specialization and to growth form of their food plants. *Ecology* 60:829–850.

Semina, H. J. 1972. The size of phytoplankton cells in the Pacific Ocean. *Internationale Revue der Gesamten Hydrobiologie* 57:177–205.

Seymour, R. S. 1979. Dinosaur eggs: Gas conductance through the shell, water loss during incubation and clutch size. *Paleobiology* 5:1–11.

Seymour, R. S., & R. A. Ackerman. 1980. Adaptations to underground nesting in birds and reptiles. *American Zoologist* 20:437–447.

Sheldon, R. W., & S. R. Kerr. 1972. The population density of monsters in Loch Ness. *Limnology and Oceanography* 17:796–798.

Sheldon, R. W., & T. R. Parsons. 1967. A continuous size spectrum for particulate matter in the sea. *Journal of the Fisheries Research Board of Canada* 24:909–915.

Sheldon, R. W., A. Prakash, & W. H. Sutcliffe, Jr. 1972. The size distribution of particles in the ocean. *Limnology and Oceanography* 17:327–340.

Sheldon, R. W., W. H. Sutcliffe Jr., & M. A. Paranjape. 1977. Structure of pelagic food chain and relationship between plankton and fish production. *Journal of the Fisheries Research Board of Canada* 34:2344–2353.

Shuter, B. J. 1978. Size dependence of phosphorus and nitrogen subsistence quotas in unicellular microorganisms. *Limnology and Oceanography* 23:1248–1255.

Silvert, W., & T. Platt. 1978. Energy flux in the pelagic ecosystem: A time-dependent equation. *Limnology and Oceanography* 23:813–816.

– 1980. Dynamic energy flow model of the particle size distribution in pelagic ecosystems. In *Evolution and Ecology of Zooplankton Communities*, W. C. Kerfoot, ed., pp. 754–763. Hanover, N. H.: University Press of New England.

Simpson, G. G. 1953. *The Major Features of Evolution*. New York: Simon & Schuster.

Sinclair, A. R. E. 1972. Long term monitoring of mammal populations in the Serengeti: Census of non migratory ungulates 1971–1972. *East African Wildlife Journal* 10:287–297.

– 1975. The resource limitation of trophic levels in tropical grassland ecosystems. *Journal of Animal Ecology* 44:497–520.

Smayda, T. J. 1980. Phytoplankton species succession. In *The Physiological Ecology of Phytoplankton*, I. Morris, ed., pp. 493–570. London: Blackwell Publisher.

Smith, A. H. & M. Kleiber. 1950. Size and oxygen consumption in fertilized eggs. *Journal of Cellular and Comparative Physiology* 35:131–140.

Smith, F. E. 1954. Quantitative aspects of population growth. In *Dynamics of Growth Processes*, E. Boell, ed., Princeton, N. J.: Princeton University Press.

Smith, R. 1981. *Phosphorus Limitation and Competition in the Phytoplankton*, Ph.D. Thesis, McGill University.

Smith, R. J. 1980. Rethinking allometry. *Journal of Theoretical Biology* 87:87–111.

Smock, L. A. 1980. Relationships between body size and biomass of aquatic insects. *Freshwater Biology* 10:375–383.

Snedecor, G. W., & W. G. Cochran. 1967. *Statistical Methods*, 6th ed. Ames: Iowa State University Press.

Spigarelli, S. A., M. M. Thommes, & T. L. Beitinger. 1977. The influence of body weight on heating and cooling of selected Lake Michigan fishes. *Comparative Biochemistry and Physiology A* 56:51–57.

Spotila, J. R., P. W. Lommen, G. S. Bakken, & D .M. Gates. 1973. A mathematical model for body temperatures of large reptiles: Implications for dinosaur ecology. *The American Naturalist* 107:391–404.

Spray, D. C., & M. L. May. 1977. Heating and cooling rates in four species of turtles. *Comparative Biochemistry and Physiology A* 41: 507–522.

Stahl, W. R. 1962. Similarity and dimensional methods in biology. *Science* 137:205–212.

– 1965. Organ weights in primates and other mammals. *Science* 150:1039–1042.

– 1967. Scaling of respiratory variables in mammals. *Journal of Applied Physiology* 22:453–460.

Stanley, S. M. 1973. An explanation for Cope's rule. *Evolution* 27:1–26.

– 1979. *Macroevolution: Pattern and Process*. San Francisco: Freeman.

Stevens, E. D., & F. E. J. Fry. 1970. The rate of thermal exchange in a teleost, *Tilapia mossambica*. *Canadian Journal of Zoology* 48:221–226.

– 1974. Heat transfer and body temperatures in non thermoregulatory teleosts. *Canadian Journal of Zoology* 52:1137–1143.

Stevens, S. S. 1946. On the theory of scales and measurement. *Science* 103: 677–680.

Straškraba, M. 1980. The effects of physical variables on freshwater production: Analyses based on models. In *The Functioning of Freshwater Ecosystems*, E. D. LeCren, & R. H. Lowe-McConnell, eds., pp. 13–84. Cambridge: Cambridge University Press.

Studiar, E. H. 1970. Evaporative water loss in bats. *Comparative Biochemistry and Physiology* 35:935–943.

Taylor, C. R. 1973. Energy cost of animal locomotion. In *Comparative Physiology*, L. Bolis, K. Schmidt–Nielsen, & S. H. P. Maddrell, eds., pp. 23–42. Amsterdam: North Holland.

– 1977. The energetics of terrestrial locomotion and body size in vertebrates. In *Scale Effects in Animal Locomotion*, T. J. Pedley, ed., pp. 127–141. New York: Academic Press.

Taylor, C. R., N. C. Heglund, T. A. McMahon, & T. R. Looney. 1980. Energetic cost of generating muscular force during running: A comparison of large and small animals. *Journal of Experimental Biology*, 86:9–18.

Taylor, C. R., G. M. O. Maloiy, E. R. Weibel, V. A. Langmon, V. M. Z. Kamau, H. J. Seeherman, & N. C. Heglund. 1981. Design of the mammalian respiratory system. III. Scaling maximum aerobic capacity to body mass: Wild and domestic mammals. *Respiration Physiology* 44:25–37.

Taylor, C. R., K. Schmidt–Nielsen, & J. L. Raab. 1970. Scaling of energetic cost of running to body size in mammals. *American Journal of Physiology* 219:1104–1107.

Taylor, St. C. S. 1965. A relation between mature weight and time taken to mature in mammals. *Animal Production* 7:203–220.

– 1968. Time taken to mature in relation to mature weight for sexes, strains and species of domesticated mammals and birds. *Animal Production* 10:157–169.

Taylor, W. D., & B. J. Shuter. 1981. Body size, genome size, and intrinsic rate of increase in ciliated Protozoa. *The American Naturalist* 118:160–172.

Tenney, S. M., & J. E. Remmers. 1963. Comparative quantitative morphology of the mammalian lung: Diffusing area. *Nature* (London) 197:54–56.

Tenney, S. M., & J. B. Tenney. 1970. Quantitative morphology of cold-blooded lungs: Amphibia and Reptilia. *Respiration Physiology* 9:197–215.

Thiel, H. 1975. The size structure of the deep-sea benthos. *Internationale Revue der Gesamten Hydrobiologie* 60:575–606.

Thomann, R. V. 1978. *Size Dependent Model of Hazardous Substances in Aquatic Food Chain*, US EPA-600/3-78-036, Ecological Research Series 47. Washington, D.C.: Government Printing Office.

– 1981. Equilibrium model of fate of microcontaminants in diverse aquatic food chains. *Canadian Journal of Fisheries and Aquatic Sciences* 38:280–296.

Thompson, D. W. 1961. *On Growth and Form*, J. T. Bonner, ed., Cambridge: Cambridge University Press.

Thorpe, J. E., & K. A. Mitchell. 1981. Stocks of Atlantic salmon (*Salmo salar*) in Britain and Ireland: Discreteness and current management. *Canadian Journal of Fisheries and Aquatic Sciences* 38:1576–1590.

Threlkeld, S. T. 1976. Starvation and the size structure of zooplankton communities. *Freshwater Biology* 6:489–496.

Tracy, C. R. 1977. Minimum size of mammalian homeotherms: Role of the thermal environment. *Science* 198:1034–1035.

Trewartha, G. T. 1969. *A Geography of Population: World Patterns*. New York: Wiley.

Trump, C. L., & W. C. Leggett. 1980. Optimum swimming speeds in fish: The problem of currents. *Canadian Journal of Fisheries and Aquatic Sciences* 37:1086–1092.

Tsutakawa, R. W., & J. E. Hewett. 1977. Quick test for comparing two populations with bivariate data. *Biometrics* 33:215–219.

Tucker, V. A. 1970. Energetic cost of locomotion in animals. *Comparative Biochemistry and Physiology* 34:841–846.

– 1973a. Aerial and terrestrial locomotion: A comparison of energetics. In *Comparative Physiology: Locomotion, Respiration, Transport and Blood*, L., Bolis, K. Schmidt–Nielsen, & S. H. P. Maddrell, eds., pp. 63–76. Amsterdam: North Holland.

– 1973b. Bird metabolism during flight: Evaluation of a theory. *Journal of Experimental Biology* 58:689–709.

– 1975. The energetic cost of moving about. *American Scientist* 63:413–419.

Tuomi, J. 1980. Mammalian reproductive strategies: A generalized relation of litter size to body size. *Oecologia* (Berlin) 45:39–44.

Turćek, F. J. 1966. On plumage quantity in birds. *Ekologia Polska A* 14:611–614.

Turner, F. B., R. I. Jennrich, & J. D. Weintraub. 1969. Home range and body size of lizards. *Ecology* 50:1076–1081.

Turpin, D. H., & P. J. Harrison. 1980. Cell size manipulation in natural marine, planktonic, diatom communities. *Canadian Journal of Fisheries and Aquatic Sciences* 37:1193–1195.

Umminger, B. L. 1975. Body size and whole blood sugar concentrations in mammals. *Comparative Biochemistry and Physiology A* 52:455–458.

– 1977. Relation of whole blood sugar concentrations in vertebrates to standard metabolic rate. *Comparative Biochemistry and Physiology A* 56:457–460.

Vácha, J., & V. Znojil. 1981. The allometric dependence of the life span of erythrocytes on body weight in mammals. *Comparative Biochemistry and Physiology A* 69:357–362.

Van Valen, L. 1973. Body size and numbers of plants and animals. *Evolution* 27:27–35.

Vinogradov, A. P. 1953. *The Elementary Chemical Composition of Marine Organisms*, Memoir No. 2, pp. 1–647. New Haven, Conn.: Sears Foundation.

Vitt, L. J., & J. D. Congdon. 1978. Body shape, reproductive effort and relative clutch mass in lizards: Resolution of a paradox. *The American Naturalist* 112:595–608.

Vleck, C. M., D. F. Hoyt, & D. Vleck. 1979. Metabolism of avian embryos: Patterns in altricial and precocial birds. *Physiological Zoology* 52:363–377.

Vleck, C. M., D. Vleck, & D. F. Hoyt. 1980. Patterns of metabolism and growth in avian embryos. *American Zoologist* 20:405–416.

Vollenweider, R. A. 1975. Input-output models with special reference to the phosphorus loading concept in limnology. *Schweizerische Zeitschrift fur Hydrologie* 37:53–84.

Walmo, O. C. 1981. *Mule and Blacktailed Deer of North America*. Lincoln: University of Nebraska Press.

Walter, A., & M. R. Hughes. 1978. Total body water volume and turnover rate in the fresh and seawater adapted glaucous winged gulls. *Comparative Biochemistry and Physiology A* 61:233–237.

Ware, D. M. 1978. Bioenergetics of pelagic fish: Theoretical change in swimming speed and ration with weight. *Journal of the Fisheries Research Board of Canada* 35:220–228.

– 1980. Bioenergetics of stock and recruitment. *Canadian Journal of Fisheries and Aquatic Sciences* 37:1012–1024.

Wasserman, S. S., & C. Mitter. 1978. The relationship of body size to breadth of diet in some Lepidoptera. *Ecological Entomology* 3:155–160.

Waters, T. F. 1977. Secondary production in inland waters. *Advances in Ecological Research* 10:91–165.

Watson, S., & J. Kalff. 1981. Relationships between nannoplankton and lake trophic status. *Canadian Journal of Fisheries and Aquatic Sciences* 38:960–967.

Weathers, W. W. 1979. Climatic adaptation in avian standard metabolic rate. *Oecologia* (Berlin) 42:81–89.

– 1981. Physiological thermoregulation in heat stressed birds: Consequences of body size. *Physiological Zoology* 54:345–361.

Weathers, W. W., & D. F. Calcomise. 1978. Seasonal acclimation to temperature in monk parakeets. *Myiopsitta monachus. Oecologia* (Berlin) 35:173–183.

Webb, P. W. 1975. Hydrodynamics and energetics of fish propulsion. *Fisheries Research Board of Canada* 190:1–158.

Webster, K. E., & R. H. Peters. 1978. Some size dependent inhibitions of larger cladoceran filters in filamentous suspensions. *Limnology and Oceanography* 23:1138–1245.

Weibel, E. R. 1973. Morphological basis of alveolar-capillary gas exchange. *Physiological Reviews* 53:419–495.

Western, D. 1979. Size, life history and ecology in mammals. *African Journal of Ecology* 17:185–204.

Weymouth, F. W., J. M. Crisman, V. E. Hall, H. S. Belding, & J. Field. 1944. Total and tissue respiration in relation to body weight. A comparison of the kelp crab with other crustaceans and with mammals. *Physiological Zoology* 17:50–71.

White, L., H. Haines, & T. Adams. 1968. Cardiac output related to body weight in small mammal. *Comparative Biochemistry and Physiology* 27:559–565.

Whitford, W. G., & V. H. Hutchison. 1967. Body size and metabolic rate in salamanders. *Physiological Zoology* 40:127–133.

Whittow, G. C. 1971. Ungulates. In *Comparative Physiology of Thermoregulation*, Vol. II *Mammals.*, G. C. Whittow, ed., pp. 192–281. New York: Academic Press.

Whittow, G. C., C. A. Scammel, M. Leong, & D. Rand. 1977. Temperature regulation in the smallest ungulate, the lesser mouse deer (*Tragulus javanicus*). *Comparative Biochemistry and Physiology A* 56:23–26.

Wigglesworth, V. B. 1974. *The Principles of Insect Physiology*, 7th ed. London: Chapman & Hall.

Wilkie, D. R. 1977. Metabolism and body size. In *Scale Effects in Animal Locomotion*, T. J. Pedley, ed., pp. 23–36. New York: Academic Press.

Williams, R. B. 1964. Division rates of salt marsh diatoms in relation to salinity and cell size. *Ecology* 45:877–880.

Wilson, D. S. 1975. The adequacy of body size as a niche difference. *The American Naturalist* 109:769–784.

– 1976. Deducing the energy available in the environment: An application of optimal foraging theory. *Biotropica* 8:96–103.

Wilson, K. J. 1974. The relationship of maximum and resting oxygen consumption and heart rates to weight in reptiles of the order Squamata. *Copeia* 74:781–784.

Winberg, G. G. 1960. Rate of metabolism and food requirement of fishes. *Fisheries Research Board Translation Services*, No. 194.

Winters, G. H. 1978. Production, mortality, and sustainable yield of North west Atlantic harp seals (*Pagophilus groenlandicus*). *Journal of the Fisheries Research Board of Canada* 35:1249–1261.

Wood, S. C., & C. J. M. Lenfant. 1976. Respiration: Mechanics, control and gas exchange. In *Biology of the Reptilia*, C. Gans & W. R. Dawson, eds., pp. 225–274. New York: Academic Press.

Wooton, R. J. 1979. Energy costs of egg production and environmental determinants of fecundity in teleost fishes. *Symposia of the Zoological Society of London* 44:133–159.

Wunder, B. A. 1975. A model for estimating metabolic rate of active or resting mammals. *Journal of Theoretical Biology* 49:345–354.

Zankai, N. P., & J. E. Ponyi. 1976. Seasonal changes in the filtering rate of *Eudiaptomus gracilis*. J. O. Sars in Lake Balaton. *Annales Instituti Biologici* (Tihany) 43:105–116.

Zar, J. H. 1968a. Calculation and miscalculation of the allometric equation as a model in biological data. *Bioscience* 18:1118–1120.

– 1968b. Standard metabolism comparisons among orders of birds. *Condor* 70:278.

– 1969. The use of the allometric model for avian standard metabolism–body weight relationships. *Comparative Biochemistry and Physiology* 29:227–234.

Zepelin, H., & A. Rechtschaffen. 1974. Mammalian sleep, longevity and energy metabolism. *Brain, Behavior and Evolution* 10:425–470.

Zotin, A. I., & V. A. Konoplev. 1978. Direction of the evolutionary progress of organisms. In *Thermodynamics of Biological Processes*, I. Lamprecht, & A. I. Zotin, eds., pp. 341–347. Berlin: deGruyter.

Index

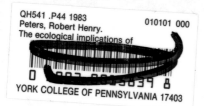